FIREGROUND STRATEGIES
SCENARIO WORKBOOK

FIREGROUND STRATEGIES
SCENARIO WORKBOOK

ANTHONY AVILLO

PennWell®

Copyright © 2003 by
PennWell Corporation
1421 South Sheridan Road
Tulsa, Oklahoma 74112-6600 USA

800.752.9764
+1.918.831.9421
sales@pennwell.com
www.FireEngineeringBooks.com
www.pennwellbooks.com
www.pennwell.com

Marketing Manager: Julie Simmons
National Account Executive: Francie Halcomb

Director: Mary McGee
Supervising Editor: Jared Wicklund
Production/Operations Manager: Traci Huntsman
Production Editor: Sue Rhodes Dodd
Cover Designer: Clark Bell
Book Designer: Wes Rowell

Library of Congress Cataloging-in-Publication Data

Avillo, Anthony.
Fireground strategies : scenarios workbook / Anthony L. Avillo.
— p. cm.
Includes index
ISBN 0-87814-841-8
ISBN 978-0-87814-841-7
1. Fire extinction. 2. Command and control at fires. I. Title.
TH9145.A97 2003   2003002018

Printed in the United States of America

2 3 4 5 5   12 11 10 09 08

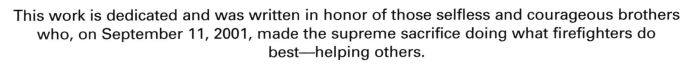

This work is dedicated and was written in honor of those selfless and courageous brothers who, on September 11, 2001, made the supreme sacrifice doing what firefighters do best—helping others.

God bless them and the loved ones they left behind.

P.T.B.

CONTENTS

Chapter Four Scenarios

Answer Section

Chapter Five Scenarios

Answer Section

Chapter Six Scenarios

Answer Section

Chapter Seven Scenarios

Answer Section

Contents

Chapter Eleven Scenarios

Answer Section

Chapter Twelve Scenarios

See Introduction

INTRODUCTION

The premise of this workbook, the companion to the *Fireground Strategies* textbook, is based on one of the fundamental principles of learning: reinforcement. One learns best when the lesson is reinforced via practical application of the materials read. Much of what we learn, if not reinforced, is forgotten as soon as a few hours afterward. The time to reinforce learning is immediately after it has been absorbed, and again on a consistent basis over time. Lessons reinforced become lessons learned. This takes effort on the part of the reader. The objective is to reinforce, through commitment, lessons learned through the textbook and workbook so that recall of the concepts becomes automatic, essentially second-nature on the fireground (see the section on "The Cycle of Competence" in the text). Remember, the only thing achieved without effort is failure.

A suggested study strategy to get the most out of this workbook is to read the textbook chapter-by-chapter, using the corresponding workbook scenarios immediately after having read the chapter. For example, read chapter 1 in the textbook, and then do the scenarios in the workbook. Then read chapter 2, followed by chapter 2 in the workbook, and so on. It will not be as effective to read the textbook first and then do the whole workbook, or read one without reading the other. They are meant to be utilized as a learning package, one reinforcing the other. Note that scenarios for chapter 3, "Building Construction," and chapter 12, "Operational Safety," are not included in this workbook. This was intentional as the information in both of these text chapters applies to every single scenario and every single fireground operation. As such, this information will permeate all the scenarios, giving the student ample opportunity to apply the information.

This concept (what I like to call challenge-based learning) allows the student to test themselves to ascertain if the information absorbed in the text has been learned and used effectively and positively in a simulated fireground situation. If mistakes (strategically or tactically) are to be made, better to make them in the classroom than on the fireground—which can be unforgiving.

CHAPTER ONE
SCENARIOS

SCENARIO 1–1
INCIDENT SIZE-UP

A fire has been reported on Central Avenue in a heavily congested section of the city.

Construction

The fire building is of four-story Class 3 construction with a flat roof.

Time and weather

The hour is 1347. The temperature is 39° F and it is lightly raining.

Area and street conditions

Central Avenue is a two-lane, two-way street.

Fire conditions

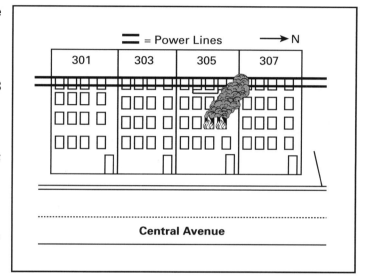

Upon your arrival, a tenant has alerted you to the possibility of a victim on the 3rd floor. Fire is showing from the 2nd floor, Side A. There is a present smoke condition from both 2nd and 3rd floors on Side A.

Exposures

This block of Central Avenue is constructed of attached four-story multiple dwellings of ordinary construction.

Water supply

The water supply is adequate and hydrants are adequately spaced.

Response

Your response is two engines and one ladder company. An officer and two firefighters staff the engine. An officer and three firefighters staff the ladder truck. You are the first officer on the scene as the company officer of Engine 1.

SHORT-ANSWER QUESTIONS

1. What would be your Preliminary Size-Up Report to dispatch regarding incident command?

_____

_____

_____

_____

_____

_____

2. What would be in your Preliminary Size-Up Report regarding the structure?

_____

_____

_____

_____

_____

_____

3. What would be in your Preliminary Size-Up Report regarding initial actions?

_____

_____

_____

_____

_____

_____

4. You are Battalion 1. You have assumed command. You have a line stretched and operating on the fire floor, a back-up line in the process of being stretched, as well as a primary search underway. It is seven minutes into the operation. What would be your Initial Progress Report?

_____

_____

_____

_____

_____

_____

MULTIPLE-CHOICE QUESTIONS

1. What would be the most important size-up factor at this incident?

    A. Street conditions
    B. Life hazard
    C. Height
    D. Apparatus and manpower
    E. Location and extent

SCENARIO 1–2
WAREHOUSE FIRE

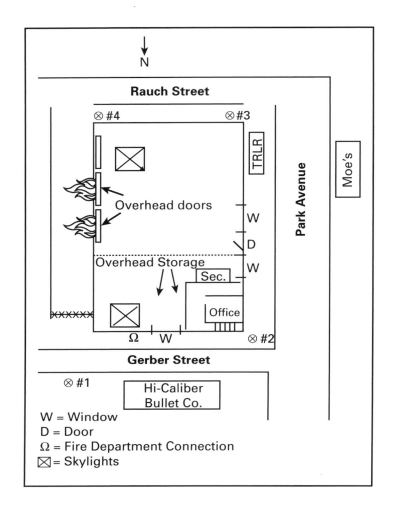

Construction

Erected in 1945 and occupying a good part of a city block, 32 Park Avenue is 200' by 100', and 25' high, of ordinary construction with a bowstring truss roof. There is a storage space between the 1st floor ceiling and the roof. This storage space is only in the north end of the building and extends toward the center of the building for 50'. The south end of this storage area is open and can be seen from the floor of the warehouse.

The storage area is filled with scrap lumber, old office furniture, and old business records. There is access to this area via a single set of stairs at the north end of the building adjacent to the offices and the security guard area.

The front entrance to the building is a large door on the west side of the building. A set of triple overhead doors is located on the southeast end of the building. There are two large windows on the Park Avenue side and one on the Gerber Street side.

Time and weather

It is Saturday, and the time is 1550. The temperature is 35° F. The wind is 25mph and is out of the northwest.

Occupancy and hours of operation

The occupancy is used to store and repair office furniture and has been undergoing massive renovation; however, the business has remained in operation. The occupancy is open Monday through Friday from 0700 to 1700, and on Saturday from 0900 to 1500. A security guard is on the premises until 2200 on weeknights and until 1600 on Saturdays. On Sundays, the structure is unoccupied. The renovation crews remain a little later than the employees each day to clean up and secure the day's work.

Water supply

The fire department connection is on the north end of the building. The sprinkler system has been temporarily placed out of service. All hydrants in the area are reliable and provide an ample water supply.

Area and street conditions

Bill's Office Supply faces Park Avenue, a two-way, two-lane street running south and north. To the north is Gerber Street, a one-way, one-lane street running east. To the south is Rauch Street, a two-way, two-lane street running east and west. A driveway leads off of Rauch Street and wraps around the rear of the building but terminates at a gate and does not go through to Gerber Street.

Fire conditions

The overhead doors are open at the time of your arrival and are showing heavy black smoke. The gate leading to the construction trailer is also open. There does not seem to be anyone around.

Exposures

Information regarding the exposures can be acquired from the accompanying scenario plot plan.

Response

You respond with two engines, one ladder company, a rescue company, and a deputy chief. An officer and two firefighters staff each engine company. An officer and three firefighters staff the ladder company and the rescue company. You are the first on the scene as the officer of Ladder 1 and will be in command.

SHORT-ANSWER QUESTIONS

1. Give the Preliminary Size-Up Report to dispatch.

_____

_____

_____

_____

_____

_____

2. As the deputy, you have arrived on the scene and have taken command. One line is stretched and operating. Additional lines are being stretched. The possibility of truss involvement is being investigated and the primary search is underway. Give an Initial Progress Report to dispatch.

_____

_____

_____

_____

_____

_____

3. What mode of operation will be employed at this fire?

_____

_____

_____

_____

_____

_____

4. How will the radio reports influence the strategy decisions of the incident commander?

_____

_____

_____

_____

_____

_____

MULTIPLE-CHOICE QUESTIONS

1. You have given your initial radio report to the dispatch center, including your incident command statement, building assessment information, and other pertinent information. What is the next action you will be required to take?
   A. Determine the whereabouts of the security guard.
   B. Position companies to match the attack strategy.
   C. Develop an attack plan and assign companies.
   D. Perform a size-up.
   E. Stretch the initial attack line.

2. What is the most important size-up factor to be considered at this fire?
   A. Weather
   B. Construction
   C. Occupancy
   D. Exposures
   E. Height

3. What are your orders for the crew of Engine 1 regarding hydrant selection and attack tactics?

   A. Establish a water supply at Hydrant #4. Position just past the overhead doors. Use the deck gun to knock down the fire from the established collapse safety zone. Also stretch a 2½" line to supplement the deck gun operation and to knock down heavy fire. No offensive operations are to take place.
   B. Establish a water supply at Hydrant #3. Position just past the front door on Park Avenue. Stretch a 2½" line through the front door to locate, confine, and extinguish fire. Protect the ladder crew attempting to locate the security guard.

C. Establish a water supply at Hydrant #4. Position just past the overhead doors. Stretch a 2½" line through the overhead door to locate, confine, and extinguish the fire. Protect the ladder crew attempting to locate the security guard.

D. Establish a water supply at Hydrant #2. Position just past the Bullet Factory on Gerber Street. Remain out of the collapse safety zone. Prepare to supply the ladder pipe operating on the north end of the building through the skylight. Operate the deck gun through the window on Gerber Street and keep Bullet Factory wet. No offensive operations are to take place.

E. Establish a water supply at Hydrant #2. Position just past the front entrance on Park Avenue. Stretch a 1¾" line through the front door to the office/security area to protect the ladder company searching for the security guard. Locate, confine, and extinguish the fire after the guard is removed.

4. What are your orders for the captain of Engine 2?

A. Secure a secondary water supply at the next most accessible hydrant. Stretch a 2½" line through the front door to the office area to back-up Engine 1's attack and to help with the protection of the ladder crew searching for the security guard.

B. Secure a secondary water supply at Hydrant #3 or #4. Stretch a booster line through the overhead doors to back-up Engine 1's attack. This line is more mobile and will be easier to maneuver through the large warehouse.

C. Secure a secondary water supply at Hydrant #3 or #4. Position on Rauch Street. Position the apparatus outside of the collapse safety zone. Utilize the ground-based deck gun to supplement Engine 1's master stream attack through the overhead doors at the rear. No offensive operations are to take place.

D. Secure a secondary water supply at Hydrant #2 or #3. Check Moe's Tavern for the security guard, report back to incident command, then search the office/security area for the missing guard.

E. Secure a secondary water supply at Hydrant #2 or #3. Stretch a 2½" line through the front entrance to back-up Engine 1's attack line and to extinguish any fire extension.

5. How would the rescue company best be directed in conducting a primary search of this building?

A. Split the company into two crews. Crew 1 enters the warehouse through the overhead doors, using a lifeline to conduct the search for the security guard and any other occupants. Crew 2 reconnoiters Moe's Tavern to see if the security guard is there, and reports to incident command the result of this. Crew 2 utilizes a second lifeline through the overhead doors, taking the opposite direction of Crew 1.

B. Split the company into two crews. Crew 1 quickly checks just inside the overhead doors for victims. Crew 2 quickly checks just inside the front doors. Then both crews withdraw and prepare for an exterior operation from outside of the collapse safety zone.

C. Keep the company intact. Utilize a lifeline to perform the primary search of interior. Utilize tethers off of the main lifeline where required. Place a large floodlight at the means of egress.

D. Split the company into two crews. Crew 1 enters the warehouse through the overhead doors, using a lifeline to conduct a search for the security guard and any other occupants. Position a large floodlight at the egress point. Crew 2 enters through the front entrance on Park Avenue. Crew 2 performs a systematic search of the office/security area.

E. Split the company into two crews. Both crews enter through the front entrance. Crew 1 searches to the right. Crew 2 searches to the left. Both crews search the office/security area first and then the warehouse area.

6. What would be the most effective position for the ladder company? How would initial ventilation of the building be accomplished?

A. Position the ladder truck on Park Avenue in front of the front door. Horizontally ventilate at the large windows, also ventilate at the large window on the Gerber Street side. No operations are to be performed on the roof.

B. Position the ladder truck on Gerber Street between the fire building and the Bullet Factory. Due to the presence of the bowstring truss, no one is to go to the roof. Allow the fire to self vent. Maintain the collapse safety zones. Prepare for exterior defensive operations.

C. Position ladder truck on Park Avenue/Gerber Street corner. Split the ladder crew. Crew 1 performs the horizontal ventilation of the large windows on the Park Avenue side. Crew 2 raises a ground ladder to the roof on the Park Avenue side. Open the skylight. Evaluate the conditions in the truss area. Report the conditions to incident command. Leave the roof after opening the skylight.

D. Position the ladder truck on the flank of Rauch Street and the rear of the building. Prop open the overhead doors to maintain the venting direction. Raise the aerial platform to the roof. Open the skylight only if it is near the seat of the fire. Prepare to cut the roof over the hot spot as per interior reports. Evaluate the conditions in the truss area. If fire is present, abandon the roof and pursue a defensive position.

E. Position the ladder truck on the flank of Gerber Street and the rear of the building. Split the ladder crew. Crew 1 horizontally vents all the large exterior windows. Crew 2 raises the aerial platform to the roof. Open the skylight from the platform. Evaluate the conditions inside the truss area. Report the conditions to incident command.

7. Would additional alarms be required at this incident?

    A. Yes, a confirmed life hazard is criteria for an additional alarm.

    B. Wait until it has been determined if the security guard is, in fact, missing.

    C. No, as there is no real fire spread potential on the leeward side of the fire.

    D. Yes, due to the inherent flaws in the construction of this building.

    E. Yes, the lack of an operational auxiliary protective system is a good indicator that an additional alarm should be sounded.

8. The fire is in a large pile of furniture and boxes. You are getting conflicting reports from your personnel. From the interior, it is reported that the firefighters can walk right up to the fire, that there is only a slight smoke and heat condition, and that control of the fire will be gained shortly. From the roof, the report is that there is a heavy fire and smoke condition at the skylight. What are your actions?

    A. Continue to operate on the interior. Reports from the roof are probably residual smoke from the blaze at the upper levels of the structure. Have the firefighters continue to monitor conditions at the roof and check for further extension at the eaves.

    B. Withdraw your companies immediately from the interior. Use the reach of the stream to sweep the underside of the truss then move back in to finish extinguishment. Have the ladder crew at the roof report on conditions as this defensive/offensive operation takes place.

    C. Withdraw firefighters from the roof. Have firefighters on the interior attack lines back out and use the reach of the stream through the windows on the Gerber Street side to knock down the fire in the truss. Provide more horizontal ventilation at the window levels.

    D. Immediately withdraw all companies from the roof and from the interior. Conduct a roll call. Use streams in a defensive manner outside established collapse safety zones.

    E. Withdraw men from the roof. Have the firefighters on the interior move back below the truss adjacent to the ones involved and use the reach of the stream to sweep the undersides of the involved truss area.

9. Your men are positioned at the Park Avenue side of the building. You are to establish a collapse safety zone at the front of the building. How large an area should this safety zone cover?

    A. 100' horizontally and 50' vertically.

    B. 200' horizontally only.

    C. 25' vertically only in the areas of the large windows.

    D. 200' horizontally and 25' vertically.

    E. 200' horizontally and 50' vertically.

ANSWER SECTION

SCENARIO 1–1
SHORT-ANSWER QUESTIONS

1. "Engine 1 is on the scene at 305 Central Avenue. Engine 1 is establishing Central Avenue command."

   This is the "C" in the C-BAR acronym and since this is not a large incident at this time, the location of the command post need not be announced. As per the routine of most departments, it will be somewhere in the proximity of the front (Operational Side A) of the building. The incident commander should take care not to block arriving apparatus or hydrant access when positioning the command vehicle.

2. "305 Central Avenue is a four-story, ordinary construction, occupied multiple dwelling, with a flat roof. Building is attached on both sides. Fire is showing from the 2nd floor, Side A. Smoke showing 2nd and 3rd floors, Side A. Trapped occupant reported on the 3rd floor."

   This "B" in C-BAR covers the most critical building information. As this is a Preliminary Size-Up Report, only the most critical building information need be provided. Using the applicable parts of the C-HOLES format, it is easy to furnish dispatch and in-coming companies with a complete and accurate description of the fire building in regard to arrival conditions. Note the use of "305" in the report. This information is culled from the diagram. To more closely examine C-HOLES here, we'll take them one at a time:

   - *C*onstruction = "ordinary (Class 3), attached."

   - *H*eight = "four-story, flat roof." Announcing the building has a flat roof during the Preliminary Size-Up Report can cue in responding ladder companies on the probable access and operational tactics required for this type roof. In contrast, a peaked roof size-up will require a different set of tactics. Ladder companies will be more effective if this information is given at the outset.

   - *O*ccupancy = "occupied multiple dwelling." There is no need to say "residential" here as it is inferred by using the terms "multiple dwelling." This fact alone will put you into the rescue mode upon arrival. (Fig. 1–1)

   Fig. 1–1 Photo by Ron Jeffers NJMFPA
   Short of total involvement of the structure, fire showing upon arrival demands that companies operate in the offensive rescue mode. All scene activities must be assigned to support the rescue (primary search) operation.

- *L*ocation & Extent = "fire showing from two windows on the 2nd floor, Side A. Smoke showing from 2nd and 3rd floors, Side A." Ensure you designate from which side fire and smoke is showing. This will help set up the direction of attack as well as concentrate primary search efforts.

- *E*xposures = providing "attached on both sides" in the Preliminary Size-Up Report gives the incoming companies as much information as they will need at this time. (This is not normally a part of the Preliminary Size-up Report, but it is a major part of the Initial Progress Report)

- *S*pecial Circumstances = "Report of person trapped on the 3rd floor." Although the fire apartment and fire floor will be the focus of the primary search, the floor above must be accessed as soon as possible. The floor above the fire will sometimes be more untenable than the fire floor. A report of a victim there will add a greater sense of urgency in not only getting a crew up there for primary search, but in the stretching of a hoseline to protect that search as well.

3. "Engine 1 will be stretching a line."

   This represents the "A" in C-BAR. This is the beginning of the initial attack plan. You must describe the initial action you will be taking to ensure those companies not yet on the scene can identify the strategic mode before their arrival, in this case, the offensive mode. This will aid them in being prepared to carry out those assigned duties they are responsible for as per their adopted SOP regarding offensive fire attacks. The entire Preliminary Size-Up Report should be as follows:

   > *"Engine 1 is on the scene at 305 Central Ave. Engine 1 is establishing Central Avenue command. We have a 4-story, ordinary construction, occupied multiple dwelling with a flat roof, attached on both sides. We have fire showing from the 2nd floor, Side A. Smoke showing from 2nd and 3rd floors, Side A. Trapped occupant reported on the 3rd floor. Engine 1 will be stretching a line."*

4. "Dispatch from Central Avenue Command. Initial Progress Report. 305 Central Avenue is a four-story, occupied multiple dwelling of ordinary construction. We have fire showing from the 2nd floor, Side A. Smoke on floors two and three. Exposures are as follows: Exposure A is a street, Exposure B is a similar occupancy of equal height and is attached, Exposure C is unknown at this time, and Exposure D is a similar occupancy of equal height and is attached. Have a report of a person trapped on the 3rd floor. We have one line working on the 2nd floor, also an additional line is being stretched at this time to the fire floor. Primary search is being conducted and is not yet complete. Dispatch a second alarm. Have all second alarm companies stage uncommitted and await orders. Have EMS and the utility company respond. Fire is doubtful at this time."

   This is the most complete picture that can be drawn at this time. All of the elements of C-BAR and C-HOLES are covered. Some of the information is identical to the Preliminary Size-Up report. In addition, other information such as exposures and resource requests are included. Also included is a very brief description of the actions taken thus far to control the fire.

In regard to exposure information, note that Exposure A is a street. If there had been a vehicle parked at the curb in front of the building, it would become Exposure A. Exposures B and D are attached structures of the same construction and height, so the term "similar" is sufficient. Using the term "attached" will tell the ladder company how to best access the roof as well as be a red flag regarding potential fire spread via the cockloft, which, until proven otherwise by direct examination, will be assumed to be open for the length of the whole row. Note that Exposure C is "unknown at this time." This is where your reconnaissance reports come into play. This information must be ascertained as soon as possible. There could be a lot going on in areas you cannot see from the command post. The sooner you know about them, the better you'll be able to address any problems presented by them.

Resources, representing the "R" in C-BAR, should be summoned early if you feel you may need them to get a handle on the situation. In this fire, the response of two engines and a ladder truck will probably not be enough to handle all the tasks that will be required to extinguish this fire, vent, search, check for extension in the fire building, and check the exposures. Directing the apparatus to stage and await orders from incident command will avoid the problem of everyone rolling into the middle of the incident and overwhelming the incident commander. This statement leads to better control on the fireground and helps eliminate freelancing. If the incident commander can state exactly where he wants these companies to stage, all the better. That way, if they are needed in a specific position or to work together in some task, they can respond from the same place. This will simplify things for all.

If the fire is quickly knocked down, these companies can be released from the scene or held in staging and used later to relieve the initial companies. If, on the other hand, the fire is already in the vertical arteries and has extended to the cockloft, the fire may burn the roof and top floor off all of the exposures. The fire situation will then require a much larger compliment of manpower than was dispatched on the first alarm.

Forecast the fire extension profile based on building construction, weather conditions (especially wind), and the location and extent of the fire on arrival. If you have additional companies on the way as soon as you realize the potential for escalation, you will be in a much better position to head it off.

Regarding outside agencies as additional resources, some of the responses should be arranged in advance through proper pre-fire planning activities. For instance, the police usually routinely respond on fire emergencies. It should also be arranged that the local EMS be dispatched as soon as a working fire is confirmed. Having this arranged in advance provides for firefighter safety in the critical, early stages of an incident where many firefighters are injured. Other special needs for outside agencies will present themselves upon arrival or as the incident unfolds. A great majority of these needs will focus on firefighter safety, such as having the utility company shut off power and gas to the structure and requesting a canteen service respond to help firefighters re-hydrate themselves. Again, experience and judgment should be a guide.

Typical incidents that will require a special response will include:

- Confirmed hazardous materials emergency

- Any situation that requires technical expertise beyond the capability of the responders on scene (i.e., confined space, trench, and high-angle rescues)

The concept to follow is, "if you think you may need an agency, have them dispatched." Reflex time for outside agencies can be quite long. Be proactive in your approach. Get the needed people to the scene and ensure that you have someone to liaison with them when they arrive. This will help you to control your incident.

Multiple-Choice Questions

NOTE: One of the keys to distinguishing between the best (+2) answer and the second best (+1) answer is that the best answer will often take the good tactics stated in the second-best answer a step further and solve more problems. Remember, as a company officer or a chief officer, your job is to solve problems. This means being thorough in your tactics and considering all of the possible aftereffects of your ordered actions.

---

1. The key to this question is the *most* important size-up factor. While most size-up factors have an impact on the fire situation, to answer successfully, you must key into what the question asks.

    A. **-1** Central Avenue is a two-lane street. There should be no problem in apparatus positioning. Due to the weather, you may get some icing, but at 39°F this should not be a problem, at least initially.

    B. **+1** Life hazard is definitely a very important factor at this incident. There is a report of a trapped person on the 3rd floor. This fact will demand that the fire be attacked using an offensive mode to protect the primary search.

    C. **0** Height should not be a problem. All areas should be accessible by ladder. The power lines may present a problem at the 4th floor and for roof access, but as these are attached buildings, roof access can best be gained via the adjoining exposure upwind. You may ask, "if a person shows at the 4th floor window, how do you get him out?" In the real world, you would not worry about this until it presented itself. If it did, a ground ladder (40' or 45' pole ladder will make it), or maybe the aerial if you can get it under the wires, or, in the pinch, a rope rescue will be required. This is not the real world. This multiple-choice question is a test question. There is no person showing at the 4th floor window. There is not even any building or fire in it. It is a test. Do not read into the question and create your own scenario. Handle the situation that is presented and you will answer it well. If you remember this in these multiple-choice questions, you will avoid this "what if" pitfall that dooms many test-takers.

    D. **+1** Manpower and apparatus are a real concern at this situation due to the potential for escalation and fire spread beyond the fire building. It is a size-up factor that must be addressed to ensure enough men will be on hand and in reserve to safely mitigate the incident.

    E. **+2** As I stated earlier, location and extent is the most important size-up factor. It will determine your life hazard, as well as dictate whether your compliment of manpower and apparatus is sufficient.

Scenario 1–2
Short-Answer Questions

1. "Ladder 1 is on the scene. Ladder 1 is establishing Park Avenue command. 32 Park Avenue is a two-story, ordinary construction, commercial occupancy, assumed occupied. Roof is of bowstring truss. Have heavy smoke showing at Side C. Also have an unaccounted-for civilian. Primary search is being initiated. Roof condition and truss involvement being investigated."

   The Preliminary Size-Up Report here, while still short and to the point, adds vital information such as the unaccounted-for civilian and the presence of the bowstring truss roof. This allows the chief officer and the other companies responding to get a better picture of the situation and allow them to formulate a projected strategy before arrival.

2. "Dispatch from Park Avenue Command. Initial Progress Report: Have a two-story, commercial occupancy of ordinary construction used as a furniture supply store, assumed occupied. Have heavy smoke showing at Side C, no visible fire. Exposure A is a construction trailer and street. Exposure B is the High Caliber Bullet Company (state height, construction, and other pertinent information if known). Exposure C is a rear driveway. Exposure D is a street. 32 Park Avenue is under renovation and has a bowstring truss roof. The auxiliary system is out-of-service. Have a security guard unaccounted-for. Have one line stretched and charged. Additional lines are being stretched at this time. A marginal primary search and attack operation is being conducted. Possibility of roof involvement is being investigated at this time. Strike a second alarm. Stage the second alarm companies at Park and Rauch and have them await orders. Have EMS and the utility company respond. Fire is doubtful, will hold at this time."

   This report includes all the pertinent information required to set the operation in motion. It also establishes the operation as being offensive but marginal due to the construction of the building. Also, the inclusion of the building being under renovation and the unusable fire department connection paint still a better picture of potential problems that may impact tremendously on the offensive/defensive decision.

3. Because there is a possibility of a person in the building, an offensive attack must be employed. This attack, depending on conditions, must be accompanied by a primary search, using guide ropes or even searching beside hoselines to attempt to find the unaccounted-for security guard and any workers who may still be in the building. This is where thermal imaging cameras are invaluable, not only to scan floor levels for victims and the fire's location, but also to determine if trusses are involved.

   Being that the roof construction is bowstring truss does not necessarily call for a complete defensive operation from the outset. The heavy smoke may be the result of a contents fire near the doors that has not involved the truss. Pulling the plug on this operation primarily because of the building's construction will cause the death of anyone inside or needless destruction of the building. Even if heavy fire is present, you may be able to perform a cautious and very marginal search of all accessible areas before withdrawing the crews. This may mean accessing all the tenable entrances to check the immediate areas near the egress points. This is often where victims are found, having been overcome while attempting to escape the fire. Also, it might be a good idea to send someone, maybe a police officer, to the tavern across the street to see if the guard is there or maybe someone from the construction crews to help account for everyone.

4. Radio reports should be the most critical deciding factor in whether an offensive mode is continued or a defensive mode is pursued. Get reports from as many places as possible to guide your decision. The most important place of reconnaissance will be the roof. Firefighters, preferably from platforms or aerials, can hopefully access natural openings, such as the skylights (See the diagram on p. 4 for skylight locations). (Fig. 1–2)

Fig. 1–2 Utilize skylights and other natural openings to ascertain if fire has involved the truss area. Strategy development will be greatly influenced by the reports generated through reconnaissance of this area.

If no fire is present in the trusses, ladder company personnel should vent, as required, and continue to monitor the truss space. The offensive operation may be allowed to continue; however, if fire is reported in the truss area, the building must be abandoned, for the roof and the front and rear walls will collapse. The walls prone to collapse are the ones on Rauch and Gerber Street. The collapsing trusses will transmit their loads to the sloping hip rafters set into these walls, causing an inward-outward collapse. This will occur without warning and cannot be predicted. Safe positions out of the collapse zone must be established and maintained. Master streams must be employed. At that point, the building is doomed.

If the interior crews report a small fire in some contents you may be able to reinforce the offensive attack and continue in this mode, but remember that the fire may still have ignited the trusses. Listen to the roof reports to guide your strategy here. Again, the use of a thermal imaging camera to scan the overhead from below or to scan the natural roof openings from above can help simplify the monitoring operation.

If the report from the interior is one of heavy fire or the fire location cannot be ascertained, put up the red flag and seriously consider withdrawing forces; for if the trusses are not involved yet, they soon will be.

Timely and accurate radio reports will be crucial to safe and effective operations at this fire. With bad or incomplete information, the incident commander can make some deadly mistakes.

Multiple-Choice Questions

1.

A. **0** If your only worry is determining the whereabouts of the security guard, you could fall victim to "tunnel-vision" as you begin to micromanage the situation. As incident commander, you should be establishing a "big picture" mentality. Without a doubt, a primary search will have to be extended to accomplish this objective, along with other support tactics such as stretching lines and performing ventilation and recon. This is known as the Incident Action Plan.

B. **0** While firefighter safety is the overriding concern at all incidents, the incident commander should not routinely be involved in apparatus positioning of initial companies. In regard to specific occupancy, this should be taken care of by SOP and/or pre-fire plans. Effective company officers should ensure their apparatus is safely positioned at all incidents, regardless of their magnitude.

C. **+2** By developing an attack plan and assigning companies, the incident commander will ensure the best control over the operational forces. The attack plan will place all operating forces on the same page, enhancing firefighter safety. Companies that are already properly positioned will be easiest to assign. Again, if the incident commander is worrying about positioning all the responding companies, he will be unable to think about an action plan, and, as a result, will be playing catch-up from the outset.

D. **-1** This has already been done. Giving an incident command statement, building assessment information, and other pertinent information to dispatch via radio can only be gleaned by performing a size-up. Picking this choice would show that the candidate did not read the question or understand what it is asking.

E. **0** This is putting the cart before the horse. To properly stretch the initial attack line, an attack plan must first be formulated, based on a sound size-up of the fire situation.

2.

A. **+1** The weather, especially the wind, is a definite factor. The wind is 25mph, but it is blowing away from the most significant exposure, the Bullet Company. Therefore, it is not the most important size-up factor.

B. **+2** The limiting factor in this situation, one that cannot be overcome, is the fact that the roof of the building is of bowstring truss construction. Bowstring trusses add significantly to the fire load of the building. This should cause the incident commander to put up a red flag of caution from the outset. If the fire involves the truss area, that factor will be the deciding factor in whether this is an offensive or defensive operation. Therefore, this is the most important size-up factor.

C. **+1** There is a heavy content fire load in this building, both on the floor of the warehouse and in the second floor storage area. This fire load will cause convection heat to rise and mushroom under the roof in the truss spaces. It will ignite these trusses unless it is properly vented, or better yet, the content fire is extinguished rapidly before the trusses become involved.

D. *0* Although the Bullet Company is an exposure, the wind is not blowing toward it. This condition decreases the significance of this factor. If the wind was blowing toward the Bullet Company, the incident commander, depending on fire conditions in the warehouse, may choose to protect this structure with the initial arriving companies and leave the warehouse to additional alarm companies.

E. *0* The building is 25' high and is reachable by ground and aerial apparatus. This is not a factor.

3.

A. *-2* The hydrant chosen is downwind of the fire and directly in line with the smoke drift. The apparatus position is also in the smoke. The tactics, strictly defensive in nature, writes off any life in the structure.

B. *+2* This is a good choice of hydrant. It can be wrapped on the way in using forward lay and the positioning will allow you to stretch a line through the front door, from the unburned side. The use of a 2½" line is necessary, as the fire conditions may be severe. The line is properly used to take care of the fire and protect personnel.

C. *+1* While the hydrant choice is not correct, and the stretch and apparatus position is in the smoke, the strategy of an offensive attack is the proper tactic as there is the possibility of life in jeopardy inside the building.

D. *-2* Again, a strictly defensive operation from the outset. A good hydrant choice, given the tactics chosen, but the wrong strategy. An attempt will have to be made to locate any victims in the building. The ladder pipe is a good idea, but much later in the operation if conditions warrant.

E. *+1* While the tactics are acceptable, concentrating on an area where the guard may be, the hoseline chosen is too small. Large area buildings demand the use of a 2½" line. A smaller line may cause a premature withdrawal, accomplishing nothing. Go for the big lines in these situations. Better to have too much than too little. (Fig. 1–3)

Fig. 1–3 Photo by Bob Scollan NJMFPA
Fires in big buildings require the stretching of large diameter lines. Think fire load and ensure your line has the best chance of making an impact on the anticipated fire condition.

4.

A. *+2* These are good tactics. At a large building where the fire may be of extensive proportions, a second, and possibly a third, water supply is mandatory. The second line is stretched through the same path of access as the first line and is the proper procedure for back-up lines. Protection of personnel engaged in the search must be a primary consideration of attack crews.

B. *-2* Stretching a booster line into a structure must be strictly forbidden. The direction of the attack also leaves much to be desired as it is where smoke, and probably heat, is venting.

C. *-2* This is again pursuing a defensive mode of attack. One of the problems of the multiple-choice scenario is that if you select one strategy in one question, there will usually be answer choices in subsequent questions that will match that strategy. This will result in deviation amplification and lead both the candidate and the fireground strategist on a downward spiral to failure in the test in the former case, which is a good thing if you don't know your business, and possible death, injury, and property loss in the latter, which is even more unacceptable. You can always take more tests. Dead firefighters, civilians, and lost buildings don't ever come back.

D. **-1** Securing a second water supply is proper, but going over to check for the security guard at the tavern is a waste of manpower. It is better to send a cop or a worker who is already out of the building. This company should be stretching a back-up line to reinforce the attack position and protect the searching personnel.

E. **+1** Proper tactics along with the proper securing of a second water source. However, if you compare this choice with choice "A," the fact that choice "A" addresses firefighter safety (protection of the searching ladder crew) makes it the +2 answer choice. Safety consideration is always a winner, both on the fireground and in the test environment.

5.

A. **0**   This is not an efficient use of manpower. While spitting the company is a good idea, especially using a lifeline in a large area, sending a crew over to the tavern to search for the guard is inefficient.

B. **-2** No search of interior. This is strictly a defensive maneuver and will cause the death of any occupants inside the building. The presence of a potential life hazard demands that ladder or rescue personnel enter the building to conduct a primary search, especially if the fire's size and extent has not yet been determined.

C. **+1** These are good tactics and will make for a safe search. Use of a lifeline with tethers and lighting at the point of egress helps ensure firefighters are able to find their way out of the structure and should be utilized at large area structures. However, in this case since there is both warehouse and office area to cover simultaneously, it may be better to split the company.

D. **+2** This is a better utilization of manpower. If there was only warehouse area and no office area, it would be best not to split the company. However, a person is reported as missing and it is best to quickly cover as much ground as possible. One crew, with a lifeline and light in place at the point of egress, searches the warehouse. This is a big job and should be reinforced by later-arriving companies. The second crew, using a standard and systematic search procedure, can cover the much smaller office and security areas. The chance of locating any victims in the shortest amount of time will be greater. If cubicles are present, a lifeline should be taken by this second crew. In either area, don't forget the thermal imaging camera.

Fig. 1–4 *Large buildings require that a lifeline be utilized to prevent companies from becoming lost in massive, open areas. A lifeline search will ensure that the safest, most effective path of least resistance out of the structure is maintained (Point of Exit Rule of Thumb).*

E. **0**   The problem here is that no lifeline is used. This may be acceptable in the office and security area, but is unacceptable in the warehouse. Staying on the walls will leave the center areas unsearched. Using a lifeline will maximize both your search endeavors and your safety. (Fig. 1–4)

6.

A. *0* The position of the ladder truck here is not the most advantageous. The power lines will block access to the roof. This information must be obtained from examining the diagram. The diagram will often give you information such as wind direction, presence of power lines, or other obstacles that the text does not give. Take a few minutes and study the diagram. If you did, you would have determined that the front of the building is not the best place for the ladder truck. Moreover, vertical ventilation and inspection of the truss area must be at least attempted. Again, looking at the diagram reveals large skylights on the Gerber Street and Rauch Street sides.

B. *-2* Here, again, is a strictly defensive posture. The apparatus positioning is upwind, however, if the ladder truck cannot be positioned outside the collapse zone it cannot be positioned in this area. Doing nothing in this case, as in most cases, is as bad as not showing up at the scene at all. Allowing the fire to self-vent will involve the trusses and cause a collapse of the walls on the Gerber and Rauch Street sides. If there are people attempting a primary search for the missing guard, ventilation must be accomplished.

C. *-1* Due to the power lines, this is not the best position for the ladder truck. Even if the positioning was conducive to raising a ground ladder to the roof, the distance that would have to be traveled to reach either skylight is an unacceptable risk. If conditions began to deteriorate, the roof firefighters might not get back to the ladder in time. There are safer and more acceptable ways of accomplishing this task.

D. *+1* This is not the best place to position the ladder truck, as it is on the downwind side and in the area of smoke drift, but it is in a flanking position. However, taking action to provide some type of vertical vent is the right idea, thus the point is given. The skylight must be opened even if it is not directly over the fire. The only way to alleviate accumulated heat at the upper levels of the building is to open the skylight. Not providing this ventilation would certainly allow the build-up of heat that will undoubtedly ignite the trusses, if it hasn't done so already. Moreover, cutting this roof as an initial tactic is not proper either. Open the skylight and evaluate conditions. There is no mention of horizontal ventilation. Providing horizontal ventilation will ease smoke and heat conditions on the interior, making searching for both the victim and the fire's location a safer task.

E. *+2* This is proper apparatus positioning. The ladder truck is on the flank, upwind, and not in the area of any power lines. Both vertical ventilation and horizontal ventilation are performed. Also, opening the skylight from the aerial platform is a safe way to evaluate conditions without actually stepping on the roof. If no fire is present at the truss level, firefighters may then recon the roof further. Also, the reporting of conditions to incident command is critical in this situation for it is by this and the interior report that incident command will base the strategy. If the report is of heavy fire (or any fire for that matter) in the truss area, companies must be withdrawn to safe areas outside of the established collapse zones. Many master streams will be required, along with additional water supplies.

If there is no fire evident in the truss area, the incident commander can reinforce the offensive operation. However, the truss area must continue to be monitored from above until the fire is well under control at the ground level.

7. Requests for additional alarms must be justified by the facts as they apply to the incident or by forecasts based on sound judgment.

   A. **0** The life hazard here is not confirmed. An additional alarm can be justified in this situation due to the potential life hazard, not a confirmed life hazard. Therein lies the difference in the point value to this question. An example of a confirmed life hazard would be someone showing at a window or many people exiting via fire escapes. Otherwise, the only way we can confirm a life hazard is by a complete and thorough primary search.

   B. **-2** "Wait and see" mentalities don't cut it on the fireground. By the time that the life hazard is confirmed it may be too late to take action to provide the proper manpower for rescue. Inaction and indecision are always losers on a test and killers on the fireground.

   C. **-1** The greatest significant exposure, the Bullet Company, is not on the leeward side of the fire. If the roof collapses, bringing down the walls on the Gerber and Rauch Street sides, there will be a more direct threat to the Bullet Company by radiant heat as the walls will release a large quantity of heat outward from the collapse. Having a tactical reserve in case the situation deteriorates will allow the incident commander to remain proactive in his approach to exposure coverage and not be outflanked.

   D. **+1** The large area of the building, the bowstring truss and its massive structural fire load, and the likelihood of subsequent roof and wall collapse are all good reasons to summon additional alarms to cope with the potential magnitude of this fire.

   E. **+2** An auxiliary protective system (such as a sprinkler system) is our biggest ally in a building of this size. Take that away and the fire forces are at a great disadvantage. Also out-of-service, along with the sprinkler system, might be any water flow alarms that alert the fire department to a fire condition in the early stages of the fire. Lacking such an operational system can allow the fire to gain great headway before it is noticed and an alarm turned in. Any out-of-service protective system is justification for additional alarms.

8. The first line in this question is the key to the whole scenario. It tells you the location and extent of the fire. Some scenarios may be vague as to the location and extent of the fire, as well as other factors. Reading all of the questions before beginning to answer them will often cue you in to the right course of action when the scenario and/or diagram does not give you all the information required. On the fireground, this lack of initial information must be filled in as soon as possible by timely reports from in and around the building. Still, it may be necessary to begin operating with less than complete information. Err on the side of safety and rely on your experience and education.

   Fig. 1–5 Photo by Bill Tompkins
   Fire involving the truss calls for immediate withdrawal of forces and the establishment of defensive positions and collapse zones. Immediate and constant recon of the truss area is critical. Do not wait for fire to break through the roof–it will be too late.

   A. **-2** This report of heavy fire at the roof level is the signal that it is time to withdraw firefighters (do not pass "go", do not collect $200, evacuate NOW!) Leaving firefighters on the interior as well as on the roof will get them seriously injured or killed. The building will collapse, often without warning. The incident commander cannot predict how long the trusses can be exposed to fire and still remain in place.

B. *-2* This defensive/offensive operation is a risk that should not be undertaken at this type of building. Leaving men on this roof is like playing Russian roulette. This building is doomed. If, as the question states, there is slight smoke and heat at the ground level, anyone missing would have been readily visible and removed. If they haven't been removed by this time, they may be an acceptable loss. The concepts the incident commander should consider are:

1. Risk a lot to save a lot (this means live people; dead people are not savable).

2. Risk a little to save a little (property).

3. Risk nothing to save nothing. Pull the men out. Burned down buildings don't come back, but can be rebuilt. Dead firefighters don't come back, period.

C. *-2* The decision to withdraw firefighters from the roof is the proper decision. Withdrawing the lines from the interior is also the right decision. However, placing these men so that they are in proximity to the walls on the Gerber Street side may be a deadly mistake. These side bearing walls can be predicted to collapse simultaneously with the roof as the failing trusses transfer their load to the sloping hip rafters tied into these walls on both the Gerber Street and Rauch Street sides. The men should operate outside the collapse zone or on the flanks.

D. *+2* This is the only acceptable tactic at this stage of the fire. After withdrawal, all men should be accounted-for via a roll call. Defensive positions should be established and maintained while all firefighting should be undertaken from outside the collapse zones. The building is surrendered at this point.

E. *0* This is an acceptable tactic at fires where the location of the involved truss is known. Men are withdrawn to a safe area behind the truss adjacent to the one involved and the reach of the stream is used to sweep the underside of the roof and extinguish the fire in the truss area. This will also limit the lateral spread of any fire under the roof. However, this is not the case here, and you cannot assume from the roof reports that only one truss is involved, therefore, the only tactic is to withdraw and pursue a defensive strategy.

Fig. 1–6 Englewood, NJ, Fire Department file photo
Where trusses are involved, the strategy is strictly defensive. Master streams should be utilized from outside collapse zones. Additional water supplies may need to be established to support this defensive strategy.

9.

A. *-1* This answer has no basis. The horizontal distance of the wall is 200'. Establishing a safety zone only half that size is unsafe.

B. *0* While the horizontal measurement of the building is being covered, the vertical height of 25' is being ignored.

C. **-1** The fact that the areas where the windows are located are the weakest points cannot be a justification for choosing where to establish a safety zone and where to ignore it. Also, no horizontal zone is mentioned.

D. **+2** This is a proper safety zone. This is the full vertical height and the full horizontal width of the wall. Some authors state that the vertical height should be equal to the height of the wall plus half that height (which would add another 12.5' to our measurement here). This is perfectly acceptable, but the minimum should be at least the vertical height of the wall. Debris may be thrown out further than this height. The best areas can therefore be positions on the flanks of the structure.

E. **+1** This is the full horizontal run of the wall plus twice the height. Erring on the side of safety will always get you a point on the test and will allow for an increased sense of safety for personnel on the fireground. Remember, that the question asks about a safety zone on Park Avenue. If it had asked about this zone on either Gerber or Rauch Street, this may have been the +2 answer, for as the roof collapses, it can collapse these walls in an inward-outward fashion, often with explosive force, which would justify the safety zone of 50', twice the vertical height of the wall. Also, don't forget that collapse can cause secondary collapses of power lines that could fall on personnel and apparatus outside the collapse zone. Take the big picture into account and everyone will survive the incident.

Passing Score for Scenario 1–2 = 12 points

Fig. 1–7 The classic hump of the bowstring truss can be seen at the roof peak. The sloping hip rafters extend downward toward the eaves. Failure of the bowstring truss will transfer the roof load to the hip rafters and cause a secondary collapse of the wall seen here.

SUMMARY

As you can see from this scenario, the radio reports from the companies located in crucial positions were critical in the establishment of the proper strategy in fighting this fire. Lacking this vital information, the incident commander cannot make the best decisions in regard to firefighter safety and incident mitigation.

CHAPTER TWO
SCENARIOS

Scenario 2–1
Dumpster Fire Exposing Structures

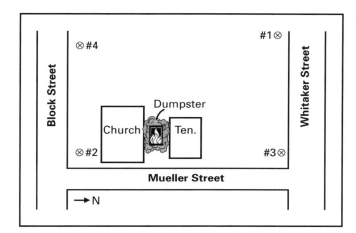
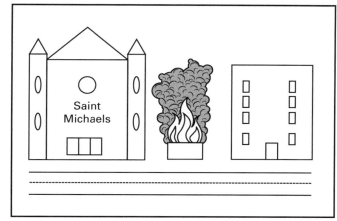

Construction

A wood-frame church built around the turn of the century is undergoing alterations. Construction debris has been placed in a large dumpster, which is located in an alleyway between the church and an adjacent structure.

Time and weather

It is a relatively warm evening. You arrive on the scene at 0300.

Area and street conditions

Mueller Street is a one-way, two-lane street running south to north. Whitaker Avenue is a one-way street running west. Block Avenue is a one-way street running east. All streets in the immediate area are one-lane streets.

Fire conditions

Heavy fire is showing in the overfilled dumpster and is seriously exposing both exposures.

Exposures

Both the church and the dwelling face Mueller Street. The church building is 50' x 90' and about 70' high. The roof of the church has no skylights or other openings. The upper 20' area comprises the cockloft space. The ceiling is of wood lath and heavy ornamental plaster. There are large stained-glass windows facing the fire. These windows do not open. At the time of the fire, the church (Exposure B) is unoccupied.

Exposure D is a 4-story, multiple dwelling of ordinary construction. Its dimensions are 25' x 60', and it is about 15' from the church. The dwelling's exterior wall facing the church is unpierced on the 1st floor, but has windows on all other floors. The stairwell in the dwelling is open and has a well hole for its entire height. The stairwell terminates at the roof with a bulkhead door. There are two apartments on each floor, one on the west, and the other on the east side of the building. The dwelling is occupied. There is a fire escape at the rear terminating at the roof via a gooseneck ladder.

Water supply

All hydrants are on separate 8" mains.

Response

Your response is three engines and one ladder company. The engine companies are staffed by an officer and two firefighters. The ladder company is staffed by an officer and three firefighters.

MULTIPLE-CHOICE QUESTIONS

1. What are your orders to Engine 1?

    A. Establish a water supply at Hydrant #3. Position in front of the dwelling. Stretch a 2½" line to knock down the main body of the fire in the dumpster, then wye off into a 1¾" line. Stretch this line to the second floor of the dwelling to knock down any fire that may have extended into the south side of the apartment. Direct water out of the window and onto the dumpster.

    B. Position in front of the dumpster. Utilize the deck gun to knock down the dumpster fire and to wet down the church. Have the crew backstretch a 5" supply line to Hydrant #2. Then take a 1¾" line into the dwelling. Stretch a line into the well up to the top floor. Check the top floor apartment on the south side for fire extension.

    C. Establish a water supply at Hydrant #2. Position just past the dumpster. Stretch a 1¾" line into the dwelling to the 2nd floor. Direct this stream out of the window onto the dumpster and onto the church. Knock down any fire extending into the apartment.

    D. Establish a water supply at Hydrant #2. Position just past the dumpster. Utilize the deck gun onto the dumpster and the church to knock down the fire and to prevent any extension to the church exterior. Stretch a 1¾" line into the dwelling to the 2nd floor to darken any fire extending into the south side of the apartment. Direct water out of the window onto the dumpster to assist in knocking down the fire.

    E. Establish a water supply at Hydrant #3. Position just past the dumpster. Stretch a 2½" line into the church. Use this to sweep the fire in the dumpster from the church window. Extinguish any fire that is extending into the church.

2. What are your orders for Engine 2?

A. Secure a second water supply at Hydrant #3. Use the deck gun to assist Engine 1 in knocking down the dumpster fire and to protect the church. Stretch a 1¾" line into the dwelling to back up Engine 1 and to cover the 3rd floor for any fire extending into the apartment.

B. Park the engine out of the way of operations. Stretch a second 1¾" line between the dumpster and the dwelling off of Engine 1's wye. Alternate the stream between the fire and the exposures.

C. Secure a second water supply at Hydrant #3. Stretch a 2½" line into the church. Use this to sweep the fire in the dumpster from the church window. Extinguish any fire extending into the church.

D. Secure a second water supply at Hydrant #3. Stretch a 1¾" line into the dwelling to the 2nd floor to back up Engine 1's attack. Then stretch up to the 3rd floor. Extinguish any extending fire in the apartment. Then, direct the stream onto the dumpster and the church.

E. Secure a second water supply at Hydrant #2. Stretch a 2½" line into the dwelling to back-up Engine 1's attack. Then stretch to the roof of the dwelling. Utilize this line to protect the church, knock down any flying brands, and assist in knocking down the dumpster fire.

3. How would you direct the ladder company to best protect the exposed structures?

A. Open the scuttles on the church roof. Open the bulkhead door on the dwelling.

B. Close the windows on the upper floors of the south side of the dwelling; remove the curtains, draperies, etc. Open the windows on the north side of the dwelling. Open the bulkhead door.

C. Take a line off of Engine 1. Use a fog pattern above the dumpster to steer combustion products up and away from the exposures.

D. Open the bulkhead door. Place a PPV fan at the entrance to the building. Pressurize the stairwell, keeping the heat and products of combustion out of the stairwell.

E. Place two PPV fans, one between the building and the dumpster, and one between the dumpster and the church, forcing the heat and products of combustion away from the exposures.

4. How would you direct the ladder company in a search and rescue operation?

A. Split the company into two crews. Crew 1 proceeds to the 2nd floor. Search the 2nd floor apartment and the 3rd floor apartment. Crew 2 proceeds to the roof via the aerial, and will descend via the rear fire escape ladder. They will search the top floor apartments and work their way down.

B. Split the company into two crews. Crew 1 proceeds into the church using a lifeline to search for any victims. Crew 2 proceeds to the fire building to perform a systematic search of the 2nd and 3rd floors.

C. Keep the company intact. Search for and remove victims from the 2nd floor and then the 3rd floor. Continue on to the top floors as conditions warrant.

D. Split the company into two crews. Crew 1 ascends the front fire escape. Perform vent operations, then enter and search the 2nd floor. Crew 2 proceeds to the rear fire escape. Perform the same function on the 3rd floor.

E. The company commander proceeds to the 2nd floor to search with hose team. The chauffeur raises the aerial to the roof. Check the bulkhead area for victims; descend via the rear fire escape. Perform vent operations, then enter and search the top floor. Two firefighters search the 3rd and 4th floors.

5. Would additional alarms be necessary at this fire?

A. No, the fire has not yet entered either exposure. A quick knockdown of the dumpster fire will eliminate all problems.

B. Yes, due to the possibility of hazardous materials from the construction site being present in the dumpster, a second alarm should be requested and a call for the Haz Mat team initiated.

C. Yes, due to the enormous exposure fire potential, strike a second alarm.

D. No, not until reconnaissance reports are received from the first crews inside the dwelling as to the conditions on the inside of the dwelling.

E. Yes, due to the time and the potential life hazard in the dwelling, a second alarm should be struck for this incident.

6. You are the ladder captain and have entered the kitchen in an apartment on the top floor of the dwelling. You observe a small fire in a set of plastic drapes. What would be your best action in regard to this fire?

A. Call for a hoseline immediately. Conduct a quick search of the apartment. Close the door until the line arrives.

B. Call for a hoseline immediately. Use water from the sink to extinguish the fire. Quickly search the apartment.

C. Call for a hoseline immediately. Use water from the sink to extinguish the fire. Then remove the drapes and close the window. Quickly search the apartment.

D Call for one of the lines operating from the window on the lower floor to quickly knock down the fire while you search the apartment. Have an additional charged line brought up the fire escape as a precautionary measure.

E. Have one of the ladder company firefighters bring up an extinguisher. Do a quick search of the apartment, then close the door and wait for the extinguisher. Call for a line if conditions worsen.

7. What is your most immediate concern regarding exposures?

    A. Flying brands will cause additional fires throughout neighborhood.

    B. The wood-frame church will readily ignite, causing a conflagration possibility.

    C. The church, being unoccupied, will present a major forcible-entry problem, thus rendering it harder to protect.

    D. Lack of operable windows on the north side of the church will make it impossible to direct lines from the inside of the church onto the dumpster, exacerbating the ignition of the wooden exterior of the church.

    E. Due to the warm weather, many open windows on the south side of the dwelling will invite fire spread to the interior.

8. Your immediate superior has arrived on the scene. As you transfer command to your superior, what are the three most vital areas of information you can supply?

    1. Strategic mode
    2. Situation status
    3. Tactical needs
    4. Identification of sectors
    5. Exposure status profile
    6. Deployment and assignment

    A. 1, 3, 4

    B. 2, 3, 6

    C. 1, 2, 3

    D. 2, 3, 5

    E. 1, 4, 5

ANSWER SECTION

SCENARIO 2–1
MULTIPLE-CHOICE QUESTIONS

1.

A. **-1** To establish a water supply at Hydrant #3, the engine would have to arrive against traffic or pass a hydrant and then hand stretch to Hydrant #3. With very few exceptions (such as a specific SOP), this is almost never a good tactic as it may cause other companies to have positioning problems. The hoseline placement is also a poor choice. The life hazard is in the dwelling and by using these tactics it would take too much time to position a line inside. As it is a warm night, many windows will be open. It must be anticipated that fire will enter the dwelling and that establishing lines in this building as soon as possible must be a priority. Also, the position in front of the dwelling should be reserved for the ladder company in case the aerial is needed for either roof access or rescue.

B. **-1** Utilizing the deck gun in a blitz attack is also a poor choice in this situation. There are hydrants available in close enough proximity to the fire that back-stretching is not necessary. It is most likely that the attack engine will run out of water, causing a delay in advancing the line to the dwelling. In addition, the top floor should not be the first choice for initial line placement in the dwelling. Although convection currents will take heat to this area rapidly, it is more likely that a fire inside the structure will first ignite on a lower floor nearer the parent body of fire in the dumpster.

C. **+1** This is the proper hydrant selection. The line is stretched to protect the most severely threatened exposure while at the same time getting water onto the fire. If fire should enter the dwelling, the lines going into the apartments on the south side will keep the fire out of the stairwell, the main artery of occupant escape.

D. **+2** The best hydrant is selected and a water supply is established. Using the deck gun to knock down the main body of fire while also sweeping the exterior of the church handles two problems, the dumpster fire and the radiant heat problem at Exposure B. Advancing a hoseline into the dwelling to the most severely exposed apartment (the 2nd floor on the south side) protects the life hazard and evacuation, handles any extension into the dwelling, and directs another stream onto the main body of fire. This is the best utilization of both manpower and equipment.

E. **-1** This is not the best hydrant due to potential positioning problems. In addition, advancing the line into the church is not as critical as protecting the dwelling. A line will eventually need to be brought into the church to protect against fire extension, but it is not an immediate priority and certainly not the place for the initial attack line.

2.

A. **+2** These are all sound tactics based on good firefighting principles. A second water supply is established, a second master stream is placed into service to help in the extinguishment of the dumpster and protection of the church. In addition, a second attack line is stretched to the next most serious area of concern (the 3rd floor), as convection currents will drive heat up and away from the parent body of fire.

B. **-2** No second water supply is established and no back-up line is stretched into the dwelling. It must be anticipated that all the floors on the south side could become involved and at least two to three lines must be placed inside the dwelling as quickly as possible.

C. **0** Here, a second water supply is established, but bringing the second line into the church will be counter-productive. The largest life hazard is in the dwelling. The loss of life in the dwelling far outweighs the loss of the unoccupied church.

D. **+1** These are also excellent tactics, but the utilization of the additional master stream in answer choice "A" makes it the better choice at this fire. Doing more with your forces without compromising safety is the mark of the great strategist.

E. **0** Secondary water is secured and a back-up line is stretched. The fact that the back-up line is a 2½" is also a good tactic, but it would be better to go to the 3rd or even the 4th floor than to go to the roof. This tactic will take a lot of time and the line can be better utilized at a lower floor where extending fire can create a larger problem than the original fire in the dumpster.

3.

A. **-1** This is a question involving critical reading. The church has no openings at the roof level. It is therefore impossible to vent the church roof in this manner. Opening the bulkhead door of the dwelling to relieve heat is a good idea only after the windows have been closed. Opening the door with the windows open, especially with the open apartment doors, could create a draft that may pull fire into the dwelling.

B. **+2** These are the best tactics. Close all exposed windows first, then remove all combustibles from the area. Opening windows on the unexposed north side and opening the bulkhead door will release any heat that may have accumulated in the building.

C. **0** A water curtain will have no effect on radiant heat and convection currents. In addition, this is not a ladder company operation. There are better ways to protect exposures in this situation that do not involve ladder companies stretching lines (typically an engine duty).

D. **-2** This tactic will create a venturi effect in the stairwell. If the apartment doors and windows on the south side are open when this foolishness is attempted, the fire will likely be pulled into the building by the high pressure created in the stairwell by the fan.

E. **-2** Another foolish move. It is folly to attempt to control the movement of an exterior fire by using positive pressure fans. I won't even attempt to speculate on what effect this tactic might have on the fire and the exposures.

4.

    A. **+2** This is the best tactic, as the most severe life hazard is in the dwelling. Apartments are usually limited in size. By splitting the crew, the job gets done in half the time.

    B. **0** In the scenario, it is stated that there are no occupants in the church. Obviously, a search must be made there, but it is not a priority at this point in time. It should not, in this case, be done by the first ladder company. The dwelling must be searched and evacuated as quickly as possible

    C. **-1** While these are good search procedures concentrating on the dwelling (and rightfully so), it is far better and more efficient to split the crew. By doing a floor-by-floor search keeping the entire ladder crew intact, deteriorating conditions may prevent upper floor access.

    D. **-1** Splitting the crew is the right idea; however, ascending the fire escape is much more dangerous than the stairway and not warranted at this fire. Moreover, most occupants will seek to escape by the way they get in—via the stairway and the front door (the paths of least resistance). Therefore, any assistance that will be required will be discovered by entering through the front door and ascending the interior stairs.

    E. **0** Splitting up the crew in this manner is done by many departments. However, on a test, and according to OSHA, the NFPA, IFSTA, and just about any text, it is far safer to operate in teams of at least two. Sending the chauffeur to the roof alone violates this guideline and is dangerous.

5. In this question, you are tested on your judgment. Some testing authorities will give the candidate a set of criteria on which to base the request for additional alarms. However, even in those cases, whatever answer you choose, if it is backed up by reasonable judgment based on a sound forecasting of events, you will usually stay out of the negative point area. In the real world, it's a little simpler. If you are not sure about requesting an additional alarm, you should request it. You can always stage the companies as tactical reserve or send them back. Notice, however, that there is no "**+2**" answer here.

    A. **+1** This is an optimistic forecast that if you put the fire out, all of your problems go away. While this is true, a prudent incident commander may strike a second alarm just in case, so that he doesn't have to play catch-up if the situation deteriorates.

    B. **0** While a dumpster can be a witch's brew of hazardous items (including chemicals from the construction site), this is not a valid reason to request a second alarm or request the Haz Mat team. If the scenario had indicated the presence of a hazardous material at the scene, the additional alarm and special call would be warranted, but this is not the case. Don't read so far into the situation.

    C. **+1** The potential for extension to exposures in this case is a real consideration. The dumpster is in close proximity to the exposures. The church is of wood-frame construction, which will not withstand ignition for a long period of time, so it must be protected. The dwelling is of ordinary construction that has non-combustible walls, but due to the fact that it is a warm night, the windows are probably open, creating a ready avenue for fire spread into the building.

D. **-1** Inaction and/or delaying a decision are never good leadership qualities. By the time the interior reports transmit to incident command that there is indeed fire in the dwelling, the reflex time for the second alarm companies may be too long to overcome the problems at hand. Again, if there is even a chance that additional manpower and equipment may be needed, request the second alarm.

E. **+1** The time of day is indeed a factor that impacts severely on the life hazard. Recognizing this, the rational incident commander orders a second alarm.

6. This is a situational question that tests your ability to make a decision given a set of conditions. This question also tests your functional fixity quotient, your ability to adapt to the situation. If you cannot make the situation better using makeshift aids, you are a victim of functional fixity.

A. **0** Here, nothing was done in regard to the fire except to call for a line. This is the top floor and the stretch will take time. All big fires start small, and by the time the line is advanced to the apartment, this fire may have substantially grown in size. The job of the company officer is to make things better and to be a problem-solver.

B. **+1** These are good tactics. Calling for the line is required in case the fire is not extinguished by the water from the sink.

C. **+2** This answer choice has all the good elements that were in answer choice "B," but they take it a step further. Prevention of a fire is always superior to suppression. By removing the drapes and closing the window, steps are being taken to remove available fuel and isolate the apartment. These tactics are excellent and thorough.

D. **-1** These tactics lack common sense. The exterior line is likely to push the fire into the apartment, causing it to spread. In addition, stretching up the fire escape is dangerous, manpower-intensive, and not necessary.

E. **-2** Calling for an extinguisher from the top floor is a poor choice. By the time it gets there, the fire will most likely be beyond its capabilities to control it. It might be a good idea to assign an extinguisher to that member working above a fire in case a situation like this arises. Also, calling for an extinguisher before calling for a line will guarantee that by the time the line gets there, the fire will also be beyond its capability. Don't take chances like this. Fires will not wait for lines to be stretched. Call for the line and in the meantime, do whatever you can to make the situation better.

7. The question here asks for the most immediate concern, which is the key to the best answer.

A. **+1** Flying brands are a definite concern, with the probability of many open windows in the area. Brand patrols may be needed to check on this, but is not the most immediate concern. Considering it got you a point, but not two.

B. **+1** The ignition of the wood-frame church is definitely a concern, for its ignition could lead to a conflagration, creating an area source out of a point source, the dumpster. While it must be protected from ignition, in this case by the deck guns, it should not be the most immediate concern.

C. **0** The first area to protect regarding the church will be the combustible exterior. Then, as soon as manpower is sufficient to address the interior of the church, it must be entered and checked for extension. Sufficient streams from the deck guns should minimize and possibly eliminate fire spread to the interior of the church.

D. *-1* The scenario states that the church windows do not open. Breaking the stained glass windows for stream access is unnecessary secondary damage. Therefore, lines from inside the church will not do as much good as the streams from the deck guns washing the facing of the exterior walls, thus minimizing the effects of both the radiant and convection heat.

E. *+2* The most significant exposure is the dwelling due to the time of day and the life hazard present. The location and extent of the dumpster fire in relation to the dwelling causes most of the problems at this incident. Fire spread into this building, especially in the initial stages of operation, will cause huge problems for incident command. Therefore, it must become the most immediate concern of the incident commander.

8. These types of questions are tough as they are filled with a substantial amount of information and sometimes hard to sort out. The easiest way to handle this is to attempt to place the answers in a chronological sequence, then find out what is absolutely essential to the answer, and then find it in the answer choices. This will help guide the way to the best point-getters. Remember the question asks for the most vital information for the oncoming incident commander. If you put yourself in his shoes, you should be able to determine, with pretty good accuracy, the information he will need to assume command of this operation.

A. *0* Strategic mode is usually not one of the pieces of information you need to give the new incident commander. Just by looking at the operation, he should be able to tell the strategic mode. If companies are outside, using exterior streams, then the mode is defensive. If lines are inside, and the sound of power saws can be heard, the mode is most likely offensive. Identification of sectors should be part of the situation status report, not included here.

B. *+2* This is the most vital information. Situation status will tell what is going on, tactical needs will tell him what resources are still required to mitigate the incident, and deployment and assignment will furnish the information about what has and is being done at this time and where. With this information, a smooth transition of command should be accomplished.

C. *+1* Again, the incoming incident commander does not usually need to be told what strategic mode the operation is in, but what is going on (situation status) and what you need (tactical needs) are good starting points for the new incident commander.

D. *+1* This is a good starting point for the incoming incident commander, but not as good as answer choice "B." Exposure status should be part of the situation status report. Deployment and assignment of personnel is critical not only for accountability, but for evaluation purposes to ascertain if current operations are effective. If the incident commander does not know who is doing what and where they're doing it, then he does not really have control over the operation.

E. *0* We've already talked about strategic mode. Both sector identification and exposure status should be part of the situation status report. The report will include these items as well as the current fire situation (escalating, stabilizing, etc.). In addition, there is no mention of tactical needs. This information will help the new incident commander determine what has been done or still needs to be done.

Passing Point Score for Scenario 2–1 = 11

SUMMARY

This was essentially a defensive-offensive operation, as the prime consideration here was the dwelling, then the church, and finally the parent body of fire. Although some of this activity occurred simultaneously, make no mistake that if only one line was available it would have been utilized to protect the dwelling and the evacuation. This strategic decision was a result of the incident commander's understanding of fire spread via heat transfer. The dwelling was subject to exposure by both radiant heat waves on the lower floors and convection currents on the lower and upper floors. Fire extension is always caused by the transfer of heat. It is this understanding that causes the incident commander to initially place his lines to head off this spread, thereby confining the fire, and finally extinguishing it.

CHAPTER FOUR

SCENARIOS

SCENARIO 4–1
OFFENSIVE RESIDENTIAL FIRE

Construction

You have responded to a fire at 14 Hetzel Street; the building is a four-bedroom cape house of wood-frame construction. The 1st floor consists of a kitchen, a bathroom, a bedroom, a living room, and a dining room. The front door is a heavy steel entrance door and the windows are of thermopane construction. There is also a wooden door side-entrance that leads into the kitchen. The 2nd floor contains three bedrooms and a bathroom. The basement, constructed of block, has a den and living space. All utilities, including the HVAC unit, are located in the basement. The basement is windowless and has no openings to the exterior.

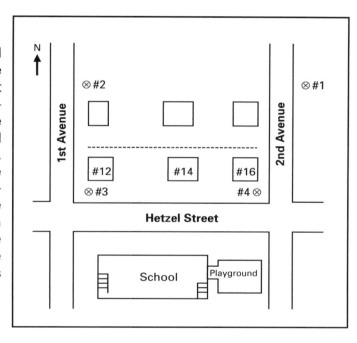

Time and weather

Central dispatch receives this alarm at 1227. The wind is 5mph out of the west. You arrive at 1230. It is 58°F.

Area and street conditions

Hetzel Street is a two-lane, two-way street with unrestricted parking.

Fire conditions

Upon arrival, you see fire through a 1st floor window of the fire building. Light gray smoke is issuing from the sides and top of the front door. You notice a car parked in the driveway.

Exposures

There is a 125' area separating 12 Hetzel Street from the fire building and a 75' area separating 16 Hetzel Street from the fire building. All buildings on Hetzel Street have rear yards that are 50' deep. Across the street there is a three-story brick school, housing elementary grades 1 through 6. Hetzel Street is 30' wide. The school is built of fire-resistive construction and has a sprinkler connection located at the rear of the building. There are enclosed stairwells on the east and west ends of the school building. On the east side of the school is a side entrance that leads out into a playground.

Water supply

A hydrant is located on the corner of Hetzel Street and Second Avenue. There are also hydrants located one block north on each corner. The hydrants are evenly spaced and each has on its own main.

Response

Your response is three engines and a ladder company. An officer and two firefighters staff each engine. An officer and three firefighters staff the ladder company.

MULTIPLE-CHOICE QUESTIONS

1. What are your orders for Engine 1?

    A. Establish a water supply at Hydrant #3 and proceed to #3 Hetzel. Stretch a preconnected 1¾" line through front steel door to locate, confine, and extinguish the fire in coordination with the ladder company engaged in the primary search.

    B. Establish a water supply at Hydrant #4 and proceed to #3 Hetzel. Stretch a preconnected 1¾" line through the wooden side door to locate, confine, and extinguish the fire in coordination with the ladder company engaged in the primary search.

    C. Stage Engine 1 in front of #1 Hetzel and stretch a preconnected 1¾ line through the wooden side door utilizing tank water. Radio Engine 2 to backstretch a 5" LDH from Hydrant #4 to provide the primary water supply to Engine #1. Locate, confine, and extinguish the fire in coordination with the ladder company engaged in the primary search.

    D. Establish a water supply at Hydrant #3, proceed to #3 Hetzel, and stretch a preconnected 1¾" line to flank the front steel door. Allow the ladder company to vertically ventilate the roof to alleviate the backdraft condition and then move in to locate, confine, and extinguish the fire. Coordinate this operation with the ladder company.

    E. Establish a water supply at Hydrant #4, proceed to #3 Hetzel, and stretch a preconnected 1¾" line to the front of #3 Hetzel. Operate this line through the 1st floor window to knock down any visible fire. Then proceed to the wooden side door to finish extinguishment in coordination with the ladder company engaged in the primary search.

2. What are your orders for Engine 2?

   A. Establish a secondary water supply. Proceed to the rear of the school. Supply the standpipe at the rear of the school. Take a standpipe kit and nozzles; connect to the standpipe on the 1st floor and use this line as needed to protect the school. Assist in the evacuation of the school

   B. Park Engine 2 out of the way. Proceed to the school and assist in the evacuation of the students. Utilize the deck gun to protect the front of the school by coating it with water. Ensure that all windows on the fire side of the school are closed. Utilize extinguishers in the school to extinguish any spot fires. Report conditions to the incident commander.

   C. Establish a secondary water supply. Stretch a 2½" preconnected line through the steel front door to back-up Engine 1's fire attack in coordination with the ladder company engaged in the primary search.

   D. Establish a secondary water supply. Stretch a 1¾" preconnected line through the wooden side door to back-up Engine 1's fire attack or to proceed to the floor above to prevent any fire extension. All operations are in coordination with the ladder company engaged in the primary search.

   E. Establish a secondary water supply. Stretch a 2½" preconnected line to flank the steel front door alongside Engine 1. Allow the ladder company to vertically ventilate the roof to alleviate the backdraft condition, then enter the building and back-up Engine 1's attack line in coordination with ladder company operations.

3. What are your orders for Ladder 1?

   A. Split the company into two crews. Crew 1 proceeds to the roof via the aerial for vertical ventilation to alleviate the backdraft condition. Crew 2 stands by to force entry at the steel front doors, and then performs a primary search in coordination with the engine company's fire attack.

   B. Split the company into two crews. Crew 1 forces entry on the wooden side door, performs a primary search of the fire floor and the floor above the fire, and removes any victims. Crew 1 also checks for fire extension and coordinates their operations with the engine company. Crew 2 performs horizontal ventilation at doors and windows of fire floor. Raise ground ladder to rear bedroom for secondary egress.

   C. Split the company into two crews. Crew 1 proceeds to the front of #3 Hetzel and forces entry on the steel front door. Crew 1 then performs a primary search of the fire floor, removes any victims, and checks for fire extension in coordination with engine company operations. Crew 2 forces entry at wooden side door, performs a primary search of the floor above the fire, and checks for fire extension. Crew 2 is to operate in coordination with the engine company.

   D. Split the company into two crews. Crew 1 forces entry on the wooden side door and performs a primary search of the fire floor and the floor above the fire. They will then remove any victims and check for fire extension in coordination with engine company operations. Crew 2 operates in the school to initiate the evacuation of the students, using fire extinguishers to knock down any spot fires. Report conditions to the command post.

   E. Send the ladder company to the school to perform the evacuation of students and work with the second engine company that is coordinating the standpipe attack/search and rescue operation. Designate the ladder company officer as the liaison with the school officials.

4. It is lunchtime and many children will be out in the street, the playground, and potentially in the area of fire department operations. How do you address this problem?

   A. Take no action; the children are not the problem, the fire is.

   B. Allow the school officials to handle the problem. This is not the fire department's problem.

   C. Assign the safety officer to liaison with the school officials regarding the safety of the children.

   D. Establish a fireground perimeter and liaison with the police to handle the problem of crowd control.

   E. Establish a fireground perimeter and assign a company to operate as "Crowd Control Group."

5. Where would you stretch a third line?

   A. To the fire department connection on the school in the event the radiant heat affects the school.

   B. To the 1st floor via the front steel door after forcible entry has been accomplished.

   C. To the floor above the fire via the wooden side door and the interior stairs.

   D. To the floor above the fire via a ground ladder.

   E. To Exposure D to check any fire extension via radiant heat.

6. If this fire were located in the basement, what would be the most effective means of ventilating the fire?

   A. Knock out the basement windows opposite the fire attack.

   B. Cut a large hole in the floor near the front windows.

   C. Breach the wall to create an opening to the exterior.

   D. Do not ventilate, evacuate the building, and flow water into the floor.

   E. Use a stream set to a wide fog pattern to create a pressure differential in front of the advancing line thereby driving the heat and smoke ahead of the attack team.

7. You are ordered to conduct post-control overhaul. The fire was confined to the living room. Where would be the first place you would make your openings to check for fire extension?

   A. The ceiling and light fixture directly above the area of most damage.

   B. Around pipe chases in the bathroom.

   C. At manmade openings in the walls, such as electrical outlets.

   D. At the highest place the HVAC duct terminates.

   E. In the attic.

8. You are in charge of salvage. The fire had extended to the 2nd floor, but was extinguished before extensive damage was incurred. You have, however, found that water has pooled in a room on the 2nd floor. What would be the most efficient way of removing this water? There is no electrical service to the structure at this time.

    A. Use apparatus current and run an extension cord into the structure. Operate a wet/dry vacuum to suck up the water; drain it into the bathroom sink.

    B. Build a catchall on the 1st floor. Punch a hole in the 2nd floor to drain the water into the catchall. Bring a generator into the 1st floor. Operate a wet/dry vacuum from this power source to suck up the water. Drain water outside the structure.

    C. Build a water chute on the 1st floor utilizing an A-frame ladder to support the chute. Place the end of the chute out a window. Punch a hole in the 2nd floor above the chute. Let water run through hole, down the chute, and out the window.

    D. Use salvage covers to construct a channel on the floor leading to the 2nd floor bathroom. Remove the toilet off its mountings. Use a squeegee to push the water into the toilet waste drain.

    E. Use mops and pails to soak up the water and then drain the pails into the toilet. Open the windows to accelerate the drying of the floor.

Scenario 4–2
Offensive/Defensive Tenement Fire

Construction

The fire building is a three-story wood-frame tenement. In all the buildings, an open stairwell runs from the 1st floor to the roof, terminating at a scuttle hatch. There are two apartments on each floor. There is a fire escape at the rear of the building.

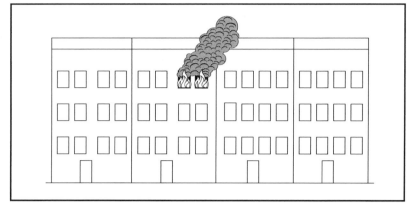

Time and weather

There is a 15mph wind blowing out of the east.

Area and street conditions

Not a significant factor at this fire.

Fire conditions

Fire has control of the west apartment on the 3rd floor and is showing out the windows on Side A. There is also smoke issuing from around the cornice at the front of the building.

Exposures

The fire building is part of a row of identical three-story wood-frame tenements. The cockloft is open over all of the buildings.

Water supply

The water supply is adequate and hydrants are appropriately spaced.

Response

Four engines and two ladder companies respond. An officer and two firefighters staff each engine company. An officer and three firefighters staff each ladder company.

SHORT-ANSWER QUESTIONS

1. What would be the orders for Engine 1?

    _____

    _____

    _____

    _____

    _____

    _____

2. What would be your orders for Engine 2?

    _____

    _____

    _____

    _____

    _____

    _____

3. What orders would you give the ladder companies?

    _____

    _____

    _____

    _____

    _____

    _____

4. Explain what would be the best way to ventilate the roof of the fire building?

_____

_____

_____

_____

_____

_____

5. What considerations should be taken into account when pulling the ceilings in the fire building during pre-control overhaul?

_____

_____

_____

_____

_____

_____

6. What would you order Engine 3 to do? What orders would you give Engine 4?

_____

_____

_____

_____

_____

_____

7. The compliment of apparatus and manpower is insufficient for this fire. Why?

_____

_____

_____

_____

_____

_____

8. Where would you assign additional ladder personnel?

_____

_____

_____

_____

_____

_____

9. Where would you assign additional engine personnel?

_____

_____

_____

_____

_____

_____

SCENARIO 4–3
DEFENSIVE/OFFENSIVE SLICK NICK'S CLEANERS FIRE

Construction

You arrive to find a fire in progress at Slick Nick's Cleaners, a one-story, wood-frame building, built at the turn of the century.

Time and weather

It is July 12, the temperature is 80°F. The time is 0400. There is a slight breeze out of the north.

Area and street conditions

The area is occupied by many turn of the century buildings. Most are attached. Those that are unattached are built closely together. Streets are narrow and double-parking is a constant problem.

Fire conditions

Slick Nick's is unoccupied and has been closed since 1700 hours the day before. Heavy fire is showing at a window on the south side.

Exposures

Severely exposed is a three-story wood-frame tenement that is occupied by six families. The building, which is separated from the fire building by a 5' wide alley, is in the process of evacuation. Windows in the alley face the cleaners on the 2nd and 3rd floors. As you look down the alley, you can see the curtains of one of the exposed apartments aflame.

Water supply

The water supply is provided by looped 8" mains that are both reliable and adequate.

Response

The first engine company to arrive on the scene stretches a line into the cleaners to locate, confine, and extinguish the fire.

SHORT-ANSWER QUESTIONS

1. Comment on the action of the first-arriving engine company.

_____

_____

_____

_____

_____

_____

2. Where should additional lines be placed?

_____

_____

_____

_____

_____

_____

3. What actions will be required of the ladder company?

_____

_____

_____

_____

_____

_____

4. State the concept that guides this action.

_____

_____

_____

_____

_____

_____

Scenario 4–4
Defensive Townhouse Fire

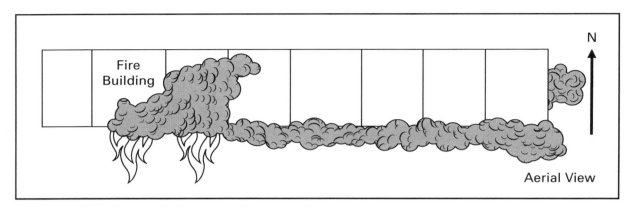

Fire Building

N

Aerial View

Construction

From your familiarization visits, you are aware of the building's construction. The walls are 2" x 4" wood, with the floors being constructed of laminated I-beams. The steeply pitched roof is constructed of lightweight wood trusses, which are held together by sheet metal surface fasteners.

Time and weather

The time is 0300 and the wind is blowing at 15mph from the west.

Area and street conditions

There is virtually no traffic in the area at this time. The streets, however, are lined with parked cars, some hanging off the corners. No hydrants appear to be blocked.

Fire conditions

Heavy fire exists on the 2nd floor of a three-story, wood-frame condominium. There is smoke showing from the top floor and the roof eaves from the fire unit to the east end.

Exposures

This condo unit is part of a complex of attached units that occupy a good deal of a city block. These condos are still in the construction phase and are not yet occupied. However, they are completely built from a structural standpoint. Installation of utilities such as water, gas, and electrical services are in the process of being installed in each dwelling unit. The roof area is open over the entire length of the row; however, there is fire-rated sheetrock between each apartment which extends to the top floor ceilings.

Water supply

The water supply is adequate.

Response

Initial response is four engines and two ladder companies. The engines are staffed by an officer and two firefighters. The ladders are staffed by an officer and three firefighters.

SHORT-ANSWER QUESTIONS

1. What would guide the actions of the incident commander here?

_____

_____

_____

_____

_____

_____

2. Where should master streams be positioned?

_____

_____

_____

_____

_____

_____

SCENARIO 4–5
NO ATTACK MODE—GAS INCIDENT

Construction

The White Rose Apartment Complex is a set of 10-story, fire-resistive high-rises. The complex consists of four identical buildings built around a quad. In the center of the quad, there is a small park-like property. Each building has an elevator, with enclosed stairwells located on the two ends of the building, both terminating at the roof.

All of the apartments have electric smoke detectors. In addition, each apartment has a balcony in the living room.

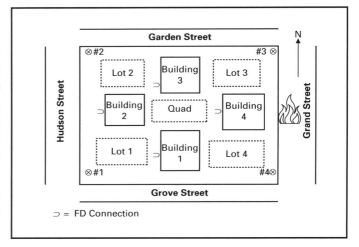

Time and weather

The time is 2335; the temperature is 40°F, with a breeze of 10mph out of the east. It is raining lightly.

Area and street conditions

All streets in the area are two-way, two-lane streets.

Fire conditions

You have been dispatched to a reported heavy odor of gas in the vicinity of the White Rose Apartment Complex. As you are responding, you are given further information by the dispatch center. An explosion has now been reported with a "fire in the street."

As you arrive, you are faced with a seven-story-high column of fire coming from a broken natural gas main on Grand Street. The sound in the area is deafening. The broken main and flame column is about 30' from the easternmost building in the complex. There are many residents on their balconies, in the lobbies, and in the area.

Exposures

There are parking lots adjacent to all of the buildings in the complex.

Water supply

There is a wet combination standpipe and sprinkler system in each of the buildings. Hydrants are plentiful and all are on their own main.

Response

Response is three engine companies and a ladder company. Each is staffed with an officer and three firefighters.

MULTIPLE-CHOICE QUESTIONS

1. What is the first priority of this incident?
   A. Notifying the utility company to respond.
   B. Shutting the gas off at the street valve.
   C. Protecting exposures.
   D. Evacuating the most seriously exposed structure.
   E. Establishing a command post and initiating an incident management system.

2. What is the most important size-up factor at this incident?
   A. Exposures
   B. Weather
   C. Location/extent
   D. Distance
   E. Apparatus/manpower

3. What are your orders for the crew of Engine Company 1?
   A. Establish a water supply at Hydrant #4. Position the apparatus in Parking Lot #4. Stretch two lengths of 2½" hose to the fire department connection and charge the system. Use the deck gun to put out the fire and protect Building #4. Stretch a 2½" handline and keep the exposed side of Building #4 wet.
   B. Establish a water supply at Hydrant #2. Position the apparatus in Parking Lot #3. Use the deck gun to keep the exposed side of Building #4 wet. Stretch two lengths of 2½" hose to the fire department connection of Building #4. Charge the system.
   C. Establish a water supply at Hydrant #3. Position on the flank of Building #4 near Parking Lot #3. Extinguish the fire using the deck gun and a 2½" line. Also stretch two lengths of 2½" line to the fire department connection of Building #4. Do not charge the system until ordered.

D. Establish a water supply at Hydrant #1. Position the apparatus in front of Building #1. Stretch two lengths of 2½" hose to the fire department connection of Building #4 and charge the system. Set up a deluge gun in Parking Lot #4 to keep the exposed side of Building #4 wet. Leave the deluge gun unmanned. Bring a high-rise pack into Building #4 to check for fire extension.

E. Establish a water supply at Hydrant #2. Position the apparatus in front of Building #3. Bring a high-rise pack into Building #4. Stretch a line to the exposed interior and extinguish any spot fires. Use the deck gun to shield the advancing companies.

4. What are your orders for Engine 2?

A. Stretch a second high-rise pack off of the same standpipe that Engine 1 utilized. Use a line out of a window onto the fire to push it away from the exposed building.

B. Set up a second unmanned deluge gun in Parking Lot #3. Stretch an additional 2½" handline to the front of Building #4 to assist in extinguishing the fire.

C. Establish a second water supply. Set up an additional unmanned deluge gun in Parking Lot #3. Bring an additional high-rise pack into Building #4. Assist Engine Company #1 in the extinguishment of spot fires. Protect the evacuation.

D. Establish a second water supply. Engine Company #2 is to enter Building #4 to assist in the evacuation of the building. Use house standpipe lines where necessary to protect the evacuation.

E. Establish a second water supply. Position the apparatus on the flank of Building #4 near Parking Lot #4. Assist Engine 1 in the extinguishment of the fire using the deck gun and a 2½" line.

5. What are your orders for Ladder 1 regarding ventilation?

A. Close all windows in the building to protect against radiant heat. Remove combustibles from the windows on the exposed side of the building.

B. Open the windows in the building on the exposed side. Send crews to the roof via the stairwells to open the bulkhead doors to establish airflow in the building to keep it cool.

C. Set up a PPV fan at the rear door of Building #4. Pressurize the building to keep heat out of the building.

D. Send crews to the roof via stairwells to open the bulkhead doors to vertically relieve any heat. Remove combustibles from the windows on the exposed sides. Ensure that all exposed windows are closed. Open the windows on the unexposed side.

E. Do not ventilate. The interior will be untenable due to the radiant heat from the fire. Evacuate the occupants rapidly and leave the building.

6. Are additional alarms required at this incident?

A. No, too many companies will congest the scene and add to the confusion that already exists.

B. Yes, a confirmed Haz Mat incident requires additional alarms be requested.

C. Wait until the arrival of the utility company. Liaison with the supervisor to assist in guiding your decision.

D. No, however, a special call for the Haz Mat unit must be made to the dispatch center.

E. Yes, an additional alarm should be sounded. Request a ladder company to assist in the evacuation of Building #4.

7. What would you do in regard to the occupants of Building #4?

A. Keep the windows closed. Keep the occupants away from the windows. They should be safe in their apartments.

B. Do not evacuate. Direct all occupants to the lobby to wait out the incident.

C. Move all occupants from the apartments on the exposed side to those on the unexposed side. Stuff towels under the doors and allow them to wait out the incident in the unexposed apartments.

D. Evacuate only occupants from the 7th to 10th floors. The radiant heat as well as the convection heat will demand this action. Occupants on the lower floors should be okay, as the radiant heat will not be so bad.

E. Evacuate all occupants from the building as this is a major emergency and the building is dangerously exposed to the incident.

8. What steps would you take in regard to gas-fed fire?

A. Attempt to shut the gas off from the street valve. Allow the fire to burn out.

B. Continue in a non-intervention mode. Allow the fire to burn while protecting exposures. Await the arrival of the utility company and let them shut off the gas.

C. Send in the Haz Mat team in proximity suits to shut off the gas line. Keep them cool with deck gun streams.

D. Apply foam to extinguish the fire and keep vapors suppressed with a foam blanket. Send in the Haz Mat team to shut off the gas flow.

E. Allow the utility company to cut off gas flow by isolating the damaged section using remote control valves. Continue to protect the exposures until the fire burns out. Check nearby buildings for accumulations of gas.

9. What would be your orders for the crew of Ladder 2?

A. Assist in the evacuation of Building #4. Remove combustibles from the exposed side windows. Vent the unexposed side.

B. Set up a ladder pipe in Parking Lot #3 or #4. Use the stream to keep the exposed side of Building #4 cool. The rest of the crew will assist in the evacuation of Building #4.

C. Check Building #3 and Building #1 for accumulations of gas. Take any actions necessary to protect the occupants.

D. Assist in moving the occupants from the exposed side of Building #4 to the unexposed side. Remove any combustibles from the windows of the exposed side.

E. Set up a ladder pipe in the quad area. Lob the stream over the roof of Building #4 to create a water curtain between the building and the fire. Also check Building #1 and Building #3 for gas accumulations. Evacuate occupants as required.

10. Considering natural gas leaks in general, which is the most dangerous type leak that the fire service will encounter?

A. Exterior leak with fire.

B. Interior leaks with fire.

C. Exterior leaks without fire.

D. Interior leaks without fire.

E. Given the right circumstances, all leaks are equally dangerous.

SCENARIO 4–6
INDIRECT METHOD OF ATTACK—MULTIPLE DWELLING APARTMENT FIRE

Construction

You have assumed command of a fire incident at a six-story multiple dwelling of ordinary construction. There are three fire escapes on the building, one at the rear and one on each side. Each floor has six apartments.

Time and weather

There is no wind to speak of. The time is now.

Area and street conditions

Street conditions and area are not significant factors at this incident.

Fire conditions

There is a slight smoke condition showing from a window on the 3rd floor, but the window is closed and stained black. The report from the interior is that the companies are outside the apartment, which has a steel door. The steel door is extremely hot to the touch. Pressurized smoke is issuing in puffs from under the tightly closed door. The report from above is that there are also pressurized puffs of smoke issuing around the baseboard moldings.

Exposures

Apartments on both the fire floor and directly above are threatened, but it does not appear that the fire has extended beyond the apartment of origin.

Water supply

Engine companies have established water supplies at the most appropriate hydrants and stretched lines up to the fire apartment.

Response

Ladder companies are located on the fire floor, the floor above, and on the roof.

SHORT-ANSWER QUESTIONS

1. What conditions can you predict in the fire apartment?

   _____

   _____

   _____

   _____

   _____

   _____

2. How should this fire apartment be ventilated?

   _____

   _____

   _____

   _____

   _____

   _____

3. What are your orders to the men on the roof and the floor above the fire?

   _____

   _____

   _____

   _____

   _____

   _____

4. What are your orders to the interior crews on the fire floor?

_____

_____

_____

_____

_____

_____

5. What other tactics can be employed at this type of fire to mitigate the backdraft condition?

_____

_____

_____

_____

_____

_____

ANSWER SECTION

Scenario 4–1
Multiple-Choice Questions

1. In this scenario, with a two-way street, the choice of hydrant (either #3 or #4) was not a factor. The differentiating issue was the tactics chosen in conjunction with the establishment of water supply.

   A. **+1** The direction of attack is proper as are the tactics. However, choosing the steel front door over the wooden side door violates the Point of Entry Rule of Thumb. The wooden door is the safest, most effective path of least resistance into the structure. The steel door will require forcible entry, causing a delay in both the attack and the primary search. Reconnaissance by the ladder company or engine company officer must find the easiest way into the structure

   B. **+2** In contrast to the steel front door, the line is stretched utilizing the Point of Entry Rule of Thumb. The line is placed between the fire and any victims, protecting the vertical artery—the stairs. This placement will not only attack from the unburned side, but will protect search crews operating on upper floors. Everything will go more smoothly once entry has been gained, and the search operation along with suppression operations are underway. In an occupied dwelling, the faster this is accomplished, the better things will turn out.

   C. **-1** No initial water supply is established here. Backstretching a supply line will take time, which can be better utilized elsewhere. In addition, placing the engine directly in front of the building will block access for the ladder company.

   D. **-1** The fact that fire can be seen through the 1st floor window indicates that a backdraft condition is not present here. If you took the light gray smoke issuing from the front door as a backdraft indicator, ignoring other information, you were a victim of tunnel vision, a very dangerous condition on the fireground.

   E. **-2** When there is a life hazard involved, it is absolutely unacceptable to operate an exterior line into an occupied space. This will write off any victims due to the expansion of steam and the fact that you probably pushed the fire into the rest of the house. The moral of this story is: Get in there!

2.

   A. **-1** In all cases, the attack must be reinforced. A back-up line is mandatory at all offensive fire operations. If there is a problem with the initial attack and there is no line to back it up, the entire operation will be in jeopardy. In regard to the initial arriving companies, the school is a distraction and is not a major priority here.

   B. **-2** If you picked this answer, you were a victim of severe tunnel vision. The presence of the school forced you to abandon the fire building and the main problem. No back-up line was stretched (see "A" above) and no secondary water supply was established, which are both critical considerations at all offensive fires, especially those on a test.

C. **+1** A back-up line is stretched and a second water supply is established. The 2½" line is a nice touch—probably not needed at this incident, but it's better than no line at all. The problem is, again, the steel front door is not the best way into the structure.

D. **+2** All the good things from answer choice "C" above, with the added bonus of the most effective path of least resistance. Also, note that not only does the line reinforce the attack, but it can also be directed to the floor above if required—additional good tactics. If you chose the right path in Question #1, you were likely to do it again in Question #2. Good for you. You are 4 points up, instead of just 2.

E. **-1** Unfortunately, if you thought this was a backdraft in Question #1, you are 2 points in the hole. More bad news is yet to come

3. Ladder company operations require many tasks to be completed. These must be coordinated with the attack operation and with each other. In this and most other small structures, it is best to split the crew into two teams of two. One crew works on the interior, while a second crew works initially on the exterior, then enters after exterior ventilation opposite the attack is complete.

A. **-1** Backdraft tactics again. You're three points down now due to deviation amplification.

B. **+2** The company is split. One crew is inside for forcible entry, primary search, fire extension checks, etc. The other crew operates on the exterior, venting for fire after the line is in place for the attack. Vent, enter, and search (VES) operations are conducted on the 2nd floor. There is plenty to do, for sure, but it should be no problem for a well-trained and disciplined ladder crew working from a solid scene-assignment SOP.

C. **0** Two crews doing the same thing at different locations. The problem is that there is no horizontal ventilation mentioned. Ventilation is as mandatory as fire attack; in fact, these two fireground operations cannot survive without one another. Most likely, neither will any victims, for that matter. Don't waste time on the steel door.

D. **0** Crew 1 does the right thing. Unfortunately Crew 2 has other ideas.

E. **-2** This is too much manpower committed to the school. According to this answer, more than half of the resources are not in or around the fire building.

4.

A. **-1** The fire department is responsible for everyone and everything in and around the fire-ground. Failure to realize this is asking for trouble.

B. **0** Again, the fire department is responsible for the area. Using the school officials to assist in handling the problem is a good idea. Ignoring them is quite another.

C. **0** If the safety officer is tending to matters at the school, who is monitoring the fire-ground for potential and present hazards? While the safety officer must be involved in the decision on how to handle this problem, they should not have to focus on it.

D. *+2* Delegation and liaison are the tools by which the incident commander accomplishes objectives on the fireground. Who better than the police department to handle this type of problem? They will more than likely have a working relationship with school officials already. Failure to delegate is an undesirable incident command trait.

E. *+1* Using a company to handle crowd control duties may not be the most efficient use of manpower, but, if there are more children than can be handled by the police and school officials (especially during the initial portion of the incident), it may be justifiable.

5.

A. *-1* The school is more than 30' away and not on the leeward side of the fire. There is little chance of the fire threatening this structure to the point where sprinkler activation would occur. If the fire escalated, supplying the standpipe would be more of a priority, but not where the third line is concerned.

B. *+1* This would be a good place for an additional back-up line, especially if the second line was stretched to the 2nd floor. Taking an alternate means of entry is mandatory as there are already two lines stretched through the wooden side door. There should never be more than two lines stretched via the same access point. However, if this line had to also go upstairs to cut off extension to the 2nd floor, this is too much hose (three lines) in any one area (the 1st floor). This will lead to entanglement and loss of line mobility.

C. *-1* This is a poor choice due to the reasons stated in the last answer. More than two lines stretched via the same access point are counter-productive and dangerous.

D. *+2* This is the best choice. Even if the second line were stretched to the 2nd floor, this line, which was stretched up a ground ladder, would not impinge on the mobility of the line already in the stairway. Again, the safest path of least resistance to accomplish the objective is what is needed.

E. *0* The downwind exposure, Exposure D is 75' away from the fire building. This is, at this time of operation, more than sufficient distance to allow the radiant heat to dissipate.

6. This question presents a problem unrelated to the fire presented in the scenario. By placing the fire in a different area, a mini-scenario is created. Answers should be based on appropriate facts presented in the scenario in relation to this new problem.

A. *-2* This is an example of critical reading. The scenario states that the basement is windowless and has no openings to the exterior. While these tactics may be the best way to operate in a basement fire where windows are present, it is a loser here. Read the scenario more carefully.

B. *+2* Given the building construction, this is the only acceptable tactic in this situation. If the basement is not vented, the heat conditions in the basement will be unbearable and the attack crews will be forced to withdraw

C. *0* This tactic will take an extraordinary amount of time and will not produce a hole of sufficient size. Again, the safest, most effective path of least resistance will be the wooden floor. Get it cut quickly and your problems will diminish, as the attack lines will be able to advance.

D. *-1* This is a last ditch effort after other efforts, such as distributors, have failed. It is a concession of the building, and at this point of the operation, is not acceptable.

E. *-2* This tactic, without sufficient ventilation, will drive the steam created by the fog nozzle right back at the attack crew, causing steam burns and chasing the crew out of the basement. A fog nozzle operated by an interior crew in any unventilated space should be strictly forbidden. The only use for a fog stream in an unventilated space is when the indirect method of attack is being deployed, and it should be applied from the exterior of the building. The unprepared test candidate may be drawn to this answer because it sounds technical and cool.

7. Post-control overhaul involves those tactics taken after a fire is under control to ensure that the fire is extinguished and fire extension is stopped. This is not to be confused with pre-control overhaul, which is utilized to expose a traveling fire, get ahead of it, and confine it. When balloon framing is suspected in a wood-frame building, opening walls and examining stud channels on the floor above the fire is an example of pre-control overhaul. Another example would be where suspended ceilings are pulled well ahead of a fire to ascertain where it has traveled and where a stand may be made to cut it off.

A. *+2* Always check for manmade openings (poke-throughs) first. This is where fire, which also follows the most effective path of least resistance, will most likely spread to the upper floor in this room. In addition, the tools and the charged line will already be in this area, simplifying the situation.

B. *-1* This is an excellent place to look if the fire was in the bathroom, but it was confined to the living room, therefore the pipe chases are not your first priority. They may be checked as part of your examination, but the question asks for the first place openings would be made. Again, critical reading leads you to the best answer.

C. *+1* Electrical outlets and other manmade openings in the wall are proper places to check, but not first. Since fire and heat will travel vertically more rapidly than horizontally, this is not the best answer, and not the first place in the room that should be opened to examine for fire extension.

D. *+1* Another good choice, but again, fire travels upward before it travels outward. HVAC ducts in dwellings are usually located at the floor level. These avenues of fire travel should be shut down as soon as possible and should definitely be checked.

E. *0* This is still another place that needs to be checked as a part of your post-fire overhaul, but since it is two floors from the fire area, it should not be the first place to be checked.

8. This question regarding salvage focuses on reducing secondary damage. There are two kinds of damage that result from a fire and the attendant fire operations. Primary damage occurs as a result of the fire (such as burned contents and structural members). Secondary damage is that damage done by the suppression forces (such as holes in the roof, broken windows, and damaged doors). Some secondary damage is necessary to reduce the effects of the primary damage. Salvage efforts should be focused on limiting both types of damage. The least damaging way of conducting salvage is often the best (and safest) way.

A. *+2* If the house current cannot be used, it will be necessary to utilize apparatus current. This is the same way that the building's interior would get lit up if the house current were cut off. The wet/dry vacuum will soak up standing water in a short amount of time. It is also effective in removing water from carpets. Using the house's drainage system or, if it is too much of a trek, a window will work well in draining the vacuum.

B. *-1* This is nowhere near as efficient as the first answer choice. First, punching a hole in the floor creates additional damage. The idea behind salvage is to reduce damage. If the wet/dry vacuum is going to be used, why not use it on the 2nd floor.

C. *0* This is a viable option if no other methods are available, but damage is again created by punching a hole in the floor. Less damaging means should be explored before using this tactic to remove the water.

D. *+1* If there was not a wet/dry vacuum available, this is the best option since it does not do any further damage to the area. The toilet can easily be removed and then replaced when the task is finished.

E. *0* While utilizing buckets and mops is not a bad idea, it is by no means the most efficient. Therefore, it garners no points.

Passing Score for This Scenario = 10 Points

SCENARIO 4–2
SHORT-ANSWER QUESTIONS

1. Engine 1 should stretch a line up to the fire floor via the front door. Line diameter here should be 1¾" or 2". Using a 2½" line in this case is not warranted. While it will provide a greater volume of water, the stretch to the third floor, up and around the stairwell, will be a formidable task and take a great deal of time. This is an offensive operation. The line should be stretched to protect the interior stairs. If there are occupants escaping, keep the apartment door closed. As soon as the stairs are clear of occupants, the apartment door can be forced if necessary and the line advanced to locate, confine, and extinguish the fire. It is usually best to first force the door to the fire apartment, and then hold it closed until the stairs are clear of occupants.

2. Engine 2 should coordinate water supply needs with Engine 1, if required. Engine 2 should assist in the stretch and operation of the initial attack line. Crews stretching second and third attack lines cannot delay this first line to the seat of the fire. SOPs on these types of buildings should assign as many men as necessary to get the first line into operation in the proper position. This usually consists of the crews of the first two arriving engines. Any delay in getting initial water on the fire will escalate the fire problem and compound all other related problems, such as the life hazard. This engine is also operating in an offensive manner.

3. At multiple dwellings of this type, ladder company duties should be split into interior and exterior duties. These duties should be assigned by SOP, and should be deviated from only under unusual or extreme circumstances. A solid SOP will place the required men with the right tools in the most crucial areas to accomplish the various support activities needed to bring the fire under control. All personnel should carry portable radios and handlights in addition to their assigned tools.

Ladder companies, in most cases, should be split. If there are four men on the ladder company, two should be assigned to the fire floor for forcible entry, primary search, horizontal ventilation, and other support activities. This interior crew should be where the officer is assigned. The results of the primary search and other applicable information regarding interior conditions should be radioed to incident command as soon as possible.

The remainder of the company should operate as per the location of the fire. In this case, a fire on the top floor, the men should work as a team and are designated as the roof division. How they get to the roof will depend on building characteristics and exposures, whether attached or unattached. Attached buildings of unequal height should be treated as unattached buildings, as the roof of the fire building cannot be accessed from them. (Fig.4–1)

Fig.4–1 Although attached, this 4-story building is unequal in height to the adjacent structures. It will be necessary to use an aerial device to reach the roof. Ladder companies must always utilize the safest, most effective path of least resistance to the roof

In this fire situation, the aerial to the roof of the upwind exposure is the best choice. Although, there is a scuttle hatch on the roof, firefighters in full gear and SCBA will have a hard time accessing the roof through this small opening. Moreover, if the roof must be abandoned in a hurry, this small hole will be even tougher to squeeze through and may lead to injuries. The aerial is much safer. The roof division must accomplish vertical ventilation of the fire building. In a top floor fire, this will include opening and checking all natural roof openings such as bulkheads, scuttles, skylights, soil pipes, dumbwaiter shafts, and the like. In a top floor fire, this will include cutting the roof as directly over the fire as is safely possible. However, it is best to first open the scuttle to alleviate heat conditions in the stairwell. The scuttle of most comparable tenements will be located over the interior stairs. Other duties will be to survey the rear and sides of the building, perform top floor horizontal ventilation, and to check the status of the cocklofts of the exposed, attached buildings.

If this fire were not on the top floor, the assignments of these two men would be different. One, usually the chauffeur, would be the roof man and perform roof function duties as outlined above. However, as the fire is not on the top floor or in the cockloft, there is no need to cut the roof. After performing vertical ventilation of natural openings such as a bulkhead door, scuttle, skylight, and soil pipes or ventilators, this firefighter should access the top floor, usually via the gooseneck ladder at the rear, to perform top floor VES.

The fourth man would operate as the OVM (Outside Vent Man). His job would be to access the area opposite the nozzle team and provide horizontal ventilation of the fire apartment/area. He may use a fire escape, ground ladder, aerial, or the building's interior to take out the windows from the floor above. (Fig.4–2). Ingenuity is the order of the day for the OVM. This firefighter must get to a position opposite the attack line by whatever means necessary. If he cannot, a report must be made to incident command so that adjustments to the plan may be made. The OVM should also perform limited VES of the fire apartment and a more thorough VES of adjacent apartments as conditions permit.

Ladder 2 should also split the company. Crew 1, with the officer, should also get to the roof in this fire situation. The roof operation at this fire will be very manpower intensive, especially if the fire extends to the cockloft. The officer should assume the duties of roof division commander and supervise all roof operations. If the fire were not on the top floor, this crew would normally be assigned an area of priority, usually the floor above the fire for primary search, ventilation, and fire extension recon.

Fig. 4–2 Photo by Ron Jeffers NJMFPA
The assignment of the OVM is to get to a position opposite the hose advancement and provide an exhaust point for the products of combustion. This ventilation must be coordinated with interior forces or control of the fire may be lost.

The remainder of the ladder company should also split up and reinforce the exterior operations of the first ladder company. Radio communication between the two ladder companies regarding access routes to assigned areas is critical. For instance, if Ladder 1's roof man accessed the roof via the adjoining building, this information must be relayed to Ladder 2's roof man. Likewise, the second OVM should ascertain the location of the first-arriving OVM. This will usually be somewhere in an exterior area opposite the attack team's advance.

If there are only three men assigned, the officer should work with the initial attack team. The remaining two members' duties would be the same as that of the first-arriving ladder companies. The only difference may be for the second-arriving officer. He may have to pair up with an engine company stretching a line to the floor above. Many texts state that all personnel should work in pairs. These ladder company assignments apparently violate this rule. The fact that there is an SOP in place where all members are assigned a specific location is one built-in safety factor when operating alone. The company officer will know where all assigned personnel are operating. The second built-in safety factor is the fact that all ladder personnel must be assigned portable radios, and as a routine, must report conditions as soon as possible. Status reports at regular intervals from these positions are also required, not only to verify the member's locations and well being, but also to inform the incident commander of fire conditions in various areas of the building.

Another alternative is to keep the three-man companies together and cover the areas of priority. A four-firefighter ladder company can also operate strictly as two two-firefighter crews in the same manner. With these alternatives, as in all cases, the areas of priority will be the fire floor, the roof, the floor above the fire, and the outside vent position. The order in which they will be accomplished will depend on such factors as the location and extent of fire, roof type (peaked or flat) and life hazard. Those areas not covered may be assigned to later-arriving ladder companies or engine company personnel.

Regardless of who is assigned to what operational area, the key to operational safety, especially in regard to ladder company operations, is to have an established plan that all members are aware of and adhere to. No plan is often equated with no safety.

As in the case of the first two engine companies, these are also offensive operations, focusing on bringing the fire quickly under control.

4. The first task of the roof team in regard to ventilation will be to open the scuttle hatch. This action will alleviate conditions on the top floor landing as well as in the stairwell. This is crucial to the life safety of evacuating occupants as well as to the ability of the attack team to make the top floor landing.

Next, it must be ascertained if the cockloft is being vented as well. Most of the time, these scuttles, and many skylights as well, will be framed-out, eliminating access to the cockloft. If this is the case, opening the scuttle will only vent the stairway and will do nothing to alleviate conditions in the cockloft. This framing, also called a "return," is often 1" x 6" boards and is easily taken out with a hand tool. Use caution as extreme heat and possibly fire is likely to vent out of this opening if the gases are above their ignition point. A long-handled tool such as a pike pole is best here. If the fire is under control quickly, and odds are that it has not spread in any major way to the cockloft, this area can be opened and examined. If the fire is advanced, it is likely that conditions in the cockloft are severe, and it will be safer to begin cutting the roof over the fire apartment. Generally, any significant fire on the top floor will require cutting the roof. This is especially true if there is a scuttle hatch present as compared to a bulkhead door and skylight. Scuttles do not provide sufficient ventilation openings in heavy fire situations. In this case, a large hole should be cut as directly over the fire as is safely possible, and the ceiling of the fire apartment pushed down. This will ventilate both the cockloft and the fire apartment. It will also serve to confine the fire, channeling it vertically instead of allowing it to spread laterally throughout the cockloft and most likely to the adjacent buildings. If this fire is not vented, it is likely that the fire building (along with the top floors and the roofs of the two leeward exposures downwind) will be lost.

5. Pulling the ceilings in the fire apartment before the roof is ventilated can have disastrous consequences. Firefighters have been killed when backdraft has occurred in this area. A serious fire that has extended to the cockloft may exhaust the oxygen supply in this confined area, resulting in a superheated, oxygen-starved atmosphere. Any pre-control overhaul to expose hidden fire must be suspended until the roof has been properly ventilated and fire is showing from the hole. If no fire shows from the hole it is unlikely that the gases were hot enough to ignite, and the ceilings can be pulled at that time.

   To safely accomplish this without getting anyone on the fire floor killed, it is imperative that the interior crew communicates with the roof division as to when the ceilings can be opened. Timing is critical here, for if there is a delay in getting a hole cut in the roof and the ceiling is opened prematurely, the missing link of oxygen is provided. This will result in an explosive return to the flaming stage, but as the fire cannot expand upward, it will expand downward into the apartment, incinerating everything in the area. For this reason, if uncontrolled fire is suspected to be in the cockloft, the ceiling below should not be pulled until the roof is ventilated above.

6. Engine 3 should be prepared to secure a second water supply upon orders. The crew of Engine 3 should stretch a back-up line to the fire floor. This line can advance in unison with the initial attack line or be utilized to cover the adjacent apartment on the fire floor

   Engine 4 should assist in the stretch of this back-up line with Engine 3. Extra personnel can assist in the searching of the fire and adjacent apartments. However, it may be necessary depending on the forecasted escalation of fire conditions, to assign Engine 4 to stretch a line into the most severe exposure, Exposure D, to cover any fire extension into the top floor and cockloft and to protect crews operating in this exposure. This exposure must be entered, evacuated, and protected as soon as possible. This exposure coverage is a defensive action.

7. This fire dictates that an additional alarm be struck. All companies are actively engaged in confining this fire to the building of origin—an offensive operation. The incident commander must realize that with all companies assigned and operating, the tactical reserve has been depleted. Striking an additional alarm early is a way of staying one step ahead of this fire. If the second alarm companies are not required, they can be staged to relieve fatigued first alarm companies. If the fire is rapidly knocked down and no extension is expected, these companies can be released from the scene.

   It must also be considered that if all the second alarm companies are going to be operating, a third alarm will have to be requested to keep the tactical reserve status at an acceptable level. At no time at a fire like this, or at any fire not under control, should the incident commander be out of personnel to send into the battle.

   A rule of thumb that I like to follow is that if I am standing at the command post and the fire situation is still escalating, I try to always have at least two companies uncommitted and ready for assignment. If I do not, I strike an additional alarm. If I am alone at the command post, this means all on-scene personnel are committed. I not only have no one to relieve them, but I have no one to give additional assignments to should something need to be done. Keep your tactical reserve well stocked and you have a better shot of staying ahead of the fire.

8. Additional ladder personnel should be assigned to Exposure D, both to the interior and to the roof. The scuttles of Exposure D and D1 will have to be continuously monitored. There is no guarantee that the cockloft and the cornices of these exposures are fire-stopped. Therefore, they must be monitored. Companies must be prepared to take whatever action is necessary to prevent involvement of these exposures.

    Exposure D (and B, at least) must be entered and evacuated. Interior crews in Exposure D will also be required to open the ceilings on the top floor and anywhere else that fire may be suspected. Don't rule out the possibility of drop-down fire around vertical shafts such as pipe chases. These will also need to be checked.

    Due to the fact that the cockloft may be open over the whole row, Exposure B will also have to be checked. Companies should be assigned here as well. A fire in this building may sneak up on you, for example, as a result of a hose stream from below the ceiling in the fire building pushing fire in that direction. For this reason, the cockloft and top floor will need to be examined.

9. Additional engine personnel should be prepared to secure additional water sources for the attack or to supply master streams should the situation deteriorate. These companies should also stretch lines into the exposure buildings or into the fire building as required. Again, if they are not immediately required, they can be staged.

Some incident command considerations

To reduce span of control and increase fireground safety, the incident commander must also consider requesting additional command officers. These officers should be placed in strategic areas to control operations in those divisions. In this fire, a chief officer (or at least a trustworthy company officer) should be assigned supervisory duties in Exposures B and D as well as inside the fire building (interior division supervisor). Decentralizing command is almost always a good idea.

Summary

As the reader can see, offensive operations are occurring in the fire building to confine the fire to the building of origin. The alert incident commander is forecasting what might happen and is taking measures to be prepared for it. This entails the striking of a second alarm (at least) and placing companies in the downwind exposure to check and cut off any fire extension, a defensive operation. The strategy, therefore, is offensive/defensive.

Scenario 4–3
Short-Answer Questions

1. The action of the first-arriving company was incorrect for this situation. The company oper- ated in a strictly offensive mode in the fire building. As heavy fire is threatening the dwelling, which is occupied, this must be the top priority here. The first water must, in this case, be used to protect the dwelling, since the life hazard is severe.

   Proper engine company positioning will allow a preconnected deck gun to be placed in service to flow the side of the exposed building in the alley as well as hit the main body of fire in the cleaners. A handline has to be stretched into the dwelling to the south side to guard against fire extension. The first line should go to the 2nd floor as the curtains are already involved and the room soon will be if no lines are placed there. If left unchecked, the fire will certainly enter the hallway and extend up the unenclosed stairway, endangering occupants. These first actions, the deck gun, and the line to the interior of the dwelling are defensive in nature, aimed at keeping the fire out of the dwelling.

2. The next line must also be placed in service in the dwelling, to the top floor south side for the same reasons as the first line. Since it is a warm July night, it is a good bet that the windows will be open. This will allow fire spread to the interior.

   Only after these two positions are covered can lines be stretched into the cleaners to locate, confine, and extinguish the fire.

3. Ladder company operations should also concentrate on the dwelling. Evacuation of all resi- dents is a top priority. Closing windows on the exposed side of the building and removing curtains and other combustibles from around the windows will slow the spread of fire to the interior. Likewise, opening windows on the unexposed side will allow any accumulated heat to be released from the building. The roof door must be opened to alleviate heat buildup in the stairwell and the roof eaves must be checked for any signs of ignition, as convection heat and flaming brands will travel upward in the alley.

   As soon as possible after these defensive tactics are completed, or at the same time if manpower permits, forcible entry, primary search, and ventilation must be performed in the cleaners.

4. This is an incident where actions will be based on priorities. If there is enough man- power on the first alarm response, then protection of the dwelling can occur simultane- ously with fire attack on the cleaners. If manpower is inadequate in the initial stages of operation, as they usually are, then the priority is placed on those actions that do the most good for the most people, in this case, the occupants of the dwelling. Therefore, the first actions in regard to the fire's location are defensive. Only after these defensive positions are covered can any offensive attack be mounted on the fire. Therefore, this strategy is defensive/offensive.

It must be stated that if the initial manpower is inadequate to cover both the dwelling and the cleaners simultaneously, then additional alarms must be struck immediately. If the parent body of fire in the cleaners goes without the quenching effects of water for a prolonged period of time, the problems will multiply. It may be very tough to keep both buildings from becoming fully involved if lines are not placed on the parent body of fire quickly. If the cleaners becomes fully involved, it may be impossible to maintain interior positions in the dwelling. This is a case where the incident commander has his work cut out for him. When a heavy body of fire threatens to extend from one structure to another, one of the major deciding factors will be the amount of manpower on hand.

Scenario 4–4
Short-Answer Questions

1. There are two factors that impact on the decision of the incident commander. The first is the fact that the buildings are unoccupied and under construction. Fire will travel rapidly throughout the voids and into the roof space that is open over the entire row.

The second is the construction of the building. Lightweight construction is a dangerous building condition. Buildings employing this type of construction can fail in as little as five minutes, often without warning. As the incident commander has no idea when the fire reached the open flaming stage, he cannot accurately predict when the building will lose the battle against gravity and collapse. Heavy fire is indicative of a fire that has already passed the flashover stage. Before flashover, only contents are involved. After flashover, the structure becomes involved. This will lead to an early collapse in this type of building. Therefore, the strategy chosen in this case is one of a strictly defensive nature. In fact, at this fire, all the buildings in the row became fully involved and collapsed within ten minutes of the arrival of the fire department. (Fig.4–3)

Fig. 4–3 Photo by Ron Jeffers NJMFPA
The fire in this condominium development under construction experienced rapid flame spread and total collapse within ten minutes after fire department arrival. Construction phase fires offer even less resistance to fire spread and collapse than the finished product.

2. Master streams must be positioned out of the collapse zone. The distance that personnel and apparatus are kept away from the building will be at least equal to the height of the building. This is known as the vertical collapse zone. This vertical collapse zone may need to be increased based on the assumption that debris will bounce farther as it hits the ground.

The horizontal collapse zone must also be considered as walls often have lateral reinforcement that will cause them to fall as a single unit for the entire length of the wall. Generally, this is not necessarily the case in this type of construction, being more prevalent in reinforced brick parapets. The prudent incident commander, however, plays it safe, and clears the entire potential collapse area in a timely manner.

If the area directly in front of the facing walls is not wide enough to accommodate a required collapse zone, then all apparatus and master streams must operate from the flanks of the building and threatened exposures and/or areas. (Fig.4–4)

Fig. 4–4 Photo by Ron Jeffers NJMFPA
Vertical collapse zones often exceed the width of urban streets. Master streams should be positioned to flank a dangerous wall. Generally, the last part of a building to fail is the corners.

Often overlooked are the areas around the building that may be struck by the collapsing wall, often causing a secondary collapse that may be more devastating and deadly than the initial collapse. Do not fall victim to tunnel vision by concentrating on just the fire building. Firefighters have been killed and apparatus destroyed by secondary collapses. Pay particular attention to power lines and utility poles, trees, and adjoining buildings of shorter height. Larger buildings collapsing on adjacent smaller buildings have caused the collapse of the smaller building. For this reason, at any serious fire, all immediate exposures and areas should be evacuated and kept clear. (Fig.4–5)

Fig. 4–5 Photo by Ron Jeffers NJMFPA
If there is any doubt as to the structural stability of a wall, apparatus must be positioned out of the collapse zone. In this case, the wall is not so high, but the possibility of the collapsing wall knocking over the utility pole necessitates a larger collapse zone.

Scenario 4–5
Multiple-Choice Questions

1.

A. **+1** Notifying the utility company is one of the primary actions you must take after establishing incident command and must be a part of the initial radio report. The scope of this incident as far as mitigation is beyond the operational domain of the fire department, so this outside agency response is critical to the safe handling of this incident.

B. **-1** This is too dangerous for fire department personnel to undertake at this incident. It is not likely to be successful. For one thing, the radiant heat from the leak will prevent a close approach. Another reason is that the valve that is chosen may not be the proper valve and may escalate the problem. For these reasons, this task is best left to the utility company who can shut the main off from a remote location. Ordering firefighters to do this will place them in unnecessary jeopardy.

C. **+1** This is also one of the first tactical objectives the incident commander must set into action, but it cannot be accomplished without a strong incident command presence to ensure the safety of responders.

D. **+1** This is also a must at this incident. However, without an incident command organization in place to coordinate this, it may cause more problems than it set out to solve.

E. **+2** Choices A, C, and D are all excellent action steps, but they will all benefit from the establishment and maintenance of a strong incident command presence. The establishment of incident command and the initiation of a management system is not only mandated by OSHA, but is also the best method of ensuring firefighter and civilian safety, especially in the hectic initial moments of operation. Without strong command and control, operations will break down and lives will be endangered.

2.

A. **+1** There are severely threatened structures on the leeward side of the flame front. As these are located only 30′ away, they are of prime importance, but the only reason this is so is due to the location and extent of the fire. This choice still gets you a point.

B. **0** While the temperature and the rain will affect your ability to keep people outdoors after they have been evacuated, the wind will be a major villain in this scenario. It will determine your protection priority as far as which structure to protect. However, it is not the most important size-up factor.

C. **+2** The only reason that the buildings, and subsequently the occupants, are endangered is because of the location and extent of the fire. This size-up factor will determine where the life hazard is most affected as well as the strategy and tactics to best protect the area. (Fig.4–6)

Fig. 4–6 Photo by Bill Tompkins
Incidents such as this ignited gas main rupture are beyond the scope of the fire department to handle. Evacuation, protection of exposures, and supply of auxiliary protective systems are among the operations that the fire department may be asked to accomplish. Do not get in over your head.

D. *-2* While the distance between the flame front and the most severely exposed structure is only 30', distance is not one of the size-up factors in the COAL WAS WEALTH size-up and does not apply here.

E. *+1* The response is totally inadequate for this incident, but again, it is a result of the location and extent of the fire. Additional manpower will be required to assist in the evacuation as well as the protection of exposures.

3.

A. *-2* To put out this gas fire and allow the presence of an unignited leak is akin to operational suicide. This fire should be allowed to burn until the utility company can shut off the valves to stop it. The hydrant selection is also poor as it is in close proximity to the fire. Also, no auxiliary systems are supplied.

B. *+1* The selection of this hydrant is good as it is not in the area of the fire. The auxiliary system of Building #4 is also being supplied. Exposure protection is the top fire department priority in this incident. The exposed surface of the building must be kept wet.

C. *-1* Again, a poor hydrant selection. The auxiliary system is hooked up to, but not charged. This is almost as bad as not hooking up to it at all. Whenever a fire situation or exposure situation exists in a building with an auxiliary system, they must be supplied and the lines charged as soon as possible. Using a deck gun to put out the fire in conjunction with a hand line is, as stated in the answer B explanation, an extremely bad move.

D. *+2* These are all good tactics. A good water supply is chosen. The auxiliary system is being supplied and the building is being kept wet with a master stream. The difference between the *+1* and the *+2* answer is that this deck gun is unmanned, which will keep personnel from being exposed to the radiant heat of the fire. It is also operating from the flanks of the building. Safety in operations is always the best tactic, both on the test and on the fireground. Also, a high-rise pack is brought into the most severely exposed building to check for and extinguish fire extension.

E. *0* Here, a good hydrant is chosen and fire extension in Building #4 is considered. However, no exposure protection is accomplished and there is no supplying of the fire department connection. Water will be the greatest defense against the ignition of exposures and the absorbency of heat in these fires. The heat from this fire can very quickly overtax the sprinkler system if fire department pumpers do not supplement it. For a fire of this magnitude, the auxiliary system will be our greatest ally in Building #4.

4.

A. *-2* There is no establishment of a second water supply which will be required at this fire for exposure protection. Also the use of a small diameter handline out of a window on the immediate leeward side of the fire will accomplish little. The area is likely to be untenable, and the best tactic would be to keep the windows closed and the exterior walls wet to protect against the effects of radiant heat.

B. *-1* Again, no second water supply is established and a line is being ordered to extinguish the fire. This is never the proper tactic for a fire fed by a gas line rupture. The unmanned deck gun is the only correct tactic being ordered here.

C. *+2* These are the best tactics and will complement the correct tactics chosen by Engine 1. Reinforcement of the master stream operation by ordering an additional unmanned master stream from the opposite flank takes firefighter safety into account. This is also evident by stretching another high-rise pack into Building #4 to assist in the spot fire problem. The securing of a second water supply will ensure a strong hydraulic reserve, always a good move in an escalating situation.

D. *+1* These tactics are somewhat acceptable. If anything, they are safe. A second water supply is established and the evacuation effort is reinforced. However, in addition to assisting in evacuation, Engine 2 should stretch an additional line off the standpipe to reinforce the spot fire extinguishment operation. An engine company should never rely on in-house standpipe lines for fire attack. This reliance may be acceptable for a ladder company operating in a search and/or evacuation mission above the fire. The house line could be used to knock down spot fires while fire department lines are being stretched, but it is never an acceptable tactic for an engine company whose objective should be water application in most cases.

E. *-2* Although a second water supply is being established, directing water on the fire for the purpose of extinguishment is the wrong tactic. The reasons have been previously stated in the "*-2*" choices in both Question #3 and #4.

5.

A. *+1* These are excellent tactics to protect against radiant heat, but will do nothing to diminish its effects. Heat will still build up in the exposed apartments. Provisions must be made for it to be released from these areas.

B. *-2* This tactic of opening windows on the exposed side and then establishing an airflow by opening the bulkhead door will pull heat into the building. The objective is to protect the building from heat. Windows must be closed and combustibles removed. This is virtually a fuel removal operation to arrest ignition.

C. *-1* This tactic will not alleviate any heat out of the building and may create a venturi effect to pull heat into the building if windows on the leeward side are opened or broken out by the heat.

D. *+2* This answer takes all the good tactics from answer A and then solves the heat build-up problem by taking steps to release it. Opening the bulkhead doors will allow any heat build-up in the stairs and halls to channel up and out of the structure via the roof. Also, after closing the windows on the exposed side, the best way to provide release of heat build-up in the room is to provide cross-ventilation by opening windows on the opposite, unexposed side. If the heat being removed is greater than the heat build-up, then unsafe temperatures can never be reached.

E. *0* This is a surrendering of the building and is not warranted at this time. This may be playing it too safe. Consider the building construction. The building is of fire-resistive construction, which is an ally to our efforts to halt any fire extension into it. If this had been a wood-frame building with asphalt siding then surrendering might have been a feasible tactic, but this was not the case.

6.

    A. *0* The evacuation and exposure protection requirements will rapidly overwhelm the forces on hand. There are more tasks than resources at this scene. Also, staging procedures established beforehand by an SOP, along with radio direction as to the location of the designated staging area, will avoid scene congestion and make incident and personnel management easier. Scene congestion will occur for the incident commander who does not make command and control a priority.

    B. *+2* This is the best action. Confirmed Haz Mat incidents will require manpower to assist in evacuation, water supply, in this case, to protect exposures, assist in decontamination if required, and a myriad other tasks which will require a strong tactical reserve. While mitigation of this incident is more the scope of the utility company than the Haz Mat team, an ignited gas main is a Haz Mat incident and should be treated as such.

    C. *0* Taking no action will never get you any points. Indecision on the fireground leads to disasters. Consulting is fine, but waiting for someone's arrival to consult him or her is a risky game; this decision is not the responsibility of anyone but yourself, the incident commander. If you even think there might be a slight chance you will need more resources, you had better call for them. It is better to have them on-scene ready to go than not to have them when the time comes that you need them. The smart incident commander will take steps to ensure he will never be "caught short."

    D. *0* The Haz Mat team may not be required to mitigate this incident, because once the utility company shuts off the flow of gas, the incident is over. As they may be a useful resource at the command post, they should not be the only additional resource summoned. The major problems facing the fire forces are ones that require a significant commitment of manpower.

    E. *+1* Evacuation will be a major reason for summoning additional assistance at this scene, but it is not the only one. Exposure protection will also require a good deal of manpower to position lines and secure water supplies. Also, summoning additional chief officers is a good idea. This incident will need to be broken down geographically to address the various problems in the different areas. Decentralization of incident command will enable the incident commander to maintain an acceptable span of control and provide for the safety of fire personnel operating in the various areas.

7.

    A. *-2* Due to the extreme heat being generated by the uncontrolled gas leak, the act of protecting the occupants in place is not advisable here. The building is too severely exposed. These occupants are best evacuated.

    B. *0* The lobby will be exposed to radiant heat and a bevy of fire operations. Having a large amount of people in the lobby of the most severe exposure will create confusion and possibly panic. The occupants should either be evacuated via the unexposed side (the best choice) or protected inside the building in a less exposed area (acceptable, but not the best choice).

C. **+1** This is better than having the occupants remain in the apartments on the most exposed side, but they may still be in danger if the fire gets into the exposed apartments. This is a calculated risk that can only be made by the incident commander who has accurate information about the degree of exposure to the building and the estimated time of shutdown by the utility company. Only with this information can this option be exercised with any level of confidence. If the shutdown will be done quickly, this may be the best option in safely protecting the occupants. However, this information is not yet available so the prudent incident commander should play it safe and evacuate the building.

D. **-1** If this statement read that "the convection heat on the lower floors will not be so bad," this might have been a viable answer and option, but it states that it is radiant heat that will not be so bad on the lower floors. This answer tests the knowledge of fire spread characteristics. Remember that convection heat will rise and affect the upper floors more than the lower floors, but it is the radiant heat that will travel in all directions and threaten occupants on all floors, including the lower floors. Knowledge of fire behavior and spread will steer the true strategist away from this answer.

E. **+2** Indeed this is a major emergency and one that warrants the removal of all persons in the immediate danger area. Building #4, especially the exposed side facing the fire plume, should be considered the "Hot Zone" and only essential personnel should operate in this area. Therefore, in the interest of safety, evacuating the building is the best option.

8.

A. **0** This is a dangerous undertaking. The valve in the street will be in close proximity to the rupture. Radiant heat will prevent the shutting of any valves in the immediate area of the incident.

B. **+1** This is about all that can be done until the gas supply is shut down. To attempt anything else would be foolish and would endanger crews.

C. **-2** Can you say "sizzle"?

D. **-1** This is impractical as there will be a substantial amount of gas escaping to think that a foam blanket can extinguish it and suppress the vapors. Most likely, there will not even be flame in the immediate area of the leak because the vapors are too rich and will not burn. The ignition point will be where the vapors are in the proper flammable range to ignite and will be somewhere above the source of the leak. Again, attempting to use a suppression agent on this leak will be unsuccessful. Personnel will never get near the area due to the incredible radiant heat in the leak area. The fact that the leak is too extensive will save the incident commander who chose this option from killing his intervention team.

E. **+2** Recognizing that remote shut-offs are the safest method of stopping the leak makes this the best choice. Shutting the gas from a distance is the wisest option and will put the least amount of responders in danger. Protecting exposures will remain a priority while this is done as considerable "line pack" will be present. Line pack is that residual gas between the shut-off point and the leak that will still need to be burned off before the incident is declared under control. However, as soon as the valve is shut down, or soon thereafter, there should be a noticeable reduction in flame intensity and pressure. Continuing to monitor nearby buildings for accumulated gas is a safe task that fire department can effectively carry out while control operations are taking place.

9.

A. **+2** Protecting life and minimizing the combustibles available to the radiant heat are top priorities in this incident. This is a large building, thus evacuation and ignition control demands are such that one ladder company cannot possibly accomplish them alone. Also, critical reading of the scenario revealed that only one ladder company responded on the initial alarm. As stated earlier, sometimes the questions will reveal answers that are not evident in the scenario. The fact that you are asked what your orders are for Ladder 2 should cue you in that additional alarms are required and should have been requested. How else could Ladder 2 get on the scene? This may cause a change in the answer choice for Question #6.

B. **+1** This action will minimize the exposure problem on the exposed side of Building #4. However, this positioning may be too close and is not as safe as the unmanned deck guns.

C. **0** These actions address lesser exposures. While these are acceptable tactics, they are not as critical at this time as the main problem, the evacuation and protection of Building #4. Ladder 2 is first due on the second alarm, and is the second ladder company on the scene. They will be required to operate at the battlefront, so to speak. Later-arriving companies can be assigned these lesser priorities.

D. **+1** This is a minimization of exposure to occupants and contents. If you chose this tactic of protecting-in-place in Question #7, you got a point then and you get a point now. Some questions will allow you to follow a line of reasoning from one answer to the next. Some may lead you into exam disaster, while others are not so bad, as is the case here.

Fig. 4–7 Photo by Ron Jeffers NJMFPA
Short of moving the exposure and increasing its distance from the heat source, the best protection is provided by coating the exposed surface with water. A single line directed in the alley between these two buildings prevented ignition of the combustible adjacent wall.

E. **-2** A water curtain is totally inadequate for exposure protection as radiant heat will pass right through it. There are only two ways to effectively protect a structure, or anything for that matter, against radiant heat. The first and best way is proper distance, but it may not be possible to achieve. In the case of a structure, it usually can't be moved. However, people and vehicles can be moved and are a preferred method of dealing with radiant heat outside of the extinguishment of the main body of fire. The second, more common approach, is the coating of the structure with a film of water. This will stop the surface from heating, releasing combustible gases, and inevitably igniting. Put the water on the exposure. (Fig.4–7)

10. This is a straight knowledge question. It is only indirectly related to the scenario. The key to the question is the phrase "most dangerous."

A. **-1** This is probably the least dangerous in terms of potential destruction. Burning gas will not usually pose an explosion problem. Protection of exposures while waiting for the gas to be shut down, either by fire department personnel in small incidents or utility personnel in larger incidents is the strategy of choice.

B. **0** The same strategy of protecting exposures while waiting for the gas to be shut down is the preferred way of handling this situation. The extension problem will be more severe in the interior, but proper coordination and controlled nozzle operation should keep things from getting out of hand. On the other hand, inadvertent extinguishment of these leaks on the interior can lead to serious problems as gas may build up inside a structure.

C. *+1* The danger of unignited gas reaching an ignition source is the primary danger, but usually outside leaks of natural gas, which is lighter than air, will dissipate easily and can be assisted by the application of fog streams to disperse the gas.

D. *+2* The danger of unignited gas inside a structure is by far the most dangerous natural gas situation that a fire department will encounter. The ignition sources must be eliminated as any spark could ignite the gas if it is in the right gas-to-air mixture. Gas detection devices are a must, as is full turnout gear and SCBA. Evacuation of the premises is also a first priority. Ventilation must occur, especially at the highest points in the building where the gas, which is lighter than air, will accumulate. The utility company must be notified immediately and steps must be taken to isolate the leak, if possible.

E. *0* This is just not true. While any situation handled poorly may cause death, injury, and destruction, unignited gas will always cause more problems than ignited gas due to the potential it presents. At least when gas is ignited it is easy to ascertain the extent of the leak, but an unignited leak may travel great distances and cause explosions that may lead to collapse of buildings.

Passing Score for Scenario 4–5 = 14

SUMMARY

It must be remembered that while this strategy of "no attack" or "non-intervention" means a "hands-off" approach to the main problem, it does not mean that the fire forces sit back and do nothing. In incidents such as these, there will be plenty of things for personnel to do, such as building evacuation, water supply, stretching of precautionary lines, exposure protection, and the like. The astute incident commander does whatever is necessary to make the situation better while still protecting the fire personnel from committing to an incident mitigation strategy that is beyond the control or scope of fire department expertise and training. A good rule of thumb to follow is, "If you can't make it better, at least make it safe."

Scenario 4–6
Short-Answer Questions

1. This apartment is showing all the signs of a backdraft potential. The sealed steel door with pressurized smoke issuing at intervals is indicative of a decay stage fire, as is the report of the same condition on the floor above. The front window, which is stained black, is another sign of an impending backdraft situation. From these indications it must be assumed that opening this door will cause a backdraft in the apartment. Therefore, the fire cannot be attacked in the conventional way. The use of the indirect attack is a viable tactic here, since opening the apartment at the highest point (an acceptable tactic at a decay stage fire) will have no impact on this condition. The highest point, the roof, is three floors away.

2. There should be no ventilation of the fire apartment at this time. Venting horizontally is likely to cause a backdraft and injure or kill anyone standing in the blast area. Venting by cutting the floor above will severely endanger the venting crew for as the superheated gases meet the ambient air, intense flaming will occur. When this happens, this apartment above the fire apartment is not exactly the best place to be as it will instantly become untenable.

   At this point the entire operations should slow down. The fire apartment is sealed and that leaves time to consider the options. The prudent incident commander will take advantage of the conditions in the fire apartment to mitigate the situation.

3. The men on the roof may open the bulkhead door to alleviate any built-up heat in the stairwell and in the bulkhead area. They can also recon any natural vertical arteries such as soil pipes for any traveling fire. Chances are, with this fire, there will be none at this time, but there may be heat that can be vented. They should then find an area of safety as the indirect attack is put into effect.

   The men on the floor above the fire should be able to quickly finish the primary search and evacuation of all floors above the fire. They should then be withdrawn to the exterior or at least to a lower floor.

4. The interior crews on the fire floor should clear away from the steel door. They should force an adjacent apartment. There are two reasons for this. The first is to establish a position of relative safety should something go wrong and the hallway becomes untenable by an explosive fire venting from the apartment. The second is to establish a position to execute the indirect attack (2½" handlines should be brought up to the fire floor if they are not there already). They will be operated on wide fog to create the required steam to successfully complete the indirect attack. The mission of the crew should be to breach the wall between the fire apartment and the adjacent apartment and apply the fog streams into the fire apartment to allow the steam to smother the fire and cool the area.

When the indications are visible that the streams have accomplished the job, the apartment should be vented as normal and the lines advanced to complete extinguishment. A thorough secondary search will need to be conducted, as due to the conditions on arrival, a primary search could not be extended.

A chief officer should be designated the operations or interior division officer and placed inside the building to supervise and coordinate the operation. This operation is dangerous and requires strict control over operating personnel. Only the essential personnel required to carry out the mission should be placed inside the building.

5. The prepared chief should always have more than one plan to achieve the desired outcomes from his objectives. Using equipment instead of manpower is a safer and possibly more effective way of applying the indirect attack on this fire. Utilizing the fog nozzle of a Telesquirt or a preconnected ladder pipe is a way of applying the attack while men are kept at a safe distance. To achieve this objective it will be necessary to withdraw all personnel from the fire floor and all floors above the fire, including the roof. The roof door should be chocked open or removed to allow the converted steam an avenue for escape. As noted above, the building should have been searched and evacuated by the interior crews. When an "all clear" is confirmed by a roll call, the operation can resume.

The Telesquirt nozzle set on wide fog is charged and extended into the window of the fire apartment. Start the stream when the nozzle is just at the window. If the glass is hot enough, the water will break the window on impact. If the window doesn't break due to the presence of thermopane windows, the boom can be extended until the nozzle breaks the window and is actually operating just inside the room. This 1000 gpm stream will have a much more violent and pronounced effect than the handlines operating through the breached walls and is safer. In this operation, no one should be inside the building as the indirect attack does its job and the desired steam is created. (Fig.4–8)

Fig. 4–8 Photo by John Hund
Although rare, a backdraft condition can exist in a relatively sealed compartment such as in an apartment. The fog nozzle from a Telesquirt, discharging 1000 gpm can be introduced through a window to produce the desired results.

CHAPTER FIVE
SCENARIOS

SCENARIO 5–1
ATTACHED GARAGE FIRE

Construction

A fire has been reported at 355 Windy Way. The dwelling is a two-story colonial residence with a central hall and an attached garage. The bedrooms are located upstairs. In the downstairs area there is a dining room, a living room, a den, a family room, and a kitchen. The attached garage, which has a peaked roof, is located just off the kitchen and is separated from the dwelling by a wooden door. There are entrance doors to the house at the front and the rear. The garage is windowless; however, there is a ceiling fan and a small vent on the roof of the garage, on the peak opposite the house. On Side B of the house, there are two second-story windows located above the roof of the garage. They are partially open with screens installed.

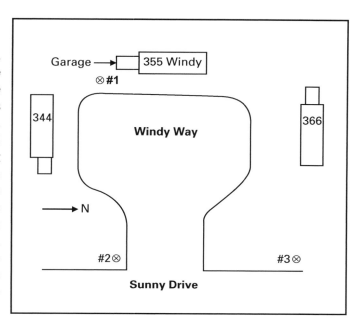

Time and weather

It is a Monday morning. The hour is 0744. The temperature is 66°F, with the wind blowing steadily out of the north at 25mph.

Area and street conditions

Windy Way is on the outskirts of the city. A long, winding hill is the only way up to this area that is located in a cul-de-sac on a ridge overlooking the rest of the city.

Fire conditions

As you are pulling out of headquarters, you can see the black smoke spiraling up from the area of Windy Way. It takes about ten minutes to arrive on the scene due to the morning rush hour traffic. Upon arrival, you see that the garage, which is closed, is showing evidence of heavy fire. It is unclear whether the fire has spread to the house. The owner of the house meets you and says that everyone has gotten out of the house. He has been burned on the face and hands and is somewhat disoriented.

Exposures

There are similar dwellings on either side of the home. They are separated by at least 50' and covered with landscaping.

Water supply

There is a hydrant located on the cul-de-sac near the fire building, and one located on the corner of Windy Way and Sunny Drive. There is also a hydrant located further down Sunny Drive that is fed from a different main than the two aforementioned hydrants.

Response

Your response is three engines and a ladder company. Each is staffed by an officer and three firefighters.

SHORT-ANSWER QUESTIONS

1. Where would you order the first line to be stretched?

2. Where would you order the second line to be stretched?

3. Where would you order a third line to be stretched?

_____

_____

_____

_____

_____

_____

4. How should search and rescue operations be conducted at this fire?

_____

_____

_____

_____

_____

_____

5. How should this fire be ventilated?

_____

_____

_____

_____

_____

_____

6. How would your water supply be established?

_____

_____

_____

_____

_____

_____

MULTIPLE-CHOICE QUESTIONS

1. What is the most important size-up factor related to this fire?

   A. Street conditions

   B. Water supply

   C. Apparatus and manpower

   D. Construction

   E. Area

2. Suppose upon arrival, the garage fire had ignited the siding of the house. The fire is extending up the wall toward the windows. What actions would you take regarding the first hoseline?

   A. Use a deck gun to knock down the fire on the siding. Stretch a line into the house to attack the fire.

   B. Use a line to sweep the siding and eaves on the house. Use a second line and enter the house to locate, confine, and extinguish the fire.

   C. Do not deviate from the placement of the first line into the structure. Raise a ground ladder and send two men up with pike poles and water extinguishers to knock down the extending fire.

   D. Before stretching the first line into the dwelling, use the line to hit the fire on the siding and to sweep the eaves. Then advance the line into the structure to locate, confine, and extinguish the fire.

   E. Place a PPV fan on the 2nd floor in the doorway opposite the exposed windows. Blow air into the room to keep the flame out of the structure at the 2nd floor level. Stretch the first line to the 2nd floor to check for fire extension.

3. How would you conduct post-control overhaul of this fire?

    A. Check around the doorframe leading into the structure as well as the ceiling and walls adjoining the garage. Completely overhaul the garage. Check the 2nd floor baseboard.

    B. Remove the siding on the side of the house facing the garage. Check the interior walls on the 2nd floor.

    C. Check the attic of the dwelling for fire extension. Completely overhaul the garage.

    D. Check the baseboards on the 2nd floor of the dwelling as well as the 1st floor ceiling. Completely overhaul the garage.

    E. Check the baseboards of the 2nd floor. Check the ceiling of the 1st floor and adjoining walls. Hydraulically overhaul the garage to ensure extinguishment. Do not pull ceilings or open walls due to questionable structural stability.

4. You have not as yet assigned a safety officer at this scene. Who would be responsible for this fireground role?

    A. The next highest superior officer that arrives.

    B. The dedicated safety officer from the previous day's shift, if on the scene.

    C. The incident commander.

    D. All company officers.

    E. Everyone on the fireground.

5. In regard to fire destruction, what is the most significant fire spread factor in wood-frame structures?

    A. The open stairway.

    B. The combustible partitions.

    C. The roof.

    D. Forcible entry difficulties in private dwellings.

    E. The entire building is made of wood, including the combustible exterior.

SCENARIO 5–2
BALLOON-FRAME DWELLING FIRE

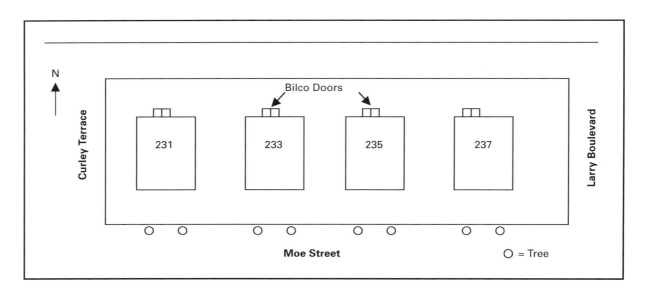

Construction

A fire has been reported at 233 Moe Street; a three-story wood frame dwelling of balloon construction with a steeply peaked hip roof. The third floor attic area is built-out with dormers and windows on each side. There is a fire escape on the building, but it is not known upon arrival whether it serves the attic or not. It is also not known whether the basement is finished or not, but there are small casement windows located on all sides of the building at the basement level. There are three electric meters on the side of the house and three doorbells on the front of the house. It is not known whether the basement or the attic is occupied. Access to the basement is via a door off the kitchen and through a set of double "Bilco" doors at the rear of the building. The "Bilco" doors lead down a few steps to another wood door that leads to the basement. (Fig. 5–1)

Fig. 5–1 These "Bilco" doors are usually located at the rear of the building and are an ideal place to vent opposite the attack line. This opening should not be your first choice for an attack point as it is possible that products of combustion may be pushed into uninvolved areas.

Time and weather

The time of arrival is 1230 and the temperature is 55°F with a calm wind.

Area and street conditions

Moe Street is heavily tree-lined with setback dwellings. There is a car in the driveway and toys and bikes strewn about the yard and driveway.

Fire conditions

Heavy black smoke is issuing from the basement windows and the first floor windows on Side A and B. There is also smoke showing lightly from the eaves all around the structure. A neighborhood man says to you that there are "a ton of people that live in that building." He does not know if they are home.

Exposures

The exposures on the right and left (Sides B and D, respectively), are similar structures and are separated from the fire building by 15' driveways.

Water supply

The water supply is adequate for the expected fire load presented by this type of structure and occupancy.

Response

Response is three engines and one ladder company. The staffing of engine companies is one officer and two firefighters. Ladder company staffing is one officer and three firefighters

SHORT-ANSWER QUESTIONS

1. Where would you assign the first hoseline?

_____

_____

_____

_____

_____

_____

2. Where would you stretch the second hoseline?

_____

_____

_____

_____

_____

_____

3. Would a third line be required and if so, where would it be stretched?

_____

_____

_____

_____

_____

_____

4. How would you order this building ventilated?

_____

_____

_____

_____

_____

_____

5. How would you order this building be searched?

_____

_____

_____

_____

_____

_____

6. Suppose during pre-control overhaul, you find that fire is traveling up the interior of the wall on the first floor. What actions will you take? You have a hose line available and standing by.

_____

_____

_____

_____

_____

_____

_____

7. In assessing the collapse profile of this building, what structural deficiency constitutes the greatest collapse danger from a firefighting point of view?

_____

_____

_____

_____

_____

_____

_____

8. Suppose this fire was in a first floor bedroom. The initial report and size-up indicate that the fire is free burning and showing signs of rollover. Your initial attack team has the line charged and is ready to go. Your ventilation is to be coordinated with your choice of attack. What type of attack would you choose for the hose team and from where?

_____

_____

_____

_____

_____

_____

9. Suppose the basement became fully involved and the first floor was becoming untenable. Your attempt at an exterior attack to knock down the heavy fire has failed due to partitions concealing the fire in the basement. What would be your next plan of attack?

_____

_____

_____

_____

_____

_____

_____

ANSWER SECTION

SCENARIO 5–1
SHORT-ANSWER QUESTIONS

1. The fire is located in the garage, possibly extending into the dwelling via the interior wood door or via auto-exposure through the windows facing on the garage. The fire will most likely take the path of least resistance into the dwelling, the wooden side door. Thus, the first line must be stretched to protect the life hazard, while at the same time confining the fire. This is a defensive/offensive strategy. Cut the fire off from the house and the greatest life hazard first, then move in and extinguish it. (Fig. 5–2)

Fig. 5–2 A fire in an attached garage will expose the entire structure, causing a severe life hazard, especially at night. There is no guarantee of a closed door between the dwelling and the garage.

On the way into the house, it will be a good idea, in this case, to give the fire and the exposed wall and window area a quick sweep with the stream to reduce the radiant heat exposure. This will work here, but in other situations, may depend on the fire's location in relation to the front door. Don't waste any time getting the line into the building.

Once ventilation has been established in the garage by opening the overhead door and propping it open, the line can advance into the garage from the kitchen door to extinguish the fire. This is the best direction of attack as it is from the unburned side and will allow the attack to proceed with the wind at the attack team's back. As there are no available vent points (windows) in the garage, using the garage door as the attack point would push the products of combustion toward the path of least resistance, that being the door between the kitchen and the garage. This direction of attack would cause the products of combustion to enter the house, as the hose stream would push them to flow in that direction.

Fig. 5–3 A line must be stretched through the dwelling to this interior garage door to protect the life hazard and the structure. Whether this line is used as an offensive attack line or in a defensive, holding position depends on variables such as available ventilation openings, and wind direction.

For a moment, let's change the ventilation profile in the garage. Say that there were some large windows at the rear or the side of the garage. These openings could be used as vent points while the kitchen door is kept closed and the attack carried out from the overhead garage door. However, this exterior attack line would still be the second line stretched. The first line, in all cases, must be stretched to the interior and placed between the fire and any possible victims. The line in the kitchen, in this instance, would operate in a holding action, ensuring no fire entered the dwelling. (Fig. 5–3)

As incident commander, keep the wind in mind (which is blowing from the house toward the garage at 25mph). This wind direction will make the advance easier, especially once the garage door (vent point) is open. However, the wind may cause flying brands to threaten nearby leeward exposures, including landscaping, and any vegetation on the hillside. Be alert to this potential. Don't forget, also, that the wind may change direction. If it does, be prepared to react accordingly. If possible, it is best to attack from a direction that causes the least discomfort to personnel.

Let's suppose now that the wind was blowing from the garage into the dwelling (from the south). In this case, the first line must still be stretched through the interior of the dwelling to the wood door between the dwelling and the garage. (Fig. 5–4) It may be smarter, however, considering the wind direction and velocity, to keep that door closed and use the interior line as a protective line at the door. The second line would then be stretched to attack the fire from the exterior, overhead garage door. This may cause some problems considering the ventilation limitations, but the dwelling and the life hazard would be better protected. The alert incident commander must be able to adapt to conditions.

Fig. 5–4 Englewood, NJ, Fire Department file photo
Regardless of the direction of the wind, the first line (shown to the extreme left) must be stretched to protect the life hazard and the threatened exposure. This line also meets the objective of confining the fire to the garage.

2. The second line must be used to back-up the initial attack line. This line can be either an additional 1³/₄" line or a 2¹/₂" line. This second line must remain at the doorway between the kitchen and the garage. In that way, no fire will extend into the dwelling via the wood kitchen door and the most severe life hazard problem is covered.

    The second line may also have to advance into the garage if the first line is having problems due to heavy fire conditions. This double-punch should do the job of knocking down the fire. If it doesn't, a change in the direction of attack may be warranted. The door to the dwelling would have to be closed and the attack switched to the overhead garage door. This, as mentioned before, is undesirable due to the fact that products of combustion would be pushed toward the dwelling, but it is an alternative, especially if the wind changes direction, complicating the attack. Always keep your options open and be one step ahead of the fire in your thinking. Plan B should always be ready, at least in your head.

    If this line were to be placed at the outside garage entrance indiscriminately, it would oppose the attack line entering the garage from the kitchen. This may cause burns and other injuries to the firefighters on the interior. Opposing attack lines is always a mistake and the incident commander must ensure this never happens.

3. As two lines are already stretched through the front door of the house, the third and any additional lines should be stretched via another access point. Any more than two lines through the same access point will cause the lines to become entangled and can possibly grind the advance to a halt, allowing the fire to grow and extend into areas remote from the origin point.

The narrative states that there is a rear door in the house. This is the ideal place to stretch this third line. The assignment for the company stretching this line would be to further protect the interior stairs and check for any fire extension into the second floor via the windows, behind the siding, or concealed spaces in the walls.

Usually all homes, and large homes especially, will have more than one access point. These should be found as part of the recon process and used as entrance points for hoselines and search teams in gaining access to less exposed areas of the building. If there are no other doors, an additional line may need to be stretched up a ladder. (Fig. 5–5)

Fig. 5–5 Ladder Company personnel on recon missions must seek alternative ways of entering the structure. Additional ventilation opportunities such as these skylights will only be found by checking the rear.

4. A primary search must be extended at every fire that is offensive in nature. This fire, while initially being defensive/offensive due to the location of the fire in relation to the life hazard, will rapidly switch to an offensive strategy as soon as the overhead garage doors are open, creating a ventilation point. The best approach would be to split the ladder company. Two members can force the overhead door to the garage, prop it open, and then enter the dwelling. These same two members can quickly search both the 1st and 2nd floors of this dwelling, especially in the initial stages of the fire when there is a good chance that the fire has not yet entered the dwelling in any extensive fashion. This crew can also recon the dwelling for fire extending into the house. Their reports to the command post will assist in determining the placement of additional lines and possibly the need for additional alarms.

The remainder of the ladder crew, assuming the staffing strength is four members, will go with the initial attack line, searching off the line as conditions dictate.

After the fire is knocked down, these crews should be switched to conduct the secondary search of both the house and the garage. It is always a good idea to have different crews conducting the secondary search than the crews conducting the primary search (or if it is the same crew, giving them a different assignment area than that of their primary search will serve the same purpose). This is because crews will tend to search the same areas they searched during the primary search. Switching the crews or using different crews will provide a different perspective.

5. The first and primary means of venting this fire would be to open the overhead garage door. Since this is a very large opening, it may be all that is needed to vent the garage. Don't forget to prop the door open with a pike pole or some other tool to prevent it from closing. Removing the ventilator on the roof may not be required, since the hole it creates will be small and not sufficient to vent the garage. If conditions deteriorate and additional ventilation is required, removing the ventilator and extending the hole by cutting the garage on the peak away from the house may be the next action taken to localize the fire in the garage.

On the interior, the windows in the dwelling on the exposed side should be closed to reduce the radiant and convection heat exposure. Curtains and other combustibles should be removed. In addition, opening windows on the unexposed side of the dwelling will help dissipate any buildup of heat.

6. Hydrants in front of or in very close proximity to the fire building should, with very few exceptions, never be used as the initial attack supply. The chances for scene congestion increase and further, the front of the building must be left open for aerial apparatus. Therefore, the hydrant next to the fire building (#1) is not to be used. It will place the water supply too close to the house. If the conditions in the house deteriorate to the point that the apparatus and crew must be moved, the initial water supply will be lost. It is also possible that Hydrant #1 is located on a dead-end main. This information should be known before the fire.

Hydrant #2 is the best choice for the attack hydrant. It will bring a sufficient amount of water to the fire scene without tying up the front of the building. Hydrant #3, as it is on a separate main, should be established as the secondary water supply.

Multiple-Choice Questions

1. Notice the absence of life hazard and location and extent in these answer choices. A decision must be made based on existing conditions and priorities.

    A. **+1** Since this is a cul-de-sac, a veritable dead-end, apparatus positioning can be a nightmare. Ensure your departmental SOP's address these situations and get the most vital pieces of equipment into position. This would be at least the first two engine companies and the first ladder company. If no SOP exists, it might be a good idea to stage other than the first-arriving companies until conditions are ascertained and a plan of action is initiated.

    B. **+1** Water supplies on the outskirts of towns may be suspect. The fact that the fire building is on a cul-de-sac may complicate the water supply problems further. As stated earlier, don't take the hydrant in front of the building and establish at least two water supplies on separate mains. The time factor in establishing a water supply on a more remote hydrant may work against the operation, causing a delay in water being applied to the fire.

    C. **0** Manpower and apparatus must be considered at every fire. Conditions will dictate to what extent their urgency is. In this instance, apparatus will not be a consideration, but manpower may be, especially because of the long response time. Forecast the needs of the incident, weigh it against the manpower on hand, and make your decision based on the balance between the two. The first alarm companies, though, should handle the situation here.

    D. **0** While the construction of the building will dictate how the fire will spread, it is not the most important size-up factor here. Due to the location and extent of the fire, well-placed, aggressively operated streams should negate any detrimental effects the construction of the building may present.

    E. **+2** The area is on the outskirts of town, the hour of the day suggests rush hour traffic, and the wind condition prevalent to the area is problematic. In addition, the cul-de-sac layout causes potential problems with water supply and apparatus positioning. All of these conditions, especially taken together, can lead to a delayed response, delayed operations, and a heavier than normal fire condition on arrival. Thus, this makes area the most important size-up factor of these answer choices. (Fig. 5–6)

Fig. 5–6 Hydrants on cul-de-sacs are often on dead-end mains and may not supply sufficient water for attack. A relay from a main road may be required.

2.

A. **+1** The deck gun is a good idea, but the fire on the side of the house should not be so extensive that a deck gun will be required. It is a huge drain on the water supply, one that can be better used to establish attack positions. It is better to use the initial attack line to sweep the fire from siding. It is quick and should be effective. (Fig. 5–7)

Fig. 5–7 Be cognizant of where venting fire from the garage may enter the dwelling. Windows above the garage should be closed, combustibles removed, and a vent established on the opposite side to alleviate any heat build-up.

B. **0** Keeping the initial attack line in an exterior position is not a good tactic. It will take time to stretch the second line into position in the house. Remember that the position inside the house is most critical for it places the line between the fire and the occupants and the stairway. Don't waste any time getting the initial attack line into the dwelling.

C. **-1** This tactic will also take a good deal of time to accomplish. By the time the ladder and manpower are in place, the fire may have already extended beyond the extinguisher's capability to control it. Also, it is better to use the stream from a handline to knock down this fire than to take a chance with an extinguisher that has limited extinguishment power.

D. **+2** This is the quickest and most effective tactic. Water will be available quickly and will do the quickest job in knocking down the extending fire. The second line can finish the extinguishment, if required, before entering the structure. (Fig. 5–8)

E. **-2** This tactic may pull the fire from the garage into the dwelling. Also, the tactic of stretching the first line to the 2nd floor will possibly allow fire to enter through the wooden kitchen door, possibly trapping men on the line on the 2nd floor. Uncontrolled PPV operations have been the cause of many building losses.

Fig. 5–8 To avoid fire extension to the upper floor via the exterior walls, use a handline to quickly sweep and knock down the extending flames before stretching into the structure. If the area of fire impingement is not close to the interior attack point, an additional line will be required.

3. Knowledge of building construction coupled with the concept of the most effective path of least resistance will lead you to the most appropriate areas to overhaul.

   A. *+2* You must think like fire to conduct a thorough job of overhaul. Complete overhaul of the garage is obvious, but forecasting fire travel will lead to a thorough examination of the other areas mentioned where fire can hide for extended periods of time. Behind exposed exterior walls, above exposed ceilings, around exposed door frames and baseboards above the fire floor are all good places to start. (Fig. 5–9)

   B. *+1* These are also good tactics and will expose hidden fire, but the most direct route for fire travel into the dwelling will be the door between the kitchen and the garage. This must be checked first. In addition, the garage must be thoroughly overhauled.

   C. *-1* While complete overhaul of the garage is crucial, the attic of the dwelling is not a top priority. The attic, while definitely on the list of areas to be checked, is not the most direct means for fire entry into the dwelling.

Fig. 5–9 There are many framed-out areas around both doors and windows for fire to burrow. Open these areas completely to ensure the fire is extinguished.

   D. *+1* The baseboards and the ceiling are definite areas of examination, as is the garage for complete overhaul. However, the area where the most fire exposure would have been, the doorframe, should also be checked. Fire can burrow in the molding and casings around the frame. This area must be opened. Remember that the fire must first pass this area before it reaches the others mentioned.

   E. *0* The overhaul of the dwelling is proper, but declaring the garage to be of questionable stability is not warranted here. Hydraulic overhaul has its place in fire service operations and a check for stability prior to overhaul is certainly warranted, but the scenario makes no mention of structural compromise. Get in there and overhaul thoroughly. If, through recon once inside the garage, the stability has been found to be compromised, then withdrawing and conducting hydraulic overhaul is the correct action. Don't make up your own scenario. It will always get you in trouble on a test and cause more problems than you need on the fireground.

4. Remember that every operation and person at a fire scene, including civilian spectators and support agencies such as the police and EMS, will automatically be the responsibility of the incident commander. He may delegate the authority to another officer to assume the duties of a position such as the safety officer or operations chief, but the responsibility of the position, as well as the obligation for every other person and operation on the fireground will be the responsibility of the incident commander. This being the case, there can only be one good answer choice for this question.

   A. *-2* The answer assumes a shirking of responsibility and poor leadership. These are not the qualities a company or chief officer should exercise.

B. *-1* If is the biggest word in the alphabet. If the previous shift is working the fire, this officer may be at your disposal, but if he is not present, expecting him to take this role is foolhardy. Know who you have on the scene, and, if necessary, choose your safety officer from them.

C. *+2* If the incident commander has not assigned a duty to another person, that duty remains his. Short and sweet.

D. *0* While all company officers should be the eyes and ears of the incident commander and are essentially partners in command, it is not their direct responsibility to assume the duties of the safety officer. To be fair to all, the position and authority must be designated by the incident commander and broadcast, via radio, to all personnel operating on the fireground.

E. *0* Again, all personnel should be the responsible for the safety of everyone else. They should protect themselves and protect each other. However, it is not their direct responsibility unless they are designated by the incident commander to hold the position of safety officer.

5. This question asks about the most significant factor relating to fire destruction in regard to fire spread. This is a stand-alone question and has nothing directly to do with the scenario. Fire destruction denotes a need for a "big picture" mentality to be formed.

A. *+1* The open stairway is one of the most significant fire spread factors in this type of construction. It will rapidly spread the products of combustion throughout the structure; however, it is not as significant in regard to fire destruction as wood-frame construction's ability to easily spread fire beyond the building of origin.

B. *+1* Partitions in this type structure may be covered with highly flammable finishes, as there is no code requirement to regulate what can be used. However, this, again, will cause more interior fire spread and not directly contribute to the widespread fire destruction that is associated with other features of this type construction.

C. *+1* If the roof becomes involved, it can spread flying brands throughout the neighborhood, igniting fires in adjacent exposures. Many conflagrations throughout history have been caused by combustible roofing materials.

D. *0* These dwellings will not exhibit the forcible entry problems found at both dwellings of other types of construction and commercial occupancies. Forcible entry at the normal means of ingress/egress that is proving difficult can often be gained at some other nearby, less fortified area.

E. *+2* The fact that the entire building is made of wood makes it and the exposures very susceptible to ignition by radiant heat and flying brands. This factor will create the largest problem relating to fire spread. This will be due to two factors that may or may not play a large part in the extension process. The first is the close proximity of exposures and the second is the material that the exterior walls are constructed of. If these factors are working against the fire forces, the exposed areas must be protected by keeping them wet.

Scenario 5–2
Short-Answer Questions

1. The first line must advance to the top of the basement stairs via the front door. This line should be a 1¾" line due to the mobility of the line and the need to get water on the fire quickly. This line placement will establish the initial attack line between the fire and any victims on the upper floors. The line is then stretched down the stairs to attack the fire. The advance must be rapid as the attack team will be descending into a chimney. A fire in the basement of any building, especially a balloon-frame dwelling, is a situation fraught with problems. The entire building is exposed to the fire, as are all the occupants. Only an aggressive interior attack executed from an interior position will solve many of the existing problems.

   Stretching via the rear Bilco doors would not only take time, but is likely to push the fire into the dwelling area. This attack position would only be used if for some reason the interior attack line cannot make it down the stairs. Then, the first floor door to the basement would be kept closed and the attack would be from the rear.

2. The second line should be advanced to the same place as the initial attack line, to the top of the interior basement stairs. This line will act to protect the initial attack team and extinguish any fire extending out of the basement. This line can also be utilized to augment the attack in the basement if the first line is stalled. For this reason, this line should be at least the same size as the initial attack line, and probably should be bigger, a 2½" with a solid bore nozzle. (Fig. 5–10)

Fig. 5–10 Photo by Bob Scollan NJMFPA
Lines must be stretched to protect the exposed floors and, if possible, attack the fire from the unburned side. Note the 2½" back-up line being stretched along with the 1¾" attack line

3. The third line is a definite requirement in this type building and should be stretched via an alternate means, definitely not through the same access point as the first two lines. This line, a 1¾", should be stretched immediately to the attic. A fire in a basement in a balloon frame building will take the path of least resistance and often show in the attic due to the open stud channels that will allow fire to spread up through the walls without any building feature to stop it. At any fire in a building that is suspected to be balloon frame, be prepared to fight fire in several areas. You may need a line on every floor to address the fire problem. Recon and pre-control overhaul are the name of the game at balloon-frame structure fires.

4. In this type of structure, ventilation will be required both from the exterior and from the interior. Personnel operating on the exterior will perform OV (Outside Vent) functions. These duties include horizontally venting basement windows opposite and in coordination with the initial attack lines. In addition, the Bilco doors at the rear must be opened to allow a larger area for the products of combustion and steam generated from the hose stream to exhaust. The outside team should then perform VES operations on the upper floors. (Fig. 5–11)

Fig. 5–11 Photo by Bob Scollan NJMFPA
Use the rear exterior "Bilco" cellar door as a vent point to push the products of combustion from the building. For the best results, remove the doors from the hinges

The interior team should enter the basement with the attack team and vent as required as the attack line advances toward the seat of the fire. A lifeline and thermal imaging camera are mandatory and should be part of an SOP-directed tool assignment.

If the situation requires additional ventilation due to an extremely difficult time advancing the attack line in the basement, a hole can be cut in the floor opposite the attack team's advance. This hole should be cut near a window and protected by a hose line.

5. The building has three doorbells and three electric meters. Therefore, it can be surmised that there is an apartment in the basement or the attic, or both. These areas must be searched. Again, the interior crew can do a quick search of the 1st floor as the line is being stretched. Then as the line advances down to the basement, the primary search of that area can be extended. It will be an excellent idea to bring a lifeline in addition to a thermal imaging camera, as there may be unusual layouts and the possibility of many mantraps in the basement.

The exterior team during their VES duties will search the upper floors and the attic. The team stretching a hoseline to the attic can also search that area and adjacent areas off the hoseline.

6. In this case of extending fire, you will have to act fast or the fire will extend past your position. You must expose the entire stud channel and adjacent ones as well. Any stud channel that shows evidence of fire travel must have water applied into them immediately. The line should be discharged upward (and downward) into each suspect channel. Then, get to the floor above, open the baseboard and make a small examination hole in the stud channel there. If you did not get ahead of the fire on the first floor, expect it to be on the second floor and on its way to the attic. You may need more than one line to control the fire. In these types of buildings, pre-control overhaul will be extensive and manpower-intensive operations. Plan for a long battle if the fire is not knocked down in the first few minutes of operation. (Fig. 5–12)

Fig. 5–12 Once fire gets into the walls of a balloon frame building, it will be demand extensive pre-control overhaul to stop it. Here, a fire that was fed by a burning gas meter in the basement followed the pipe chase and stud channels up the front of the building. It was necessary to practically rip the whole front off the building to expose the hidden fire.

Expect to find heavy smoke conditions if fire has entered the concealed spaces. Smoke in these spaces will be oxygen deficient and create poor visibility conditions. If you find a heavy smoke condition inside the building and little evident fire, you must begin extensive pre-control overhaul before you are pushed out of the building.

7. The primary collapse danger in wood frame buildings is the small vertical members holding up larger horizontal members. Sometimes 2" x 4" studs will be holding up 3" x 10" roof and floor joists. Water application will add still more weight to these joists, which may have been deteriorating from the fire attack for quite some time.

    In balloon-frame buildings, the entire exterior of the building (stud channels) will be exposed to fire spread that will feed on floor and roof joists, spreading fire both horizontally and vertically.

8. This question stands alone and changes the whole scenario from a basement fire to a 1st floor bedroom fire. Since the fire is showing signs of rollover, water must be applied quickly to the fire area to relieve the flashover potential. This must be done from a protected area outside the room. In this case, applying water from the bedroom doorway is the best choice. The line, in a tight fog, should be directed at the ceiling to create steam, thereby cooling the overhead and possibly smothering the fire.

    It is critical that ventilation be coordinated here. The OV must take out the rear bedroom windows opposite the attack as soon as the hose team confirms they are in position to attack and they have water at the nozzle. This will allow the steam and the products of combustion to vent opposite the attack line, allowing the line to advance to the seat of the fire. Without proper ventilation, the attack can backfire as the steam will choose the path of least resistance, the doorway from where the attack is being made, and cause burns to the attack team. Proper coordination of attack line operations with ventilation opposite the stream is probably the most critical factor in safe interior fire attack.

9. This is a sequence question. The key is to determine the first action to take in the event the basement becomes untenable. This would be a great multiple-choice question as there are many tactics that may be attempted at untenable basement fires.

The first tactic, as stated in the question would be to withdraw the interior forces and attempt an exterior attack through the windows. If this is successful in knocking down the heavy body of fire, an attempt to once again advance into and gain control of the basement must be made. As this tactic has failed according to the question, the next tactic to try would be to cut a hole in the floor and insert a cellar pipe or Bresnan distributor. The choice between the two will depend on how close over the seat of the fire you can get. While the cellar pipe has a longer reach, about 50', its stream operates in only two directions and must be manned. A Bresnan distributor, on the other hand, is multi-directional in its water application, but has a limited stream reach, only about 15 or 20 feet. However, it does not have to be manned and may be controlled by a gate valve placed a length or two behind the nozzle in a safe position. Conditions will dictate which appliance to use.

If this fails to work, high expansion foam may do the trick. Foam, being mostly water, will deliver both a cooling effect in addition to a smothering effect on the inaccessible fire. However, in order to be effective, high expansion foam must be used as soon as the need arises as extreme heat and fire will eat up the blanket and negate its smothering and cooling effect. For this reason, there must be sufficient quantity of foam on hand to accomplish the task. Not having enough high expansion foam is like not applying it at all. It must be of sufficient quantity to be effective.

Another alternative is to set up a deck gun at the front door and "flow the floor." This may create an inefficient sprinkler system that may find its way to the seat of the fire. This is almost a last ditch effort to save the basement and the building.

The last alternative would be to flood the basement with many master streams through as many openings as possible, attempting to fill it up like a swimming pool. While this may precipitate a collapse of the first floor into the basement, it is, at this point, the safest and most desirable method of applying water to the cellar. At this juncture, the building is doomed anyway.

It is crucial to be cognizant of the fact that while all these operations are underway, the fire, which is not being quenched, is burning away the floor supports. These floor supports will not last forever. A good rule of thumb to follow is if you haven't succeeded in getting to the seat of the fire in the first fifteen minutes or so of operation, a defensive strategy may be the only safe way to operate at a well-advanced basement fire.

CHAPTER SIX
SCENARIOS

SCENARIO 6-1
OLD-LAW TENEMENT CELLAR FIRE

Construction

A fire is in progress in the cellar of 105 West Street, a four-story residential building. 105 West Street is an old-law Class 3 building, with dimensions of 25' x 75'.

The interior cellar stairway is directly beneath the main stairs and is protected by a solid, wooden door on the first floor. The only other entrance to the cellar is in the rear yard, but the cellar door is eight steps below grade and is heavily padlocked.

Time and weather

The time is 0430 and the temperature is 61°F.

Area and street conditions

The streets are virtually deserted at this hour. The area is dominated by buildings similar in construction and occupancy to the fire building.

Fire conditions

On arrival, heat conditions in the cellar are severe, but the first floor is tenable. There are occupants attempting to escape the fire building via the rear fire escape. There are also many people exiting the building via the main stairwell.

Exposures

There are exposure buildings of similar type and construction attached on either side. The rear yards are fenced.

Water supply

The water supply is adequate for the fire load and construction.

Response

Your first alarm assignment is three engine companies and one ladder company. Engine staffing is an officer and two firefighters. Ladder staffing is an officer and three firefighters. Appropriate engine companies have secured their own water supply and the first due ladder truck has been placed directly in front of the fire building.

Short-Answer Questions

1. Where is the first line placed and why?

_____

_____

_____

_____

_____

_____

2. Where is the second line placed and why?

_____

_____

_____

_____

_____

_____

3. What would be the duties of the first-arriving ladder company?

_____

_____

_____

_____

_____

_____

4. Would additional alarms be necessary at this fire?

_____

_____

_____

_____

_____

_____

5. Where would you position a second ladder company and what would be their duties?

_____

_____

_____

_____

_____

_____

6. Suppose there is an exterior basement entrance located at the front of the building. To access the basement from the front, you must descend three steps to the exterior door. What changes to hose line placement and attack operations would be warranted in this situation?

_____

_____

_____

_____

_____

Multiple-Choice Questions

7. Regarding old-law Class 3 construction, what would be your major concern?

    A. Collapse of the parapet wall in the initial stages of the fire.

    B. Vertical extension via the interior stairs.

    C. Vertical extension via the pipe chases.

    D. Horizontal extension across the basement ceiling and vertical extension via the exterior wall studs.

    E. Auto-exposure into the 1st floor windows.

8. Regarding the life hazard, what would be your major concern at this incident?

    A. Protect the interior stairs.

    B. Assist occupants down the fire escape.

    C. Extinguish the fire.

    D. Control panicking occupants.

    E. Spandrel wall collapse due to fire exposure.

9. Regarding the location and extent of the fire, what major problem will be encountered by firefighters?

    A. Difficulty in advancing downstairs to attack the fire.

    B. There should be no problems at all in this type of construction and occupancy.

    C. Difficulty in locating the fire.

    D. Difficulty with ventilation.

    E. Difficulty with forcible entry.

10. How would you order a further increase in ventilation, if it were required at this fire to advance the attack line?

    A. Use fog streams behind the attack team.

    B. Use positive pressure fans at the top of the interior stairs.

    C. Use positive pressure fans at the rear exterior entrance to the basement.

    D. Use a sledgehammer to breach the cellar wall opposite the direction of the advance.

    E. Cut a hole in the rear on the 1st floor near a window opposite the direction of the advance.

Scenario 6–2
New-Law Building Basement Fire

Construction

In this scenario, we have 38 Main Street; a new-law Class 3 apartment building that is six stories tall. The open interior stairway terminates at a bulkhead door on the roof. There are four fire escapes. There is also a grade-level passageway that runs through the building from the front, leading to a courtyard, the basement entrance, and the rear yard. There are eight apartments per floor.

Time and weather

It is now 1705; the temperature is 80°F and the humidity is 70%. The wind is blowing towards the west at 10mph.

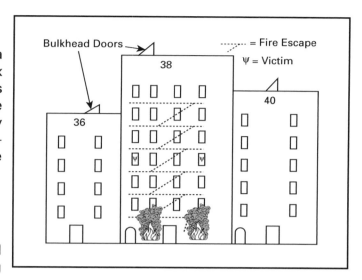

Area and street conditions

Rush hour has begun and the area is busy with afternoon traffic.

Fire conditions

The fire is located in the basement of 38 Main Street and smoke and flame issuing from the front windows facing the street. The building is occupied. There are people showing at two 3rd floor windows. It is unknown whether or not the basement is occupied.

Exposures

Exposure B is four stories tall and exposure D is five stories tall. Both exposures have bulkhead doors on their roofs. These buildings abut 38 Main Street and are of identical construction.

Water supply

The water supply is adequate.

Response

Your response is three engines and one ladder company. The engine companies are staffed by one officer and two firefighters. The ladder company is staffed by an officer and three firefighters.

SHORT-ANSWER QUESTIONS

1. Properly position the first line.

_____

_____

_____

_____

_____

_____

2. Properly position the second line.

_____

_____

_____

_____

_____

_____

3. Properly position the third line.

_____

_____

_____

_____

_____

_____

4. Discuss proper ladder company operations at this fire.

_____

_____

_____

_____

_____

_____

5. You have been assigned to check for fire extension on the 1st floor. Where would you expect the most likely places for extension to occur?

_____

_____

_____

_____

_____

_____

6. In buildings of this type (and in consideration of the location of the victims in relation to the fire) what would be the safest way to remove these people from the building?

_____

_____

_____

_____

_____

_____

SCENARIO 6–3
TOP FLOOR MULTIPLE DWELLING FIRE

Construction

The fire building, 225 South Street, is a six-story apartment building of ordinary construction. The open interior stairway terminates at a bulkhead door on the roof. There are two fire escapes; one at the front and one at the rear. There is also a passageway that runs under the building from the front leading to the rear yard and the basement. There are four apartments per floor.

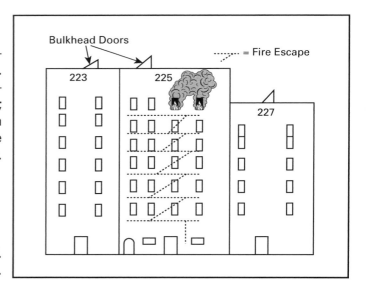

Time and weather

The time is 1515 and the temperature is 40°F. The wind is blowing towards the east at 15mph.

Area and street conditions

It is Sunday. Traffic is light, but many cars are parked and hanging off street corners, which may impede aerial access into the block.

Fire conditions

Fire is located on the 6th floor of 225 South Street. Upon arrival, no smoke alarms are sounding. Smoke is issuing from front windows on the top floor.

Exposures

Exposure B is six stories tall and Exposure D is five stories tall. There are bulkhead doors on the roofs of both exposures. These buildings abut 225 South Street and are of ordinary construction.

Water supply

The building has no auxiliary protective systems.

Response

Three engine companies, one ladder company, and a rescue company respond. An officer and two firefighters staff the engine companies. An officer and three firefighters staff each of the ladder and the rescue companies.

Short-Answer Questions

1. Compare the problems of top floor fires in these buildings to basement or cellar fires.

_____

_____

_____

_____

_____

_____

2. Where should the first line be positioned?

_____

_____

_____

_____

_____

_____

3. What factors would come into play regarding hose diameter and length of stretch?

_____

_____

_____

_____

_____

_____

4. Where would the second line be positioned?

_____

_____

_____

_____

_____

_____

5. How should the rescue company be assigned?

_____

_____

_____

_____

_____

_____

6. What ladder company operations should take place inside the fire and adjacent apartments?

_____

_____

_____

_____

_____

_____

7. What operations should take place on the roof?

_____

_____

_____

_____

_____

_____

Scenario 6–4
Braced-Frame Multiple Dwelling Fire

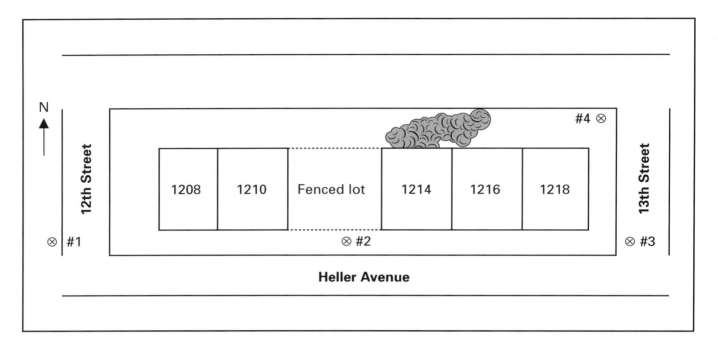

Construction

The fire has been reported at 1214 Heller Avenue. The building at this address is a three-story building of wooden braced-frame construction. Open wooden stairways run from the 1st to the 3rd floor. The roof is flat. There is a fire escape at the rear of the building that does not serve the roof. There is a scuttle hatch on the roof over the interior stairs. The front of the building is covered with vinyl siding, while the other three sides are covered in asphalt shingles. The cellar is not finished and there are small windows located on all sides of the building at the cellar level. There are six electric meters on the side of the building and six doorbells on the front of the building.

Time and weather

At the time of arrival, 0125, the temperature is 5°F with a 15mph wind.

Area and street conditions

Heller Avenue is a two-lane, one-way street running from west to east. 12th Street is a two-lane, two-way street. 13th Street is a one-lane, one-way street running from north to south. As you pull into the block, there is no one around.

Fire conditions

Upon arrival you find heavy smoke issuing from the 2nd floor windows on Side C. There is also smoke issuing from the 3d floor windows on Side C, but of a lesser volume.

Exposures

The fire building is attached on Side D. Exposure D is identical to the fire building.

Water supply

Water supply is adequate; however, frozen hydrants will be a concern due to the weather.

Response

Two engine companies and one ladder company respond to this alarm. An officer and two firefighters staff each engine. An officer and three firefighters staff the ladder company.

MULTIPLE-CHOICE QUESTIONS

1. What are your orders for Engine 1?

    A. Establish a water supply at Hydrant #1. Position the apparatus just past the fire building. Stretch a preconnected 1³/₄" line through the front door. Advance up the stairs to the 2nd floor to locate, confine, and extinguish the fire after the stairwell has been cleared of occupants.

    B. Establish a water supply at Hydrant #3. Position the apparatus just past the fire building. Stretch a preconnected 1³/₄" line through front door to the top of the 2nd floor stairs. Use this line in a holding position to allow a primary search to be completed and to protect egress from the 2nd floor. Attack the fire after any victims have been removed.

    C. Establish a water supply at Hydrant #2. Position the apparatus just past the fire building. To keep the interior stairway free for occupant egress, stretch a preconnected 1³/₄" line to the 2nd floor via a ground ladder to locate, confine, and extinguish fire.

    D. Establish a water supply at Hydrant #1. Position the apparatus just past the fire building. Stretch a preconnected 1³/₄" to the 3rd floor via the interior stairs. Locate, confine, and extinguish the fire.

    E. Establish a water supply at Hydrant #2. Position the apparatus just past the fire building. Stretch a preconnected 1³/₄" line to the 2nd floor via the interior stairs. Locate, confine, and extinguish the fire.

2. What are your orders for Engine 2?

   A. Establish a second water supply at Hydrant #2. Stretch a 1¾" line into the dwelling to the 2nd floor to back-up Engine Company 1's attack; extinguish any fire extending into the 3rd floor.

   B. Establish a second water supply at Hydrant #3. Stretch a preconnected 1¾" line through front door to the 2nd floor. Back-up Engine 1's attack line and also stretch to 3rd floor via interior stair to check for and extinguish any concealed fire.

   C. Establish a second water supply at Hydrant #3. Stretch a 1¾" line up the fire escape to the 2nd floor to supplement Engine 1's fire attack; utilize a pincer tactic to extinguish the fire.

   D. Stretch a preconnected 2½" line into the lot to the rear of the building. Utilize the hoseline to knock down any fire spreading via the exterior wall and protect the rear of the exposure.

   E. Establish second water supply at Hydrant #3. Stretch a preconnected 1¾" line to the 3rd floor to cut off any extending fire. Conduct a primary search off of this line. Horizontally ventilate at windows as required.

3. Would additional alarms be required at this incident?

   A. No, the fire area is relatively small. The on-hand complement of men and apparatus will be sufficient to handle this incident.

   B. Yes, strike a second alarm just in case fire is inaccessible. Stage apparatus and men on 12th Street. As they may not be needed for this fire, leave them on an "in-service at the scene" status in case other alarms are received in the district.

   C. Yes, strike a second alarm because of the possibility of a conflagration due to the presence of highly combustible asphalt siding.

   D. Yes, strike a second alarm due to the severe weather conditions.

   E. Contact your superior via radio. Advise your superior of the situation and current conditions. Advise that you feel an additional alarm is necessary. Ascertain their time to arrival.

4. How would this fire best be ventilated by the ladder company?

   A. Provide horizontal ventilation of the 2nd and 3rd floor windows from the exterior via the fire escape and ground ladders. Vent from the interior as conditions allow. Vertically vent the building by opening the scuttle to clear the stairwell. Check the cockloft for any presence of fire.

   B. Provide exterior horizontal ventilation of windows on the 2nd floor opposite the attack. Perform interior horizontal ventilation of the windows on the 3rd floor.

   C. Open the scuttle to vent over the stairs. Cut an appropriately sized hole in the roof. Push down the top floor ceiling. Horizontally vent windows on the 2nd and 3rd floors from the interior.

   D. Use ground ladders to knock out the 2nd and 3rd floor windows. Open the scuttle to vent over the stairs.

   E. Perform vent, enter, and search on the 2nd and 3rd floors via the rear fire escape. Access the 2nd floor via the interior and horizontally ventilate the windows opposite the attack.

5. Regarding search and rescue, what orders would you give to the ladder company?

   A. Keep the company intact. Perform a systematic primary search of all floors. Then perform a secondary search.

   B. Split the company into crews. Crew 1 conducts a primary search of the 3rd floor. Crew 2 searches the remainder of the structure, including the cellar.

   C. Split the company into crews. Crew 1 searches the 2nd floor and Crew 2 searches the 3rd floor. The remainder of the building will be searched when time and conditions allow.

   D. Keep the company intact. All ladder company members perform a primary search of the 3rd floor. Split the company into crews and then search the other floors in a systematic manner.

   E. Split the company into crews. Crew 1 searches the 1st floor and the cellar. Crew 2 searches the 2nd and 3rd floors.

6. The interior companies are reporting that progress is being made on the fire. Where would you stretch a third line and how would you get it there?

   A. Stretch to the 3rd floor via the fire escape to cover any extension.

   B. Stretch to the 1st floor stair landing to cover the occupant evacuation.

   C. Stretch to the exterior rear. Use the line to keep fire from auto-exposing into the 3rd floor via the window. Use the stream to wash the wall above the window. Keep water out of the window.

   D. Stretch to the top floor of Exposure D to guard against fire extension via the cockloft.

   E. Have the line stand by until its required location is ascertained by the ladder company performing checks for fire extension.

7. Suppose the fire had extended to seriously involve the 3rd floor. What would be the best way of protecting the adjacent exposure?

   A. Make a trench cut between the fire building and the leeward exposure. Stretch lines to the roof and the top floor of the exposure. Use the lines as a defensive fire barrier to prevent involvement of the exposure.

   B. Stretch an additional line to the top floor of the fire building to provide more fire flow to control the fire. The additional water will control the fire and protect the exposure

   C. Evacuate the fire building. Use a master stream to knock down the heavy body of fire on the 3rd floor. Stretch lines to the top floor of the exposure. Pull the ceilings and direct streams into the cockloft space.

   D. Cut a large hole in the roof of the fire building as safely over the fire as is possible. Stretch lines to the leeward exposure. Pull the ceilings on the top floor to expose any fire traveling in the cockloft. Direct streams into the cockloft to prevent lateral extension. Monitor the exposure cockloft from the roof by cutting examination holes.

   E. Cut a large hole in the roof of the fire building over the heaviest concentration of fire in the cockloft. Push the ceiling down into the fire apartment. Cut examination holes in the roof of the leeward exposure.

8. In assessing the collapse profile of the building's walls, what type of collapse could you expect from this type of structure?

   A. Lean-over collapse
   B. Inward-outward collapse
   C. Curtain fall collapse
   D. 90° angle collapse
   E. Tent collapse

9. What type of building failure would most likely cause this type of collapse?

   A. Failure of the front or rear bearing wall
   B. Fire escape collapse
   C. Failure of a side bearing wall
   D. Failure of the mortise and tenon joints
   E. Partition failure at the center of the dwelling

10. You are supervising the overhaul operation on the 2nd floor. The incident commander has cautioned against causing unnecessary water damage to the dwelling. How can you best accomplish this?

    A. Leave the charged line outside the building until it is determined that it is needed.
    B. Leave the nozzle hanging outside the window in the overhaul area until it is needed.
    C. Use an extinguisher instead of a hoseline.
    D. Do not open the nozzle until it is absolutely required for fire extinguishment.
    E. Spread salvage covers across the floor and over all furniture before beginning overhaul.

Scenario 6–5
Fire-Resistive Multiple Dwelling Fire

Construction

The fire building, 633 Pacific Avenue, is a fire-resistive multiple dwelling built in the late 1960s, and is six stories tall. There are no elevators or automatic fire suppression systems in these buildings. There is a compactor chute that pierces all floors and terminates at the roof. There are also no fire escapes on the building. All stairways are enclosed. Access to the roof is via the enclosed stairwells that terminate at the roof at a bulkhead door on both the north and south ends of the building. There are self-closing fire doors on each end of the hallways.

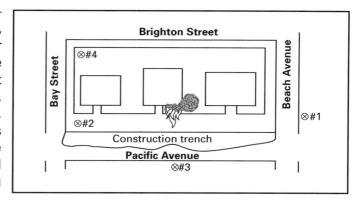

Time and weather

The time is 1230. The temperature is in the low 70s and there is no breeze to speak of.

Area and street conditions

Your company is "on the air" when you notice a heavy volume of black smoke issuing from a six-story building. In front of the fire building and for the entire block to the corner is a construction trench that has been dug to replace the aging sewer system in the area. The trench is about 10' wide for the length of the block. As it is lunchtime, there are no construction workers in the area.

Fire conditions

The smoke seems to be emanating from the 3rd floor, south side. The fire alarms begin to sound as you arrive.

Exposures

It is part of a three building project complex of identical construction that stretches out for a full city block.

Water supply

Hydrants in the area are spaced well and are all on separate mains. The hydrants on the construction side of Atlantic Avenue are usually not in service during the construction period each day, but there have been no notifications of out-of-service hydrants so far this day. This is not unusual, as there have been many days when the construction companies have failed to notify the dispatch center of out-of-service hydrants.

Response

Three engines and one ladder respond. Each is staffed by an officer and three firefighters

MULTIPLE-CHOICE QUESTIONS

1. What is the most important size-up factor at this fire?
   A. Construction
   B. Street Conditions
   C. Water Supply
   D. Life Hazard
   E. Weather

2. What is your first action at this fire?
   A. Request a full alarm assignment.
   B. Establish a command post and give a size-up report to the dispatch center.
   C. Stretch a line of appropriate size into the building.
   D. Secure a water supply at the appropriate hydrant.
   E. Use the deck gun to knock down heavy fire on the fire floor.

3. What are your orders to the crew of Engine 1?
   A. Secure a water supply at Hydrant #1. Position the apparatus just past the fire building. Stretch a preconnected 1¾" attack line to the fire building after a ground ladder has been laid across the trench. Advance via the interior stairs to locate, confine, and extinguish the fire.
   B. Secure a water supply at Hydrant #1. Position the apparatus on the corner of Beach Street. Stretch a 1¾" attack line down the sidewalk on the building side of the construction trench. Advance into the building to the 3rd floor via the interior stairs. Locate, confine, and extinguish the fire.
   C. Establish a water supply at Hydrant #2. Position the apparatus on the corner of Bay Street. Stretch a 1¾" line down the sidewalk on the building side of the construction

trench. Advance into the building to the 3rd floor via the interior stairs. Locate, confine, and extinguish the fire.

D. Establish a water supply at Hydrant #3. Position the apparatus just past the fire building. Place a ground ladder across the trench. Advance a preconnected $1^3/_4$" attack line via the interior stairs to locate, confine, and extinguish the fire.

E. Establish a water supply at Hydrant #1. Leave a crew at the corner of Beach Street and Pacific Avenue. Have the crew walk down the sidewalk on the building side of the trench with a rope. Position the apparatus just past the fire building. Use the rope to pull a preconnected $1^3/_4$" attack line across the trench to the front of the building. Advance a preconnected $1^3/_4$" attack line via the interior stairs to locate, confine, and extinguish the fire.

4. What are your orders for Engine 2?

A. Secure a secondary water supply at Hydrant #3. Assist with the initial stretch, then stretch a second attack line from Engine 1 into the building to assist and back-up Engine 1's attack line and/or advance to the floor above.

B. Secure a second water supply at Hydrant #2. Have the crew stretch an attack line down the sidewalk on the building side of the trench. Advance this line into the building to back-up Engine 1's attack line and/or advance to the floor above.

C. Secure a secondary water supply at Hydrant #3. While assisting in the initial attack line advancement, use the deck gun to knock down heavy fire as the line is being advanced into the interior. Shut down the deck gun when ordered by the company on the interior.

D. Secure a secondary water supply at Hydrant #2. Place a ground ladder across the trench. Stretch a second attack line into the building to assist and back-up Engine 1's attack line and/or advance to the floor above

E. Secure a secondary water supply at Hydrant #4. Stretch a back-up line through the rear yard to the front of the fire building. Advance up the interior stairs to back-up Engine 1's attack line and/or advance to the floor above.

5. What are your orders for Ladder 1?

A. Position the apparatus on Bay Street. Split the company into crews. Crew 1 performs a primary search and horizontal ventilation of the fire apartment and the fire floor. Crew 2 performs a primary search and horizontal ventilation of the floors above the fire.

B. Position the apparatus on Pacific Avenue. Raise the aerial ladder to the roof of the fire building. The company ascends via the aerial and descends via the opposite stairwells to the fire floor to execute a primary search and horizontal ventilation of the fire floor.

C. Position the apparatus on Pacific Avenue. Split the company into crews. Crew 1 places a ground ladder across the trench. Enter the building via the front entrance. Perform a primary search and horizontal ventilation of the fire floor and the floor above. Crew 2 ascends via the aerial to the roof of the fire building. Open the bulkhead doors and all natural openings. Leave the roof via the aerial and enter the building via the front to assist in search and vent operations of upper floors.

D. Position the apparatus on Pacific Avenue. Entire company places a ground ladder across the trench. Then, split the company into crews. Crew 1 performs a primary search of the fire floor and the floor above. Crew 1 also performs a horizontal ventilation of the fire area. Crew 2 accesses the roof via the interior stairs furthest from the fire. Open all natural openings as well as the bulkhead stairwell door of the stairwell closest to fire. Then, descend through the safe stairwell to search and vent the top floor.

E. Position the apparatus on Pacific Avenue. Split the company into crews. Crew 1 places the aerial to the roof of 625. Use a ground ladder to bridge the roof. Perform a vertical ventilation of the building by opening both bulkhead doors and all natural openings. Crew 2 places a ground ladder across the trench and enters the fire building via the front entrance. Crew 2 performs primary search and horizontal ventilation operations on the fire floor, the floor above, and the top floor.

6. If you were assigned as the captain of Ladder 1, what tool(s) would you carry in regard to forcible entry in this type of building?
   A. Rabbit tool and flathead axe
   B. K-tool and flathead axe
   C. Maul and flathead axe
   D. Halligan tool and flathead axe
   E. Pike pole and flathead axe

7. At structures of this type, what is your most critical concern regarding floor to floor fire spread?
   A. Concealed spaces
   B. Common cockloft
   C. Combustible exterior walls
   D. Open stairwells
   E. Autoexposure

8. The RIT or FAST team has arrived on the scene and is awaiting your orders. What action do you take regarding this team?
   A. Stretch a line to the front door to place between the endangered firefighters and their exit.
   B. Perform a walk-around of the structure to size-up access/egress points.
   C. Report to the interior division chief.
   D. Set up the command board and study the tactical worksheet to familiarize yourself with the operating companies' locations and tasks.
   E. Stand by and await orders.

9. Suppose on your arrival, fire was blowing out the windows of the 3rd floor, severely threatening the 4th floor. Attack crews are in the process of stretching attack lines into the

building. What would be your best tactic regarding the abatement of the threat of fire extending into the floor above?

A. Radio advancing crews to take refuge in the stairwell. Use the deck gun to knock down heavy fire. Then, contact interior crews and tell them to advance and finish the extinguishment of the fire.

B. Put the fire out and all other problems will go away.

C. Request a progress report from the interior crews. Base your next move on their report.

D. Stretch a line to the floor above the fire. Operate this line to keep fire out of the apartment. Remove any combustibles from the windows. Ventilate opposite of the exposure.

E. Use the deck gun or a handline to wash the spandrel wall above the fire floor and allow water to flow down the exposed wall. Do not put water into fire floor/apartment/floor above. Once crews are in position and ready to advance, shut this line down.

10. Suppose the fire on the 3rd floor was a small fire and easily extinguished. Your task is to vent the floors above. The smoke condition in the hallway is light to moderate on the uninvolved floors above the fire. How would you vent these floors?

A. Block open the compactor door on the affected floors. Ensure that the roof above the shaft is open.

B. Place a PPV fan on the affected floors. Open the doors at the end of the hall and the bulkhead doors. Blow products of combustion off the floor and out of the vertical artery.

C. Open the doors to all of the apartments on the affected floors.

D. Operate a fog nozzle out of the stairwell window to pull the smoke from the affected floors

E. Leave the bulkhead doors open. Allow the floor to air out naturally.

ANSWER SECTION

SCENARIO 6–1
SHORT-ANSWER QUESTIONS

1. The first line must be placed to protect the vertical artery, the open interior stairs. Placing this line at the top of the stairs to the cellar positions the line between the fire and the occupants. The door to the cellar should be kept closed until all occupants have been evacuated down the stairs and a back-up line is in place. Once these tasks have been accomplished, the line is rapidly stretched down the cellar stairs to locate, confine, and extinguish the fire. This will usually take the manpower of the first two engine companies. As the top of the stairs will be most discomforting for the attack team, the descent must be made quickly. Once at the floor level of the cellar, the conditions are not likely to be as severe. This is not to say that conditions in the cellar will be pleasant, but unless the fire is right at the cellar stair door, there will generally be less heat at the floor level than at the door as heat will rise up the stairwell.

    Generally, cellar fires in this type of building are usually tough and hot. Aggressive fire attack along with aggressive and sometimes creative ventilation operations will make the difference between the successful operation and the withdrawal of personnel.

2. The second line should be stretched by the third engine company via the same access point as the first line, the front door. This is, of course, the safest, most effective, path of least resistance to accomplish the objective desired (in this case, the placing of a line between the fire and the victims in order to protect the most significant vertical artery, the open stairway). This second line must remain at the top of the stairs and extinguish any fire extending into the 1st floor by way of the cellar stairs. If this line must descend the stairs to assist in the extinguishment operation, a third line must be stretched to protect the vertical cellar opening.

3. The first-arriving ladder company will have numerous duties to perform to accomplish life protection and fire confinement operations. Ladder company SOPs should address the critical tactical areas in the building. These are the interior, the rear, and the roof. It is no coincidence that these are also the areas that are not directly visible from the command post. Proper and timely recon reports from these areas are imperative as they will greatly assist the incident commander in the development and maintenance of an effective action plan.

    A primary search will need to be conducted in the cellar, which will include locating the fire and removing obstacles to the hoseline. If the crew descends the stairs prior to hoseline advancement, it will be necessary to use a lifeline as well as a thermal imaging camera. Cellars, even small ones, are relatively easy to get lost in. If possible, horizontal ventilation of the cellar should be performed, if the fire will not be spread by this action. The primary search and fire extension check must be extended to the upper floors as well, especially the 1st floor

(floor above) and the top floor. One of the most important tasks the team searching the 1st and upper floors can do is to ensure the cellar door in the 1st floor hallway is closed. If it is left open, fire can roar out of the cellar, trapping them on upper floors. Firefighters have been killed and severely burned by fire roaring up the open stairway via an open cellar door.

Horizontal ventilation of the door at the rear is critical for safe cellar operations and hose-line advancement at this incident. Any windows found in this area should also be vented. After this is accomplished and water is being applied to the fire, any other windows found should be vented. The name of the game in cellar fires is total, albeit controlled, ventilation. Recon of the rear for victims and fire extension is also a priority. Members operating at the rear should assist occupants down the fire escape.

Roof operations will include accessing the roof via the upwind exposure (preferred) or the aerial. The bulkhead door must be opened and removed from its hinges to prevent reclosing. The area immediately inside the bulkhead opening must also be probed for victims. Any other natural openings that are found on the roof must be checked for heat and opened as required. (Fig. 6–1) These buildings some-times have scuttles or skylights instead of bulkheads. While the ventilation will not be as effective, these must be opened nonetheless. It is also a good idea, if there is an access door, to check the condition of the cockloft. This is most safely accomplished from the

Fig. 6–1 Natural openings must be vented at the roof level. These will include sky-lights, scuttle hatches, and bulkheads. Note the narrow shaft located between these attached buildings just behind the sloping bulkhead door area.

bulkhead stairwell of the attached exposure. At this time, it will most likely not be involved, but it is better to be safe than sorry. As the fire is in the cellar, there is no need to cut the roof.

After roof operations are complete, the team should descend the rear fire escape via the gooseneck ladder and conduct VES operations on the top floor. (Fig. 6–2) Finally, the floors between the top and the first floor should be searched and evacuated.

Ladder company members operating in different parts of the building must report results of searches and fire extension checks as soon as possible to incident command. Any area that cannot be accessed must also be reported so that the incident commander can assign another company to that task.

Fig. 6–2 After completing roof duties and reporting roof conditions to Command, the roof team must descend the rear gooseneck ladder and conduct a VES (Vent, enter, search) operation of the top floor apartments. This fire escape serves two attached buildings.

4. The time is 0430, and the building is occupied. The fact that there are occupants attempting to evacuate via the interior stairs and the rear fire escape denotes a severe life hazard. Remember that these are just the people you can see. There may be a great many more still trying to get out, trapped in illegal cellar apartments or unaware that a fire is in progress. In addition, the building is

attached, which may mean spread to the exposures via the adjoining cellars. Checking all these areas requires a large commitment of manpower. An additional alarm must be requested and possibly a third alarm if the fire in the cellar proves extensive or difficult to locate.

5. This question may cause you to deduce that a second alarm is required, especially if only one ladder is on the initial response card. That is why it is important to read all the questions before attempting to answer.

    The second-arriving ladder, in this situation, should position in front of the exposure building. This way, the ladder is available for equipment and aerial operations, if required.

    The duties of the second-arriving ladder company will be to reinforce or accomplish any unfinished or difficult operations that were assigned to the first ladder company. This may be to supplement the search, check for fire extension, and assist in ventilation efforts in the fire building. In addition, the second ladder company should examine the rear of the building and assist in fire escape evacuation. Do not forget the duty of utility control as well (this also could be assigned to the second-arriving ladder company).

    The second ladder company should also examine the adjoining cellars for fire extension and request line support if required.

6. It is much easier to attack a fire from the same level or below than it is to descend a veritable chimney to fight it. Therefore, the attack should now be made from this front, exterior cellar entrance. This is the safest, most effective path of least resistance to accomplish the objective of fire extinguishment. However, even though this is the best path of attack, this is not the position of the first line stretched. The placement of the first line does not change. It must, without exception, still be positioned at the top of the cellar stairs to protect the vertical artery. In this case, however, the first line remains at the top of the stairs while the attack line and back-up line are advanced from the front, exterior entrance. To assist in the advance, the crews at the interior first floor cellar door may, after the evacuation has been completed, intermittently open the cellar door to aid in the vertical ventilation of the cellar. Their tasks will include extinguishing any fire extending out of the cellar. It must be remembered that there are crews operating below. For this reason, under no circumstance, should water be applied down the cellar stairwell. This will endanger the attack crews in the cellar.

    As mentioned earlier, the cellar door on the 1st floor must be closed. Usually the ladder crew entering to search and evacuate the 1st and upper floors will be in the building before the line is stretched. Sometimes the lines into the 1st floor and the cellar are stretched simultaneously, especially if manpower is sufficient. However, strict control must be exercised over the operation of these lines. If the line is operated from the cellar before the 1st floor line is in place and/or the cellar door is closed, it can cause a blast of heat and fire to erupt from the cellar. This blast of heat can immediately pyrolyze the hallway and stairwell walls, causing instantaneous ignition all the way up the stairs as the heat and fire from below feed on the combustible gases liberated from the walls. This is especially true if flammable or combustible varnishes were used to seal the walls. The flame plume will rise until the gas/air mixture is too lean to cause flaming, however, the heat will travel further. This ignition is called a traveling flashover and has caused death and injury to firefighters operating on the floors above the fire floor. If the door is closed and a line is in place at the 1st floor cellar door, chances are this will not happen. The door will act as a barrier to the heat and the line will protect the 1st floor.

Multiple-Choice Questions

7.

A. **0** Collapse of the parapet wall is more of a concern for new-law construction, and is usually not present in old-law construction, especially at the front. Usually found at the front is the decorative metal cornice, although parapets may be found at the sides and rear if the building is unattached. Regardless, these cornices or parapets are not usually prone to collapse in the initial stages of the fire, unless there is significant direct flame exposure, the building feature is in a deteriorated condition, or there have been previous fires that have weakened it.

B. **+2** This is the primary means by which fire will spread to upper floors and endanger life. Recognition of this concern will prompt the incident commander to commit a major portion of manpower to keep the fire from controlling this artery. The victor of this battle for control of the stairwell will likely win the war.

C. **+1** While fire travel in these structures will certainly be influenced by pipe chases, this will not usually be the major contributing factor to the problems involved in fighting a fire in this type building.

D. **-1** Horizontal fire travel will almost always be slower than vertical fire travel. It is certainly not your major concern. In addition, as the walls are brick, there will be very minimal, if any, fire extension via vertical channels at the exterior walls. These buildings, unlike wood-frame buildings, and especially balloon frames, are simply not built this way.

E. **+1** Auto-exposure into the 1st floor windows in this type of construction is unlikely, but if the fire is blowing out of a rear cellar window, it is a possibility. Recon efforts must ensure that any extension problems are addressed immediately. (Fig. 6–3) The auto-exposure hazard will be more acute at frame tenements as the fire will likely feed on the combustible exterior en route to the floor above.

Fig. 6–3 Photo by Bob Scollan NJMFPA
Reconnaissance of the rear of the building is essential. Reports of conditions must be made as soon as possible so that Command can take action to address these areas.

8.

A. **+2** All of the apartments will open into the interior stairs. As such, most people will attempt to flee the building by this route. Protecting the interior stairs by properly placing a hoseline there is the best way to safeguard the life hazard.

It has been said that the more deviation from the normal means of exiting the building the occupants take, the more time it will eat up and the more danger is involved. In addition, more manpower will be required. This is why the interior stairs, which are the quickest, normally safest, and the most direct way out of the building must be maintained. Sounds a little like safest, most effective, path of least resistance, doesn't it? This little rule of thumb applies to a multitude of situations.

B. **0** As there is only one fire escape at the rear, there should only be two apartments per floor in a railroad configuration (where apartments that occupy one half of the floor have the rooms arranged from front to rear to be similar to cars in a train).

This is definitely a concern, and any person on a fire escape at a fire, especially children, occupants with infants, and the elderly, must be considered a life hazard. However, these people are already on the exterior of the structure. The real life hazard and the major concern will be those occupants still on the inside trying to escape via the interior stairs.

C. *+1* This is the safe answer. Obviously, extinguishing the fire will make all the other problems go away and should be accomplished as soon as possible, but the protection of the interior stairs will be the major concern regarding life hazard. If there are any problems with the successful and rapid extinguishment of this fire, having the stairs protected will often make the difference between evacuation and entrapment.

D. *+1* Panic is contagious and will cause people to act irrationally. The best way to prevent panic of occupants is to control the situation. Access point control will allow the occupants to quickly evacuate the building in a timely manner. Protection of the stairway is pivotal to avoiding widespread panic.

E. *-2* The spandrel wall is that part of the wall between the top (lintel) of one window in a building and the bottom (sill) of the one above it. This will not be a major concern as far as life hazard is concerned. If this were to become a concern, interior firefighting will have already shifted to a defensive strategy and collapse zones will have been established.

9.

A *+1* The stairway to the cellar may be a chimney of venting smoke and hot gases. If the area is tenable, the descent should be made as rapidly as possible. However, any ventilation in the fire area is likely to move the products of combustion to the exterior quickly, making the trip to the cellar via the interior stairs more bearable.

B. *-2* If there were no problems, there would be no need for this question or any of the others as well. The mission of the incident commander and line officers is to forecast the problems likely to be faced and develop solutions based on sound judgment. An action should then be developed to overcome them.

C. *+1* Many cellars are divided into partitions and conducive to unusual layouts. However, the primary reason that the fire will be difficult to locate is due to the fact that visibility will be severely restricted.

D. *+2* The visibility in the basement will be severely restricted due to the limited opportunities for effective ventilation. Usually, in old-law buildings, the entire cellar is below grade, with only small windows at the rear or in a grate near the building's wall. If lines are unable to advance, creative ventilation tactics may be required. If you can't properly vent, the line can't advance. If the line can't advance, the seat of the fire cannot be located or attacked. If this is not done, the interior crews will be forced to withdraw and change to a defensive strategy. If the situation deteriorates to this point, the tactics regarding operations in a well-advanced cellar fire discussed in the private dwelling chapter may be useful.

E. *0* Forcible entry is usually not a major problem in these structures. There will be times when the lock is more sophisticated than the entire door. The damage created by forcible entry should be directly proportional to the emergency at hand. Minor emergencies necessitate minor damage or looking for an easier way in. In major emergencies where life is at stake and fire spread potential is great, even substantial damage is justified.

10.

A. *-2* Fog streams will create steam and cause the firefighters operating the attack line to be burned. If anything, a 2½" back-up line with a solid bore nozzle may provide reach and penetration into an untenable area.

B. *-1* Positive pressure fans are unpredictable during fire attack and, in this case, may cause turbulence in the cellar, whipping fire around. If the seat of the fire cannot be located, use of the PPV fan may unnecessarily endanger the attack team.

C. *-2* This will negate the benefit of the only natural vent point available. It will also push the products of combustion opposite the nozzle back at the attack team. This is a dangerous option and one that should be forbidden.

D. *0* This will take an awful long time to do as the walls are usually below grade and constructed of brick or something stronger like bluestone or solid rock. There are other alternatives that may accomplish the task easier.

E. *+2* This is a quick and effective method of supplementing the available ventilation openings. A charged line must be in place on the 1st floor to ensure that fire does not extend in a major way to the 1st floor. This stream should not be applied into the hole or the vent will be negated.

Scenario 6–2
Short-Answer Questions

1. The manpower of the first two engine companies should stretch the first line. Once the line is in place, the manpower of the second engine may assist on this line or be assigned to stretch the second line. Stretch the 1¾" line through the front, grade-level passageway opening to the basement door in the courtyard. The more mobile, smaller diameter 1¾" line will get to the basement the quickest. If faced with heavy fire on arrival, it may be wise to stretch a 2½" line from the outset. Advance the line through the courtyard basement door to locate, confine, and extinguish the fire.

   In some new-law multiple dwellings, the basement entrance may be located in an enclosed courtyard. In this case, the only access to the basement will be via the interior hall to a door located a half-landing beneath the main stairs. Other times, the grade-level passageway may be heavily secured. This alternate route to the basement is longer than stretching through from the front passageway and may impede evacuation from the building via the main hallway. Ensure enough hose is stretched to reach the basement with line to spare for the advance to the seat of the fire. Remember that in a new-law building, the basement is isolated from the rest of the building, unlike the basement as found in old-law construction. Thus, a line stretched to protect the interior stairs is the wrong tactic and will accomplish nothing.

   Chief officers and companies should be thoroughly familiar with the best attack routes to the buildings in their area of responsibility. A hoseline stretched to the wrong place will waste time, allow the fire to intensify, and place occupants and firefighters in needless danger. In addition, if the first line is improperly placed, the primary search will receive no protection.

2. The second line must also be stretched into the basement. It may be stretched by the second engine if they are not needed on the first line or by the third engine company. The line should be stretched the same way as the first line unless the first line encountered an obstacle or other problem. In this case, the officer of the first attack team should radio the second attack team with any information that is required to reinforce the attack in the safest, most efficient manner possible.

   If nothing out of the ordinary is occurring, the second line should back-up the first line and, if necessary due to heavy fire and heat conditions, advance in unison with them to confine and extinguish the fire.

3. The third line, unless unusual circumstances dictate otherwise, should stretch to the first floor as close to directly over the fire as possible. Areas of least resistance regarding vertical extension out of the basement include the kitchen and bathroom. If the third engine crew is not needed for the second line, they may stretch this third line. Otherwise, additional resources may be needed to properly position all required lines. The third line should protect against fire extension via vertical arteries and support the ladder company operations working to reveal extending fire.

4. Ladder company operations will focus on the fire area, specifically the primary search, and attempt to locate the fire and make efforts to confine it to its area of origin. As there may be illegal apartments in the basement (some in areas that were originally intended for storage), all accessible areas must be examined as quickly as possible. This will necessitate the use of a lifeline and thermal imaging camera to aid in the search of a large, irregular area where visibility may be reduced.

Forcible entry should not be a problem, except in the areas where illegal living spaces may be. They may be heavily fortified. Ladder companies should be carrying forcible entry tools at all times when on the interior of a fire building. A ladder member without a tool is useless.

The basement must be ventilated, preferably opposite the hoseline advancing toward the fire. This is best done by exterior crews in coordination with interior crews. Interior crews may vent horizontally provided it does not spread the fire.

For this fire, the best access route to the roof will be via aerial device. (Fig. 6–4) It is usually taboo to reach the roof via interior stairs; however, being that the basement in new-law construction is isolated from the rest of the building by a fire-rated ceiling, roof access via the interior stairs may be acceptable in this type of structure. Note that this is permissible only if the fire is in the basement and only in multiple dwellings of new-law construction. If the fire is anywhere else in the structure or if the building is of old-law construction, using the interior stairs as a route to the roof is definitely not an option! When in doubt as to the fire's extent, find another way to the roof and stay out of the vertical artery.

Fig. 6–4 Photo courtesy of Newark NJ Fire If the building is unattached or attached, but of different height than the attached exposures, the aerial is the preferred way to the roof. Ladder company personnel must waste no time in getting to the roof to ventilate at the natural openings.

If the building is attached or the aerial is not being used for rescue, accessing the roof by way of the adjoining building first, and by aerial second, are still the two preferred ways of reaching the roof of any multiple dwelling, either new-law or old-law, regardless of the location of the fire. The previously mentioned route is an alternative used only under specific circumstances.

The roof team, upon reaching the roof of the fire building, must also recon all sides of the building from the roof, including any light and air shafts. The basement door from where the attack is being made is usually located in a shaft on one of the sides of the building. This shaft may create a strong updraft of venting smoke and heat, exposing apartments on upper floors, especially in warm weather where windows may be open and easily ignited curtains may be flapping out of them. Firefighters should make note of and examine, or request via incident command that another team examine, any exposed apartments for fire extension and close windows bordering on any exposed shafts.

The roof door must be opened and examined for victims and smoke conditions. (Fig. 6–5) Then, the top floor must be accessed by way of the gooseneck ladder for search and evacuation operations. Make sure the gooseneck ladder is checked for stability before climbing down it. Attempt to pull it out of its mounts. If you can't, it's likely to be okay. On the way down to check other floors, the windows in the stairwell can also be opened to add to the ventilation effort.

Fig. 6–5 Regardless of the location of the fire in the building, the bulkhead door must be opened to clear the stairwell of the products of combustion. This is venting for life and may provide the required relief of smoke conditions that allow a great majority of occupants to evacuate.

Utility control is another duty of the ladder crew. Gas and electric to affected areas should always be shut down as a matter of routine at all fires. Unless the fire extends in a major way to the upper floors, it should only be necessary to disconnect utilities to areas affected by the fire. Leave the rest of the building in service, as the building power can be used to run lights, salvage equipment, or smoke ejectors. If in doubt, however, shut the utilities down to the entire building.

In a large building, the electric shutoffs are usually on the wall where the service connection enters the building. The gas shutoffs may be located anywhere, but are usually found near a front wall. The location of these items should be noted in pre-fire plan visits.

5. In addition to a complete overhaul of the fire area in the basement, the floor above the fire and all shafts in the area of the fire must be examined. Being that the basement ceiling is protected, the examination on the floor above the fire should concentrate on those vertical arteries that originate in the basement. Manmade openings such as pipe chases located in both the kitchen and bathroom must be examined as soon as possible. Any protrusions on the 1st floor that could hold a steel I-beam and act as a channel rail should also be examined. Also, any wiring run from the basement to the 1st floor, such as cable television and phone, should be located and examined. An excellent idea would be to get a look at the basement layout and building features in the area of the fire to aid in locating the most likely paths of vertical fire travel. (Fig. 6–6)

Fig. 6–6 These wires are run blindly to the first floor and possibly to all floors. This artery alone may be the cause of vertical fire spread right to the cockloft. Take note of these channels and ensure fire is not spreading through them.

Baseboards on the floor above the fire, especially in partition walls in the apartments, must be examined. Those baseboards located near exterior walls are not as likely to be an avenue for fire spread due to the presence of brick, but interior partitions may be suspect. All openings must be justifiable and accomplished so that repairs may be easily made.

6. The fire is in the basement. The people are showing at the 3rd floor windows. On a test, this is known as a distracter to get the candidate to shift focus of the fire to this non-problem. These people can be easily guided down the interior stairs to the front door for a safe exit. If the fire is quickly controlled, protection-in-place may also be an option here.

Remember that panic of some people in remote areas from the fire may cause you to lose your focus. Consider the location of the victims relative to the fire's location and act accordingly. The more you vary from the normal path of egress from the building, the more manpower, time, and danger will be involved in the rescue. Tunnel vision will make the presence of these victims a problem for the incident commander if he or she lets it. A reassuring word from firefighters on the street may calm the individuals. If it doesn't, then more direct means of intervention may have to be performed immediately. In all cases, if it is not a problem, don't make it one.

Scenario 6–3
Short-Answer Questions

1. In contrast to a basement or cellar fire, a fire on the top floor of these types of buildings will exhibit a different set of problems. In a cellar fire, the entire interior of the building will be exposed to fire and the products of combustion. Control of the stairwell is critical. In a top floor fire, the problems encountered in the stairwell will more likely be due to the human load, namely the mass evacuation. This can cause a delay in personnel and equipment accessing the top floor. In addition, a cellar fire will have a relatively rapid hoseline stretch to the fire. A fire on the top floor will require a longer and therefore, a more time-consuming stretch. This will result in fatigue of companies and possibly more advanced fire conditions upon arrival at the fire floor. In comparison with a basement fire, ventilation of the fire area will not be as much of a problem on the top floor due to the availability of windows and the proximity of the fire floor to the roof and bulkhead door. However, the fire may extend into the cockloft, which may cause it to spread to a larger area of the fire building and even to adjoining buildings. Steps must be immediately taken to confine the fire. (Fig. 6–7)

Fig. 6–7 Photo courtesy of Newark, NJ, Fire Department. Top floor fires in large multiple dwellings present the problem of reflex time regarding hose stretch, roof ventilation, and sufficient manpower. Well-advanced fires of this type often extend beyond the structure of origin

2. The first line should not deviate from the rule of thumb of using the safest, most effective path of least resistance. The objective to be accomplished is the placing of a line between the fire and the victims to maintain a viable path of egress from the building. This would also position the attack line to advance and operate from the unburned side of the building. Thus, the line should be stretched as quickly as possible to the fire apartment on the 6th floor via the main interior stairs. The manpower of at least two engine companies and possibly a third should be pressed into service to accomplish this task. The reason for the additional manpower is that if the initial hose line is not placed to protect the stairway and attack the fire as quickly as possible, other lines may be useless. This is why it is important to use as much manpower as required to get this initial line in position as possible. (Fig. 6–8)

Fig. 6–8 Doors must be chocked open to allow for line advancement. Failure to secure the door in the open position will result in reduced fire stream volume and has led to unnecessary firefighter casualties and burned down buildings.

Once at the door to the fire apartment, the hose team should ensure that the stairway is clear of occupants. This will not be as critical as when placing a line at a lower floor fire where more people are exposed and are subsequently evacuating down the stairs, but the stairs must remain as free of the products of combustion as is possible until the civilian life hazard is cleared from the area. Then the line is aggressively advanced into the fire apartment to locate, confine, and extinguish the fire. This advancement must be coordinated with the outside vent team. In this situation, as it is a top floor fire, window venting opposite the attack will most likely be accomplished from the roof with either a Halligan tool on a rope or a Halligan hook or pike pole from the roof. If ventilation opposite the attack is not accomplished, line advancement may be difficult, stalled, or doomed to failure.

3. In any fire in an elevated area, the efficiency of the length of the hose stretch and the diameter of hose must be addressed. There will be a point where specific diameters of hose will reach their absolute maximum usefulness, and the stretching of any more lengths beyond that is counterproductive. Whereas high-rises and even some newly-renovated multiple dwellings will be equipped with a standpipe, most three- to six-story multiple dwellings will not. Therefore, the line must be stretched by hand to the fire floor.

In many buildings with an open stairwell from the 1st floor to the top floor, one length of hose stretched in the stairwell will be sufficient to reach from the ground floor level to the top floor, plus at least one length to attack. Many times, this will be the case in old-law wood frame and Class 3 tenements where the stairs are wide open from the ground floor to the top floor. One can drop a ball (or spit) from the top floor and hit someone on the ground floor. This will not be the case in new-law buildings. In these structures, the hose must be laid on the stairs and the rule of thumb will be one length per floor plus at least one to operate on the fire floor.

Knowing which of these two distinctly different stairwell layouts is present is the responsibility of the company officer, as well as all firefighters and chief officers. If only one length is needed to stretch from the 1st floor to the 5th, then dragging six lengths into the building will create an extreme excess of hose and cause a delay in applying water to the fire. The opposite is also true, as the stretch will be about three or four lengths short of the objective. Preplanning or at least checking the stairwell profile before any hose is stretched will save time, embarrassment, and possibly the building. (Fig. 6–9)

Fig. 6–9 Stretching the attack line up this stairwell will result in couplings being snagged where the two stairways meet. These stairs will require the attack line be laid on the stairs. In addition, someone has placed a stove at the foot of the stairs, which will further complicate the advance.

As far as hose diameter, 1³/₄" hose is the minimum size hose that should be used to attack fires in these buildings and is, in fact, the preferred attack diameter due to its mobility and sufficient discharge. As a rule, about 180 gpm is the maximum gallonage discharge from these lines and should be adequate to extinguish most fires in multiple dwellings as well as in private dwellings and small stores. Room size is relatively small and the fire load is usually light.

Length of stretch is, however, another ballgame. I have seen departments who insist on stretching ten lengths of 1³/₄" hose with a fog nozzle and pump at 125 psi to start or 150 psi if the nozzle operator asks for an increase. With ten or even more lengths pumped at these pressures, it is a surprise any water comes out of the nozzle at all. I would be expecting dust. Using these pressures and lengths is dangerous and a waste of operational time. I have heard at times when buildings were lost that there was too much fire for the initial line to handle. More than likely, there was not enough water. If you'playing the game any other way, you will be a loser most of the time. What is required is a sufficient punch behind the attack, and that punch is water of sufficient volume (gpm) to overcome the heat release from the fire (Btu).

When using a fog nozzle at 100 psi nozzle pressure, four lengths of 1³/₄" hose is about the maximum stretch possible where you can still get 150 gpm at relatively low pump pressures. When using a solid bore tip, which requires 50 psi nozzle pressure, up to seven lengths can be stretched flowing the same gallonage and similar pressure.

4. The second line should be placed on the fire floor to operate as a back-up line for the initial attack line. This line may reinforce the attack in the fire apartment or provide coverage to adjacent apartments where fire extension examinations are taking place. The route this second line takes should be the same route as the first line. If a 2¹/₂" line connected to a gated wye supplied the first line, the second line can easily be attached to the gated wye and stretched much more quickly than having to stretch seven lengths from the street.

   Any fire extension examinations performed in adjacent apartments must have line support in the immediate area so that if fire is found traveling in the cockloft or in any wall for that matter, steps can be immediately taken to cut off its spread.

5. The duties of the rescue company will vary from department to department, but they usually support and reinforce ladder company operations. An effective SOP addressing specific occupancies will guide most operations.

   A rescue company with four or more members assigned can be split just like a ladder company. In this scenario, half the rescue company should reinforce the primary search on the fire floor. The other half should reinforce the operations on the roof. It may be just as effective and acceptable to have the entire rescue company at these type fires conduct the primary search on the fire floor while the entire ladder crew handles roof operations. As long as a sound plan is in place and enforced, with members assigned to critical areas of operation, it doesn't really matter who is responsible to accomplish the various tasks. They should be accomplished as a matter of routine.

6. As at all interior, offensive operations, a primary search must be extended on the fire floor and the floor above. However, since this is a top floor fire, this simplifies matters a little as far as the primary search is concerned; there is no floor above. Access to the fire floor should be made via the interior stairs. Search teams may also access the top floor via the rear gooseneck ladder. In this way, like in private dwellings, search teams will converge on the critical areas from a number of access routes. Horizontal ventilation will also need to be accomplished, usually via a tool from the roof or a fire escape. This must be coordinated

with interior teams. The conditions in the cockloft must be checked early to ascertain if the fire has reached that area. Any cockloft examinations conducted from the interior must be coordinated with the roof team in case a backdraft condition exists in the cockloft.

In the adjoining apartments, the same tactics as in the fire apartment should be performed. These include forcible entry, a primary search, horizontal ventilation to release any accumulated heat and smoke, and controlling fire extension. The ceilings in the adjacent apartment, especially the downwind (leeward) apartment must be pulled as quickly as possible. If fire is found traveling in this area, a hose stream must be applied immediately, and the ceiling must be further opened to determine the extent of the fire in the cockloft. The incident commander as well as the roof team must be made aware of conditions on the fire floor, both in the fire apartment and in adjacent apartments.

Fig. 6–10 When multiple dwellings are attached and bulkhead doors are present, it is safest to access the roof of the fire building via the bulkhead door of the attached building. Be cautious of shafts between the buildings.

7. Roof operations will be dependent on the location and extent of the fire. Reaching the roof, in this case, will be by climbing the interior stairway of the adjoining, attached building and crossing over to the roof of the fire building. Be cognizant of shafts when traversing from one building to the next, especially at night or if a heavy smoke condition is present. (Fig. 6–10) If the building is unattached, the aerial will be the next best way of reaching the roof. There will be times when overhead wires and setbacks prevent aerial use. In this case, using the rear fire escape will be the only choice in these apartment buildings. It may be easier to access the rear fire escape from an apartment on the floor below the fire or the fire floor in a safe area than to ascend the fire escape from the ground level. (Fig. 6–11)

Fig. 6–11 Photo by Bill Tompkins
When a building is unattached, the aerial is the preferred way to access the roof. Note the position of the aerial in relation to the power lines. Always access the roof from the safest area possible.

Note that attached buildings of unequal height should be treated as unattached buildings in regard to roof access. It may also be possible, in unattached buildings, to access the roof via interior stairs of the fire building. This can be done only if the stairs to roof are in separate wing of building from fire. This is usually not the case and must be known by the firefighter assigned to the roof beforehand. Absent this condition, it should be absolutely forbidden in a working fire situation to access the roof via the interior stairs and bulkhead of the fire building. (Fig. 6–12)

Fig. 6–12 Photo by Bob Scollan NJMFPA
Attached buildings of unequal height must be treated as unattached buildings when accessing the roof. The preferred route in this case is the aerial device.

Upon reaching the roof, at least one member must perform a walk-around to survey conditions on all sides of the building, including any light and airshafts. A survey for fire conditions and trapped victims at inaccessible windows is essential. In some instances, the roof may be the only place to observe these conditions. A report of conditions must be made to the command post immediately after this is done.

Regardless of the fire's location, the bulkhead door must be opened and checked for victims. If there is a glass skylight on the bulkhead on the side opposite the door, this must also be vented. (Fig. 6–13) If a skylight is available, it might be easier to initially vent the skylight before working on the door, which may be heavily secured. It may also be easier to chop or cut a hole in the wall opposite the door if no skylight is available to alleviate the heat and smoke while the door is being worked on. Codes require that these doors be operable from the inside. This may mean that there is panic hardware installed on the inside of the door, which will make for a difficult forcible entry, but victims seeking refuge may still be able to get out. It is not uncommon to find that the door is chained closed and secured with casehardened padlocks. This will not only make it harder for firefighters to force the door, but will also make it impossible for occupants to escape via the stairway. As the heat and smoke will be accumulating at the door, these occupants will become victims in a very short time. In these cases, it will be easiest to attack the hinges. The hinges will be the weakest link. If worse comes to worse, as a last resort, fire up the saw and cut a "door in the door." If there is a skylight located at the top of the bulkhead, once the door is removed, it can be used as a makeshift ramp to gain access to the top of the bulkhead to break out the skylight. (Fig. 6–14)

Fig. 6–13 There is a skylight on the rear of this bulkhead opening. In addition, there are also shaft termination points for the dumbwaiter (right) and the elevator (left). These natural openings must be vented as early as possible.

Fig. 6–14 After this bulkhead door is opened, tear the door off and use it as a ramp to access the top-mounted skylight. Bury an axe in the door to use as a handhold on the way up. Also, wedge a tool against the butt of the door at the roof level to keep it from sliding.

Natural openings on the roof must also be opened and checked. These will include soil pipes, vent pipes, any skylights and scuttles, and any other vertical channels that may be present such as dumbwaiter and elevator shaft penthouses. Any openings hot to the touch must be opened and their location communicated to the interior teams for further examination.

On top floor fires, it will easier and safer to provide top floor horizontal ventilation from the roof. A heavy tool such as a Halligan attached to a rope via a snap clip can easily be launched over the side of the building to swing into and break the windows opposite the hoseline attack. If this is not available, a Halligan hook or a pike pole can be used to take out the windows, thus providing horizontal ventilation of the fire apartment. The Halligan hook is preferable to the pike pole as it has a heavier head that is better suited for breaking windows in this manner.

Once horizontal ventilation has been accomplished and the bulkhead door has been opened to clear the stairway, the roof over the fire area must be cut. If you are lucky and manpower permits, the roof cut can begin while the aforementioned tasks are being completed. This cut is an attempt to localize the fire and keep it from spreading throughout the cockloft and burning the roof off the building and possibly spreading to the cocklofts of exposures. The cut should be made as directly over the fire as is safely possible. Cutting these roofs is no easy task, as they may be very thick from multiple renovations and the saw may bind up in the tar, which turns to a sticky goop from the heat of the fire.

Regarding the size of the roof cut, a 4' x 8' hole is the best size, but is better to first get a 4' x 4' hole cut first. Even the four by four cut should still be sliced to make a two by four cut so it is easier to pull the roof boards up. Before the initial vent hole is opened, extend the upwind side by cutting "legs" that extends from the hole toward the upwind side. (Fig. 6–15) This way, once the initial hole is pulled open, it will be easy to cut the last side of the hole which will enlarge the hole to 4' x 8', doubling the area of the vent opening. Knockouts must be provided in order to get a bite on the roofing material to pry it out of the opening. Always work with the wind at your back. Once the cuts have been made and the initial hole is open, use the back end of a pike pole or Halligan hook to push the ceiling down in the fire apartment. Otherwise, only the cockloft will be vented, which is fine, but provides no relief to the crews operating on the fire floor and no ventilation of the main body of fire in the fire apartment.

At least two saws should be on the roof to make cuts, especially if multiple cuts are to be made. There must also be small exam holes made in adjacent areas to determine if the fire is traveling through the cockloft. These may be small plunges of the blade into the roof called "Kerf" cuts, which work better at night due to the illuminating quality of the fire beneath the slice cut. During the day, this slice may not be large enough to see if fire is located below the cut. If the Kerf cut is not sufficient in ascertaining the extent of the fire, triangular cuts can be made by cutting three Kerf cuts that overlap each other at the edges. This causes the center material to either fall into the hole or it might have to be pushed in with a tool. In any case, these are initial attempts that must be made to locate the fire and then to localize it. (Fig. 6–16)

Fig. 6–15 The standard 4' x 4' roof cut should be made to be enlarged. Here, the cut was made, the roof boards pulled, and the ceiling pushed down, but the fire had not penetrated into the cockloft. The "legs" on the left side of the hole did not have to be utilized to cut the 8' hole. Note also the angled knockout cut on the removed roofing.

Enlarging a 4' X 4' vent hole to a 4' X 8' hole.

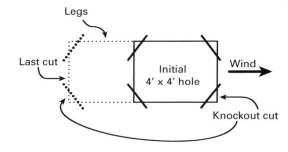

Fig. 6–16 A triangular examination hole is an effective method of determining where to cut the main vent hole. If fire shows, make the cut. Note the thickness of the tar on this roof. If a fire below was melting this tar, it may bind the saw, making operations more difficult.

The trench cut is usually not appropriate in these buildings unless there is a "throat" which effectively cuts off the burned and conceded sections from the unburned areas. A throat is an area over a hallway that has adjacent airshafts on each side. A trench cut is a homemade fire stop that is made by cutting the roof from one side to the other in a trench-like fashion about 3' wide. Fires burning on top floors and cockloft areas in large multiple dwellings can benefit from the trench cut if a throat, which separates the fire side from the unburned side, is present. In this case, a line can be placed in each apartment on the safe side of the throat to operate a stream across the shaft and one line can be positioned in the hallway on the safe side of the trench. The ceilings on the safe side of the trench are pulled and streams operated to keep the fire away from the side where the "stop" will be made. (Fig. 6–17)

Fig. 6–17 *This trench cut was positioned in anticipation of rapid fire spread in a one-story U-shaped strip of buildings. It would have been better to use the skylight as part of the trench. At this fire, there was no need to open this trench as the fire was confined before the trenched area was reached.*

Operations between the interior team on the safe side of the trench, the roof division commander, if assigned, or the roof team making the cut, and the incident commander in the street is critical.

I once operated on a mutual aid assignment at a multiple alarm fire at a large multiple dwelling that had been renovated so that the walls of several buildings were literally breached and the hallways were attached, making it a block-long structure. The fire was in the end of the building on the top floor and in the cockloft, having spread up a shaft from the basement. I was placed in charge of operations on the top floor with orders to "hold" that floor. Although I had some companies from the home jurisdiction in the area, I did not have direct contact with command. I had the department frequency on my radio, but unknown to me, there was a "private line" block on the system, preventing me from breaking in. I could hear incoming transmissions on the fireground, but my outgoing communications was blocked.

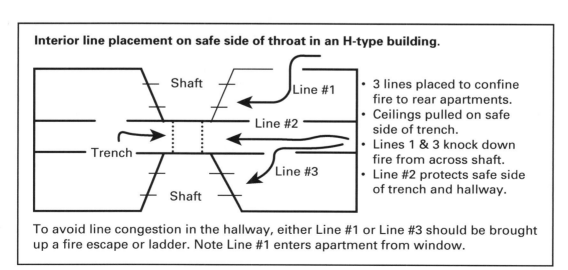

Interior line placement on safe side of throat in an H-type building.

- 3 lines placed to confine fire to rear apartments.
- Ceilings pulled on safe side of trench.
- Lines 1 & 3 knock down fire from across shaft.
- Line #2 protects safe side of trench and hallway.

To avoid line congestion in the hallway, either Line #1 or Line #3 should be brought up a fire escape or ladder. Note Line #1 enters apartment from window.

As quickly as possible, three lines were stretched to the top floor to operate in a defensive manner to keep the fire from extending down the long hall and to the rest of the roof area via the cockloft. As there were shafts on each side of the throat, the lines were placed as mentioned above and shown in the diagram (see p.139). The plan was to confine the fire and then switch to an offensive mode to extinguish it once the main body of fire was knocked down by the lines operating across the shaft. The strategy was progressing smoothly until the incident commander, witnessing fire venting from the now burned-through roof of the surrendered area, ordered the operation of a ladder pipe into the vent hole. This subsequently sent the interior teams into a quick retreat down the hall as the top floor conditions rapidly deteriorated. The fire took place in a jurisdiction that was on a different frequency than the North Hudson companies, so we had no direct contact with the command post, nor he with us. As a result,

we were never advised of the ladder pipe operation, nor could we stop it. I quickly liberated a portable radio from a firefighter from the "home team" and explained our dilemma and how unappreciated we were now feeling. The ladder pipe was shut down and the lines were once again advanced into position where the fire was soon placed under control. Luckily, the ladder pipe did not push the fire past the trench or injure anyone on the interior. The importance of communication between the interior and command post is crucial to coordinate operations. Mutual aid radios are critical to firefighter safety. Incidentally, this communication problem was almost immediately rectified after this incident.

Fig. 6–18 Photo by Bob Scollan NJMFPA
The fire has already entered the throat. A well-defended trench cut may have stopped this fire as it is not yet evidenced in the right (D side) wing.

If the building has no throat, it is better to cut large vent holes over the fire because it will take too long to cut a trench across a large roof area. Taxpayers, strip malls, and row houses are occupancies where trench cuts are not usually effective without surrendering a large amount of area to the fire while the trench is being cut.

The roof operations mentioned above, forcing and checking the bulkhead, horizontally venting the fire apartment from the roof, cutting the main vent hole, and other support operations should be completed simultaneously if enough manpower is available. If it is not, it is best accomplished in the order stated. If only one man or team is initially available, it will be wise to request additional alarms to reinforce the roof operations. If roof ventilation, both horizontally and vertically is not accomplished in a timely manner, the interior forces will most likely be driven off the fire floor and the fire will burn through the roof and possibly spread to adjacent exposures. The key in the control of all fires, but especially in top floor fires, is to "vent early and often."

Fig. 6–19 Photo by Bob Scollan NJMFPA
The fire has spread past the throat and into the adjacent wing of the building. The entire roof and top floor will now be lost.

SCENARIO 6–4
MULTIPLE-CHOICE QUESTIONS

1.

A. *+2* The best hydrant and the most effective tactics. Where open, interior stairs are present, it may be necessary to hold the line at the door to the fire apartment only until the stairs have been cleared. This will likely be the decision of the officer on the landing. This decision may be influenced by reports from the ladder company on the floor above the fire. Once the door to the fire apartment has been opened, the stairwell may turn into a flue, funneling heat and smoke upward toward unprotected escaping occupants. Take the no-nonsense approach to aggressive interior attacks and you will be a winner a good majority of the time.

B. *-1* To use this hydrant, the engine would have to enter the street against the traffic flow or hand stretch an unreasonable distance to the hydrant chosen. This hydrant is clearly not the best choice for these reasons. In addition, using this line in a holding position will allow the fire to spread. Water must be applied to the seat of the fire as soon as possible as the building is old and will not withstand the ravages of fire for an extended period of time. Attacking the fire while at the same time placing the line between the fire and the stairway and hall is the best action for the initial attack line.

C. *-2* This tactic is completely unacceptable. Recalling the point of entry rule of thumb, it states that the line must be stretched via the safest, most effective, path of least resistance to, in this situation, place the line between the fire and the victims and/or the vertical arteries. This means that the attack line in any offensive operations must be stretched through the interior to achieve the objective of location, confinement, and extinguishment. Stretching via the exterior will take time and endanger both occupants and firefighters engaged in primary search. In addition, the hydrant chosen is too close to the front of the building and will cause apparatus congestion, possibly blocking the ladder company from the most effective laddering and support position.

D. *0* There is an old axiom in the fire service: Never pass fire. Stretching the line above the fire without first at least attempting to knock down the fire on the initial fire floor is dangerous and could cause the attack team to be trapped above a fast-spreading fire. The floor above the fire is, in most cases, the place for the second line or even the third line.

Fig. 6–21 Photo by Bob Scollan NJMFPA
The first-arriving engine company must leave the front of the building open to allow for placement of an aerial device. Blocking the front of the building with an engine may hamper rescue and support operations.

E. *+1* These are basically the same tactics as answer choice "A," but the hydrant chosen may be too close to the fire building. This position should be reserved for the ladder company. (Fig. 6–21) The point is given for the proper direction and location of attack, however, it must be remembered that the stairs above must be evaluated for life hazard before the apartment door is opened and the attack is made.

2.

A. **+1** The only reason that the hydrant choice here is acceptable is if the first engine took Hydrant #1. In this case, although the hydrant chosen is not the best, the tactics are sound. The back-up line is properly advanced into the structure to provide reinforcement of the initial attack line and protection against fire extension on the floor above.

B. **+2** If the best hydrant was chosen in Question #1, then the next best should be chosen in Question #2, following the proper path of logic. Hydrant #3 is a better choice than Hydrant #2 as the street is wide enough for the engine to pass through. Perhaps a required supply line or needed equipment can be left at the front of the building before the engine proceeds to Hydrant #3. Also, positioning at Hydrant #3 will not cause the same scene congestion as positioning at Hydrant #1. The line placement and action chosen, as in answer choice "A." are proper firefighting tactics for this type of structure given the fire's location and extent.

C. **-2** This would cause opposing lines to be operated. A "pincer" movement means to "work in opposition" or "in a simultaneous flanking movement." This may work as a viable strategy in a large fire such as at a lumberyard or large warehouse, but utilizing this strategy in a small area such as an apartment can result in steam burns and an unintended channeling of fire gases into undesirable areas. In just about all cases, the back-up line should enter the building through the same access point as the initial attack line. This way, it can provide adequate coverage as a back-up line and advance to adjacent areas or the floor above as required. The most significant feature of a back-up line is that it must be there to protect and, if required, reinforce the initial attack operation.

D. **-1** Autoexposure via the exterior siding to both the floor above and to the adjoining, leeward exposure is a genuine concern at fires in these types of buildings. The scenario makes no mention of this spread. Do not make it up. In almost all cases, the second line must back up the first. The concern for firefighter safety demands this action. If the fire is beyond the control of the first line or if the first line bursts, the second line will be available to either supplement the initial attack line in the former case or to take over the job of fire attack in the latter. Another crucial tactical action is also omitted here. No secondary water supply is established.

E. **0** The tactics used here of stretching a line and providing horizontal ventilation and search of the 3rd floor are all good firefighting tactics. Even though they are typically a ladder company responsibility, the engine company advancing a line may still perform them. However, there is no mention here of a back-up line. This is crucial both on the fireground and in a test. There's another old axiom that states, "If you didn't say it, then you didn't do it." The moral here is to make sure you back-up the attack line in all situations.

3. As stated before, questions based on your willingness or reluctance to request additional alarms will test your judgment regarding existing conditions and circumstances of this incident.

A. **+1** Although the fire does appear to be relatively small, it may not stay that way. However, with a well-coordinated and aggressive interior attack, the fire should be contained.

B. **-1** Requesting and staging an additional alarm for a tactical reserve is a good idea in the field, but on a test, there should be a good reason to back up this request. In addition, leaving the companies in staging on an "in-service" status is not a good policy. If you find you need them and they are out answering another alarm, the benefit of the additional alarm is negated.

Staged companies at multiple alarms should be considered "committed" resources, and available only to the emergency at hand until the incident de-escalates to a point that the danger of running short of manpower and apparatus at the scene has been reduced.

Furthermore, on a test, you are being graded on your ability to make decisions. This answer is sort of a wishy-washy way to request an additional alarm and not really commit them to the incident if you're not sure they are required. Be decisive. Indecision will always bite you in the behind sooner or later. Here, your indecision cost you a point. On the fireground, it may cost you more.

C. **+1** Asphalt siding has caused the destruction of many attached and closely spaced structures. Fire spread across its surface can be extremely rapid. A well-placed exterior stream that does not hinder the interior operation may be the best defense against this threat. Remember, however, that fires of this type must be fought from the interior. Streams operated exclusively from the exterior will burn down the building from the inside out.

D. **+2** The temperature is freezing, the wind chill is worse. The firefighters will be wet and need relief, if not in fighting the fire, then in overhauling and picking up afterward. In cold weather, the potential for frozen hydrants, hose, and malfunctioning apparatus can lead to delayed operations that can negatively influence fire conditions. Requesting an additional alarm on the side of firefighter safety is the best choice here, and the best reason for making the decision. (Fig. 6–22)

Fig. 6–22 Photo by Ron Jeffers NJMFPA
Weather extremes are a sound justification for additional alarms. Only a sound fireground organization with an adequate tactical reserve will allow for an effective rotation and relief of fatigued and frozen firefighters.

E. **0** This is a no-decision answer and, as such, displays poor leadership and decision-making qualities.

4. Ventilation to confine the fire first to the immediate area and then to the building of origin is of the utmost importance in this situation. Failure to ventilate will result in withdrawal from the fire floor and the eventual surrendering of the building. Vent aggressively if you want to keep this fire manageable.

A. **+2** These are excellent tactics at this type building. The exterior team, either working from the fire escape or a ground ladder opposite the attack line may best accomplish horizontal ventilation of the fire floor. This is ventilation for fire and must be coordinated with and performed when the attack line is at the apartment door and has water. Proper communications between the outside vent (OV) and the attack crew is critical. Personnel operating on the interior may accomplish additional ventilation on the floor above the fire.

Ventilation of the floor above the fire may be better accomplished from the interior. In this case, it may be wiser to open windows instead of breaking them. This will depend on conditions. If you can stand and open the window, do so. If the heat is so severe that you have to stay low, break the glass. Glass is the cheapest part of the building.

Opening the scuttle can be described as venting for life and will, in effect, burp the stairway, where occupants may be trying to escape and from where the attack is being mounted. The cockloft must be checked for involvement. If the fire has spread to the cockloft, steps must be immediately taken to confine the fire in the cockloft by cutting the roof over the area of most involvement or the safest area closest to it. If this is not done, the fire can spread laterally to the adjacent buildings via the cockloft.

B. *0* This method of horizontal ventilation, with the exterior team venting opposite the attack line, is the safest course of action to take in the ventilation of the fire floor. Third floor windows may be opened from the interior if fire has not yet extended to that area. This set of actions fails to provide relief on the interior stairs by opening the scuttle. This is one of the most important actions that can be taken at a fire with an open stairwell. Whether the top of the building contains a bulkhead door, a skylight, or a scuttle, opening it will relieve conditions in the area of initial operation and occupant escape.

C. *-1* Here, the scuttle is properly opened to vent the stairs, but, given the current conditions, it is not necessary to cut the roof at this point. If the fire is confined to the 2nd floor, then the roof cut will be unnecessary secondary damage. Instead, vent through natural openings, examine the cockloft via the scuttle area, and monitor fire conditions and extension reports from the interior. The practice of exclusively venting windows from the interior may be dangerous and jeopardize firefighters if they find themselves opposite the attack line when ventilating. Ventilation opposite the attack line is best executed from the exterior.

D. *+1* The scuttle is properly opened to provide a vertical vent. Using ground ladders to knock out windows may quickly achieve a horizontal vent opposite the attack. This will work well on the fire floor, but it may be easier, if conditions warrant, to vent the floor above from the interior. Providing ventilation from both the interior and the exterior will give the attack the best chances of success.

E. *-1* Vent, enter, and search tactics will be effective at this fire and should be a primary method of both venting and providing quick primary search at this fire. Personnel operating in this manner should close doors between them and the fire. This action will not only deter fire spread, but also create a barrier between themselves and the fire. Venting opposite the attack from the interior will place those firefighters accomplishing this task in an unforgiving position as the line blasts the products of combustion in their direction. This action is best left to the exterior teams on ground ladders or the fire escape. In addition, no vertical vent is provided.

5. It is always a better plan to split up the ladder company. This covers the building twice as fast and gets things done in half the time.

A. *0* This will take a long time to accomplish. The rooms in this type of dwelling are small and do not warrant more than two crew members each to search. In fact, members operating in close proximity to each other can search two rooms at once if the smoke and heat conditions are not too severe. If the conditions are questionable due to severely reduced visibility, the partners should stay together, entering and exiting the individual rooms and the structure together.

B. *+1* While this is an effective manner in which to search, Crew 1 should probably go to the fire floor as they will probably be the interior team and can be most effective in the area of greatest danger.

C. **+2** Concentrate on the danger areas, the fire floor and the floor above, in this case, the 2nd and 3rd floors, respectively. The remainder areas must indeed be searched, but will receive a lower priority being that these areas are below the fire floor. Priorities are based on the most endangered areas of the building.

D. **-1** The ladder crew should be divided from the beginning. Searching the 3rd floor first, and then splitting to search the remainder of the building will neglect the most dangerous area from an initial search, the fire floor. By the time the fire floor is searched using this approach, it may be too late.

E. **0** Without a specific objective like a bedridden occupant, don't waste time on the 1st floor and cellar initially. The most severe life hazard will be on the upper floors where the fire and the rising products of combustion will be located. The time of day suggests that most people will be asleep, which demands as rapid and thorough of a search as is possible.

6. It is important to remember that there are already two lines stretched into the structure. Since tactics on the fireground must be flexible and there are many ways to address a situation, there are two **+2** answer choices here.

A. **+2** This is a good tactic and one that will forecast any future fire travel. There are already two lines operating on the fire floor. Stretching up the fire escape to the 3rd floor to cover any extension will avert line congestion and properly reinforce fire attack and extension control operations. (Fig. 6–23) Be aware, however, that before this line attempts to make the floor via the fire escape, that the officer on this line coordinate operations with the interior lines so as not to cause an opposing line scenario. In addition, the fire escape must be quickly inspected and tested for strength. If there is any doubt as to its stability, the line must be stretched another way.

B. **-1** This will cause congestion in the stairwell and hinder line movement. By this time, occupants should have been removed, and the third line can do nothing but harm.

C. **+1** This is a good strategy to prevent autoexposure by keeping the exterior wall wet. This operation, however, requires discipline on the part of the nozzle operator. It is critical that no water enter the 2nd or 3rd floor window as this will change the direction of the vent flow and push the products of combustion toward advancing interior attack teams.

D. **+2** Whether it be the third or fourth line, a line must be stretched into the adjacent, attached exposure building. The most likely path of fire extension will be via the cockloft. Even if it is believed that firewalls are present between the buildings, remember that buildings of braced frame construction will

Fig. 6–23 Photo by Ron Jeffers NJMFPA
Once an attack line and a backup line have been stretched through the front door, all other lines must enter the building via alternate means. This includes fire escapes, ground ladders, aerial devices, or hauling via rope. Ensure operations are well coordinated to avoid opposing lines.

be at least one hundred years old and older. The incident commander who believes that all firewalls are compromised until proven otherwise will take steps to cover the potential extension and not be caught off guard by a sneakily extending fire. (Fig. 6–24)

Companies performing recon in the exposure should be on the lookout for any other easy avenues of fire travel between the fire building and the exposure. Renovations over the years as well as poor building maintenance may make for virtually effortless lateral fire travel.

Fig. 6–24 Photo by Louis "Gino" Esposito
These attached braced frame dwellings will require the incident commander commit a task force of companies to exposure protection. Attached building operations will include evacuation, line placement, and extensive pre-control overhaul on the top floor as well as the roof level.

E. **0** By the time the ladder company ascertains this information, it may be too late, and the fire may have already extended in a major way to the 3rd floor. If you think you will need it, begin the stretch as soon as the potential for the need arises. This will allow you to stay one step ahead of the fire and possibly head it off.

7. The incident commander who, on the fireground in real time, has waited until the fire had seriously involved the 3rd floor to take action in regard to the exposure, will almost certainly lose both the fire building, the exposure, and maybe the rest of the leeward buildings. The only saving grace for the unprepared is the presence of a parking lot on the B-side.

A. **-1** The trench cut can be used as a homemade fire barrier in certain instances on the fireground. Unfortunately, this is not one of them. These braced frame buildings may be as much as 80' deep. To cut a trench will require at least twice that length to cut both sides as well as the extra time it will take to slice up the trench, pull it, and push down the top floor ceiling. Cutting a trench on this size roof will be very time-consuming, manpower-intensive, and doomed to failure. The roof crews as well as the interior crews below will most likely be driven from the building before the task is complete. While not advisable, if a trench must be cut, it may be necessary to surrender the roof of both the fire building and the immediate leeward exposure and cut the trench on exposure B1, the corner building. It should not be necessary to have to resort to such tactics if you are proactive in your approach.

B. **-1** This is classic "candle-moth" syndrome. While the fire fight must continue to be aggressively fought on the top floor of the fire building, requiring additional lines and roof support, using these tactics as the best way to protect the adjacent exposure, will very likely be unsuccessful. A strategy is needed to cut the fire off in addition to sustaining the attack on the main body of fire. Here, only the fire building problem is being addressed.

C. **+1** This may be necessary if the 3rd floor becomes untenable. The scenario states no such thing. It is not yet time to surrender the fire building. Instead, reinforce the attack. The point is given, however, because lines are being stretched into and used to defend the exposure. This fire condition warrants aggressively fought confinement tactics to keep the fire confined to the building of origin. Get the lines there as early in the attack as manpower warrants. Don't wait until fire is already showing from the top floor of the exposure or the cornice covering the cockloft to commit lines to the exposure.

D. **+2** The act of cutting the roof of the fire building will help localize the fire and may buy some time for the incident commander in amassing troops to protect the exposure. It is best to use a task force approach and assign a division commander to oversee and coordinate operations in the exposure. Get at least two lines to the top floor of the exposure and get the ceilings down well ahead of the fire. It is best to pull the entire ceiling of the exposed side apartments. In this way, it will be easier to direct the streams from a safe distance if necessary. (Fig. 6–25) If the ceilings are tin, a more time-consuming operation may be required. For this reason, it is necessary to get a recon crew in the exposure as soon as possible to help the incident commander forecast any problems that may hinder the protection of the exposure. The roof team must also monitor the cockloft of the exposure. One way would be to use examination holes on the exposure roof or in the combustible cornice. (Fig. 6–26) Another, possibly better way, would be to open the scuttle and punch out the cockloft enclosure. Thermal imaging equipment can also be of assistance here. This will allow better examination and be a possible vantage point for stream penetration prior to the ceilings being pulled. Knowledge of building construction will always assist in best protecting exposed buildings.

Fig. 6–25 Pull the entire ceiling in the exposed top floor apartments as soon as possible. If the ceiling is tin, get there as early as possible due to the amount of time and effort required to open this area. Having charged lines in the area is a must.

Fig. 6–26 Photo by Louis "Gino" Esposito
Don't ignore signs that the fire has spread to the cockloft. The color and volume of the smoke here suggests that superheated gases have permeated the cockloft. Take steps to localize horizontal spread by opening the roof of the fire building.

E. **0** While cutting the roof of the fire building will help stall the lateral movement of the fire, it is imperative to get lines stretched to the exposed areas as soon as possible. If they are not required, they can be repacked. However, if they are required and are not there, the entire block may be jeopardized.

8. This is a straight stand-alone question on your knowledge of the structural stability of braced-frame dwellings.

A. **+1** Since this is a building that is not abutted by any buildings on the Exposure B side, a lean-over collapse is a definite possibility and must be monitored. If the building is seen to be slowly listing to one side, prepare for a lean-over collapse of the structure. Lean-over collapses often exhibit 90° angle collapse characteristics, falling outward for a distance equal to the full height of the wall.

B. **+2** The inward-outward collapse is the most prevalent and deadly collapse of braced-frame buildings. Often giving no warning, all four walls can simultaneously collapse, with the upper two floors collapsing inward and the bottom floor collapsing outward. If the fire conditions are heavy for an extended period of time, anticipate this type collapse. (Fig. 6–27)

Fig. 6–27 Photo by Bob Scollan NJMFPA
This braced frame building failed in an inward-outward manner. Heavy fire was present on the first floor when a sudden collapse occurred. Note the lack of fire damage to what were once the upper floors. The lower floor fire destroyed the integrity of the supports that held up the upper floors

C. *0* The curtain fall collapse is more prevalent in ordinary or heavy timber construction, however, if the walls of this building were covered with a decorative stone or masonry veneer, there can be a potential for a curtain fall collapse. Do not get caught off guard.

D. *+1* This type collapse may occur as the walls are non-bearing and may fall straight out as one piece. However, in this type construction, the walls will usually crack in two in an inward-outward fashion. It is critical that all incident commanders, when establishing collapse zones, anticipate to protect men and apparatus against the occurrence of a 90° angle collapse, which is will usually cover the most area with collapse debris.

E. *-1* A tent collapse is a type of floor collapse and does not apply here.

9. The answer to this question is based upon your answer for the last question. If you had to guess on #8, you'll have to guess on #9. We have already established the inward-outward collapse as the prevalent type of collapse in braced-frame buildings.

A. *-1* The walls in the braced-frame building are not bearing, as the load rests on the other bearing members for structural stability. These load-bearing members are the vertical posts and the horizontal beams making up the outside walls. For this reason, braced-frame construction is sometimes called "post and beam" construction.

B. *0* A fire escape will certainly put an eccentric load on the wall it is anchored into, but it would not be the most likely cause of the failure of this building.

C. *-1* As stated above, none of the walls in braced frame construction are load bearing.

D. *+2* The mortise and tenon joints are the critical connection points between the posts and beams that support the structural load of the building. As such, a failure of any one of these connections will likely result in the failure of the building—possibly all four walls at one time and without warning. (Fig. 6–28)

E. *-1* This partition failure is the cause of the tent collapse, which we have established as a type of floor collapse.

Fig. 6–28 The weight of the braced frame structure depends on the stability of the mortise and tenon connection. Note how the end of the tenon has been whittled down in comparison with the rest of the joist. The mortise also has less mass where the tenon is connected. Less mass equals earlier failure.

10. This is a situation question relating to your knowledge of salvage operations and your ability to think like water.

   A. *-1* A charged line in the area of operation is an absolute necessity when conducting over-haul operations. If it is left outside, by the time its need is realized, the fire may spread beyond its control. There are better ways to reduce water damage.

   B. *+2* This is an excellent technique as even the best intentions when attempting to limit the use of water can cause water damage. Nozzles invariably leak, can be inadvertently charged, or dropped and opened unintentionally. Once the nozzle is out the window, the water will not damage the interior. Of course, due to the temperatures, it will be necessary to use caution or cordon off the area outside the building where water and ice may accumulate.

   C. *0* Extinguishers are limited in many ways. The line is already on the fire floor. To remove the line and bring up and extinguisher to accomplish the same objective is a waste of time.

   D. *+1* Leaving the nozzle closed until absolutely necessary is an acceptable tactic and is a no-brainer here. But, if the nozzle leaks, which nozzles usually do, then a better plan is required. This is why answer choice "B" is the *+2*, and this is the *+1*. Functional fixity; don't be a victim. Effective officers offer creative solutions to routine problems.

   E. *+1* Water damage is a direct result of uncontrolled and overzealous hoseline operation. The best way to reduce this condition is not to cover items so they don't get wet, but to take measures to keep water away from them altogether. This can be accomplished in a proactive manner by isolating the nozzle until it is required.

Passing Score for Scenario 6–4 = 14 Points

SCENARIO 6–5
MULTIPLE-CHOICE QUESTIONS

1.

A. **+1** The construction of this building will act as an ally to firefighting in that the fire should be confined to the area of origin by the inherent structural features of the building. The stairs are enclosed and the fire spread problem created by structural deficiencies such as the open stairs found in non-fire-resistive multiple dwellings will not be present. One construction-related problem that may be encountered by the attack teams is that the fire area may be extremely hot and may retain this heat for a long period of time after the fire is knocked down. It may be necessary, if the attack plan is not having the desired effect, to use a bigger line and rotate companies frequently.

B. **+2** All of the tactical problems that will be encountered in the initial operation at this incident will be directly related to the problems presented by the excavation trench at the front of the building. This street condition will cause a delay in operations and warrant an additional alarm.

C. **+1** There is no mention of a hydrant being out of service, but due to the fact that some hydrants in the construction area are placed out of service during the construction period, it is a size-up factor that must receive serious consideration. Engine pump operators must test the hydrant before connecting to ensure the water supply is adequate. This problem is caused by the street condition size-up factor.

D. **+1** Life hazard is a consideration at every fire. The time of day, early afternoon, does not mean that all the occupants are awake, although a majority of them probably are. A primary search and evacuation must be conducted. The life hazard problem here is nevertheless compounded by the operational delays caused by the street conditions in the area.

E. **-1** The weather is probably the only relatively favorable aspect at this fire.

2. Recalling the acronym C-BAR will guide you to the best answer choice.

A. **+1** This is actually the second action you will take. Command must be established first to initiate incident management functions and begin command and control actions at the scene.

B. **+2** Command and control are the top priority of any incident and the first action taken upon arrival at the scene. Establishing command will provide accountability and responsibility for the incident from the outset. In fact, in some states, it is mandatory to establish command and initiate an incident management system. Establishing command and furnishing dispatch with a size-up report will give responding companies a chance to conduct their own pre-arrival size-up and formulate a plan of action in their heads. Information is power and vital to all fireground operations. (Fig. 6–29)

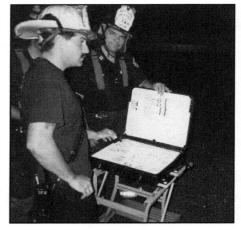

Fig. 6–29 Command establishment is the first step in organizing and maintaining a safe and effective fireground. A solid command SOP along with disciplined officers and firefighters will create an organized and structured approach to mitigating any incident.

C. **0** This will be one of the last things you will do as far as initiating action to organize the fireground. This is done after you give a size-up report and indicates which action you are taking. You must set up the organization before you can take action to mitigate the incident. In fact, establishing incident command is the most important step in mitigating the incident.

D. **0** This again is an action step taken after setting up the incident command organization, but before stretching a line to locate, confine, and extinguish the fire.

E. **-2** While this is an action step and not a command action, it gets a **-2** point value because this is assumed to be an occupied building and an offensive strategy must be pursued. The use of a deck gun into an assumed occupied building is a defensive action and must be forbidden as it will create havoc and push fire toward victims.

3. This scenario challenges the ability of the fire officer to adapt to the situation and overcome the obstacle of the construction trench with creative tactics.

A. **+1** This is the best hydrant to choose given the circumstances, but the ground ladder placement will take a lot of time and stretching across the homemade "bridge" will be dangerous. There is a better way.

B. **+2** This is a much safer method of stretching the attack line in this instance. The trench is avoided altogether, averting any chance of stretching-related injuries. This will be the quickest way of getting the initial line in place to attack the fire. Again, the Point of Entry Rule of Thumb leads to the best answer.

C. **-1** This is a suspect hydrant. Don't take a chance on a hydrant that may be out of service. Planning prior to the fire should dictate which hydrant to use.

D. **-1** The hydrant in front of the building should be avoided if at all possible as a water supply for the initial attack engine. It may be used by a later-arriving engine that uses LDH if no other hydrant is available or close by, but the first engine should avoid it if at all possible. This front of the building position should be reserved for the ladder company. If Hydrant #1 was not available, Hydrant #3 would be the best hydrant due to the service status of Hydrant #2. However, this is not the case. Also, it has already been stated that stretching the initial attack line via the ladder bridge is dangerous, will take time, and should be avoided if possible.

E. **0** Dropping off the crew, having them walk down the block on the fire side of the trench, and then using them to pull the line across the trench with a length of rope is a waste of time and an almost unbelievable display of tunnel vision. This is like turning the building to unscrew a light bulb instead of the other way around.

4. There is no mention of the size of attack line in the question. This decision will have to be made by the incident commander based on fire conditions and reports from the interior. If companies are having a hard time reaching the fire, stretch the 2½" as a back-up. If the attack is making progress, it may be easier and quicker to back-up with a 1¾" line. As always, when in doubt, stretch the bigger line.

A. **+2** Second water supply is established, the first line is in place and operating, and a back-up line is stretched. This is reinforcement of the initial attack position using the proper methods.

It is also a good idea to take the second line from Engine 1 as the second team will already be on the building side of the trench and this will be a safer, more expedient way to accomplish the task.

B. *0* The tactics are proper, but the hydrant may be out-of-service. Hooking up to a dry hydrant is like not hooking up at all. Hydrant #3 or #4 is a better choice. As stated before, Hydrant #3 is an acceptable choice as a secondary hydrant here because Hydrant #2 is suspect. As long as the engine position does not impede the positioning of the ladder, this hydrant may be used as a back-up.

C. *-2* This requires an extraordinary amount of close coordination between the interior and exterior companies. However, it is not warranted here as the stream will probably not reach and the extent of the fire is not known. Only smoke is issuing from the windows and a cardinal rule of firefighting is not to open a line onto smoke unless heat is hindering your advance. This is not the case here. In addition, the use of the exterior stream while companies are advancing from the interior as well as the high probability of occupants still in the building make this tactic unacceptable and dangerous.

D. *-1* The hydrant choice is poor and the tactic chosen for the hose stretch is dangerous. There are better ways.

E. *+1* While this will take longer, the hydrant is probably a good one and will not cause congestion at the front of the building. In addition, the stretch will also avoid the trench and still get to the fire building and provide reinforcement to the initial attack team. The only problem with these tactics is that the second engine company will probably not be available to assist in the stretch of the initial attack line. This may cause a delay in getting initial water on the fire.

5.

A. *0* The ladder truck is nothing more than a rolling toolbox and for that reason, must be positioned at the front of the building if at all possible. It is the ladder company's duty to get across the trench. Positioning on a side street leaves tools, lights, and equipment far from the building. Also, no one in this choice goes to the roof to provide vertical ventilation of the stairwells. This is imperative, especially ventilation of the chosen attack stairway. As soon as the hall door is opened and then the apartment door, the smoke will choose the path of least resistance and rapidly fill the stairwell. This must be vented.

B. *-2* It is easier to bridge the trench. For both crews to ascend the aerial and then descend via the north and south stairwells may be dangerous as the products of combustion may vent if the stairway door and apartment door is left open or the attack has already begun and the fire apartment has been opened. Stairwells could rapidly turn into chimneys. In this building, if you are in the wrong stairwell at the wrong time, there are few places on the face of the earth that will be more unforgiving.

C. *+2* These are sufficient tactics if the aerial reaches the roof. If it does not, Plan B must be ready to go. The stairs will then be the only route to the roof. The attack will have to wait behind a closed stairwell door until ladder crews assigned the responsibility of providing vertical ventilation access the roof via the interior stair. Proper and timely coordination between the fire floor and the roof team is critical to firefighter safety. In regard to leaving the roof via aerial, even though it is time-consuming, it is a safe method of egress.

D. *+2* These are also safe tactics and are completely acceptable given the construction of the building. In ladder company work, there are many ways to skin a cat as long as it is done in the safest manner possible. There is no room for victims of functional fixity in ladder work. However, before descending the "safe" stairwell, it might be a good idea to get a report from someone on the fire floor as to the status of the stairwell farthest from the fire.

E. *0* These are poor tactics in regard to the bridging operation at the roof level. The diagram shows the buildings to be far apart, making this operation extremely hazardous.

6. This question tests your knowledge of door construction in regard to fire-resistive multiple dwellings and how to best go about forcing these doors.

A. *+2* These doors will most likely be inward-swinging metal doors with a metal frame set in a block or double-thickness fire-rated sheetrock. These are very substantial doors. The rabbit tool is made to order for these sturdy door assemblies. In fact, the more substantial the assembly, the more effective the tool is. The tool only works on inward-swinging doors. The flat head axe is used to tap the jaws of the tool into the doorframe. Used correctly, the rabbit tool can open doors quickly and is effective under heavy smoke conditions. Do not forget to attach something to the doorknob to control the door swing.

B. *0* The K-tool works only on cylinder locks where you can see both the lock cylinder and the guts of the lock to duplicate the action of the key. It works great on raised commercial exterior locks where visibility is usually good. Under heavy smoke conditions that are likely to be found in apartment hallways, the tool will not be as effective. There are better ways.

C. *+1* This is the modified brute force method to "gorilla" the door. The maul is used to strike the door and force it open. It is extremely important in this case to control the door as the maul will blast the door open, often knocking it off its hinges. This will not maintain the integrity of the door and may allow fire, smoke, and heat into the hallway. While this method of forcing steel doors is safer in adjoining apartments and other proximal apartments where evacuation must be accomplished, extreme caution must be used when employing this method on the fire apartment door.

D. *-1* These tools are useful on wood doors set in wood frames because they are able to remove the frame from the door, compromising the integrity of the doorframe. However, the sturdier door assemblies in fire-resistive multiple dwellings will render these tools next to useless. It is just not possible to get the proper leverage. The doorframe offers no "give." Sometimes, if the right tools are not on hand, it may be quicker and easier to breach the wall next to the knob, reach in, and open the door. Hopefully, the apartments are not set up so that your breach next to the knob will access the adjacent apartment. It might not be a bad idea to breach a wall away from the danger area to ascertain the door-to-separating wall relationship. Otherwise, you just have to be lucky.

E. *-2* By the time these tools force this door, it will be the only thing left standing. The fire will have consumed everything else around the door. Functional fixity is one thing, but no matter how ingenious you may be, if the tool and the method chosen are unsound, the job will never be accomplished.

7. Again, a question regarding your knowledge of building construction and the areas and conditions that will promote fire spread.

   A. **+1** While these buildings will not contain the same concealed spaces inherent in multiple dwellings of ordinary construction such as a cockloft, there will still be areas of potential vertical and horizontal fire spread such as pipe chases and other utility conduits. These will most likely be found in bathrooms and kitchens in the areas of the soil pipes. While many of these pokethroughs will be properly firestopped, others may not and may have been compromised over the years. These areas must be checked on the floor above the fire apartment , as this is a ready way for fire to spread into these areas.

   B. **-2** Fire-resistive multiple dwellings do not have a cockloft, common or otherwise.

   C. **-1** The walls of these structures will be constructed of brick or some type of block. They are not combustible and will not be an avenue of fire spread in this structure.

   D. **-1** The stairways will be enclosed, and thus should not spread fire from floor to floor. If the fire originates in the stairway, it may be critical to the ability of occupants to escape. Fortunately, these type buildings usually have more than one stairwell. Coordinated operations should make use of unexposed stairways as evacuation routes. In addition, if the fire doors between the hallways and the enclosed stairways have been blocked open by tenants (see Granton Avenue fire in chapter 3 of the text), deadly products of combustion could permeate the stairways and upper floors.

   E. **+2** The most common avenue for floor-to-floor fire spread will be via auto-exposure. In fire-resistive structures, the fire is usually contained to the apartment of origin, however, heavy fire venting out a window can blow into a window on the floor above, spreading the fire. This will be more prevalent in warm weather when windows are open and curtains may be flapping out in the breeze. These are easily ignited.

8. NFPA 1500 Fire Department Occupational Safety and Health Programs sets the standards for Rapid Intervention Crews (RICs) (also known as RIT [Rapid Intervention Team] or FAST [Firefighter Assist and Search Team]). The duties of these firefighter rescue teams is quite clear. They are not to be assigned to any function other than firefighter rescue. They cannot be deployed to fight a fire in lieu of calling another team, as this would create a time lag where no team is available. They must be made available at every working fire.

   A. **0** This is a duty that may be performed if firefighters are cut off by an extending fire. This is not the situation here. Therefore, this is not a duty given to them as they arrive.

   B. **+1** This is not a bad idea, however, the whole crew should not be involved in this. Only one member should perform this size-up. If you have the manpower, you can send two members. It doesn't even have to be the company officer, but may be one of his crew with a portable radio. The rest of the crew stands by, properly equipped at the command post. Any problems found by the recon team on the walk-around should be communicated to the FAST team commander immediately. The member should take no action until the team is at his location.

   If interior members are searching upper floors of a building, the FAST team, if applicable to the building, can place ground or aerial ladders at strategic points to ensure a secondary egress is secured for interior teams. The team placing the ladder must communicate its position to the interior teams

C. **-2** This is incorrect at this type of fire. The team should report to and remain at the command post. In a high-rise fire, they would report to the operations chief at the operations post where they will be closer to the action, but at a structure fire other than a high-rise, they should routinely report to and remain at the command post.

D. **+2** Operational familiarization is one of the keys to a safe fireground. If this task can be done while remaining at the command post and not cause an unnecessary occupation of the FAST team, it will do no harm to firefighter safety, and may enhance it. By studying and being actively involved in command board and tactical worksheet operations, the FAST team will have the best up-to-the-minute information on the probable whereabouts of all personnel.

E. **+1** This is probably the NFPA 1500 answer. The team is standing by at the ready. They are committed to firefighter rescue and well-being. According to the standard, this is their only job.

9. Situation questions often throw a snag into the operation and provide a test of your tactical knowledge and your ability to think on your feet and be flexible.

A. **-1** This is unacceptable as it would probably cause the death of any victims in the apartment as well as forcing the fire into uninvolved areas.

B. **0** This is great, but offers no solution to the problem. Chief officers are only worth their salt if they can find solutions to problems in a quick and efficient manner.

C. **0** The interior crew will most likely not be able to see the fire exposing the floor above from the exterior. In addition, this will waste time. The time to make a decision is now. Putting the decision in a subordinate's hands is the sign of an ineffective leader. Indecision is always a loser.

D. **+1** This is a good option, but it will take time and manpower, something which may not be in abundance at the time of arrival. By the time the line is stretched up four floors, the fire will most likely have extended into the apartment.

E. **+2** The spandrel wall is that space between the top of the window on one floor and the bottom of the sill on the floor above. The stream should be used to sweep the wall in a back and forth motion. Direct application may cause an unstable wall to collapse. Using this tactic will arrest flame spread into the interior by cooling the Btus in the form of convection currents being generated by the fire blowing out the window. Heat is the most common cause of fire spread in a building, and convection currents are the most common cause of heat spread. Stream application discipline is crucial here. Avoiding the fire apartment with the stream will not cause occupants to be endangered by hydraulically driven fire. In addition, when the main body of fire is about to be attacked, shutting the line down is the safest move. Hitting the seat of the fire from the interior will cause the exterior flame and heat production to abate, minimizing the auto-exposure threat.

10. Another situation problem that tests your ability to operate effectively in a given situation. The mark of a good officer is one who can adapt and make the environment work in his favor.

   A. *+2* This is the most direct means of removing the smoke. The compactor chute pierces the entire building where it terminates at the roof. It is the most effective path of least resistance to remove the problem.

   B. *0* Most PPV fans are gas-powered. Unless the fan is electric, this will cause a build-up of gas fumes on the floor. If the fan is electric, this is an acceptable tactic, but it must be controlled to ensure that no other floors in the building are contaminated.

   C. *-1* This will contaminate the apartments that would have otherwise remained unaffected by the incident. Don't create a larger problem than the one that already exists.

   D. *-1* This may be effective, but will cause water damage. There are better ways. Salvage is about minimizing damage, not creating more of it.

   E. *+1* This will take time and will work, but it is not the most effective. A more effective way to accomplish this would be to place the PPV fan at the bottom stairwell door if there is an exterior one and blow the air up toward the bulkhead door. Then, opening the stairwell door will create a venturi effect to pull the smoke out of the hallway and into the stairwell, where it will be exhausted out of the bulkhead door.

Passing Score for Scenario 6–5 = 14 Points

CHAPTER SEVEN
SCENARIOS

SCENARIO 7–1
HIGH-RISE OFFICE BUILDING FIRE

Construction

The fire building is a 28-year-old, ten-story, fire-resistive building that was remodeled eight years ago. It is occupied by Desort's Sports, an international sports clothing manufacturer. In order to accommodate all of the renovations, drop ceilings of acoustical tile were installed. Most of the floors have different floor layouts. The 1st floor is actually two floors and is used as a computer center. Floors 3 through 7 are used for offices and showrooms. Floors 8 through 10 are used for shipping, receiving, and unused merchandise. All elevators were replaced, and a central air system with a rooftop unit was added. The roof is flat and is poured concrete. There is an elevator bulkhead on the roof.

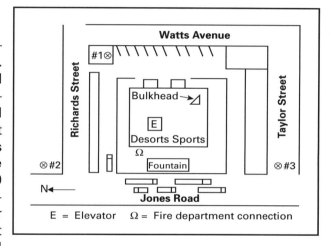

The fire building is 105' across the street front and 125' deep. There are large double-pane windows on all sides of the building. There is a steel door in the rear of the building that leads to the basement and rear staircase. A double steel door serves as a delivery entrance on the south side of the building. There are two interior staircases, one in the rear and one in the front. The stairwells are enclosed.

Time and weather

The report to central dispatch came in at 1437 on an overcast day in April. The temperature is 40°F, and the wind is 15mph from a southerly direction.

Area and street conditions

The fire building's address is 728 Jones Road. Jones Road is a heavily trafficked four-lane thoroughfare. Richards Street and Taylor Street are both two-lane, one-way streets. They run in alternate directions; Taylor Street runs east. The sidewalk in front of the building is 25' wide. This part of the city is mostly commercial in nature, and trucks will double-park if there are no available parking spaces.

Fire conditions

The caller stated that black smoke was coming from a locked door in a 7th floor storage room. The building's fire safety director informs you that people are trapped on the upper floors. The elevators have been returned to the lobby and smoke is beginning to fill the stairwells as it is evident that stairwell doors have been left open.

Upon arrival, heavy black smoke and flames are coming out of two front windows on the south side of the seventh floor.

Exposures

The building sits on a 160' x 210' lot. The building is situated so that there is a 25' area between the northern side of the building and the street. There is a 15' area between the building and the driveway. The front of the building faces Richards Terrace; the area in the rear faces Watts Avenue. There are driveways on the south side and north side of the building. Large dumpsters abut the rear of the building.

Water supply

Standpipes are located in the stairwells. The standpipe connection is located on the west side of the building.

Response

Four engines, two ladders, a rescue company, a battalion chief and a deputy chief are on the response. The staffing for all apparatus is one officer and three firefighters. A safety officer is also on the response.

SHORT-ANSWER QUESTIONS

1. Discuss the incident command organization.

   _____
   _____
   _____
   _____
   _____
   _____

2. Where would the first line be positioned? What are the objectives of this line?

   _____
   _____

_____

_____

_____

_____

3. Where would the second line be positioned? What are the objectives of this line?

_____

_____

_____

_____

_____

_____

4. Where would a third line be positioned? What are the objectives of this line?

_____

_____

_____

_____

_____

_____

5. Discuss ventilation operations at this fire.

_____

_____

_____

_____

_____

SCENARIO 7–2
RESIDENTIAL HIGH-RISE FIRE

Construction

A fire has been reported in a 36-story fire-resistive high-rise. The building is circular with the apartments located on the perimeter. The utilities, stairwells, a compactor chute, and elevators are located in the center core. There are two elevator banks in this building, each in its own fire-rated enclosure. One bank serves floors 1 through 20. The other serves floors 21 to the penthouse. The elevators are not equipped with fire service operation. There are also two enclosed stairwells. Both stairwells serve the roof. All the apartments have balconies that overlook either a large park on the west side of the building and a river on the east side. The interior of the apartments is of masonry walls, ceilings, and floors. There are steel doors set in steel frames that are mounted in the masonry hall wall. Each unit has its own HVAC unit.

There is a 24-hour doorman and a staff of porters and maintenance personnel. The alarm panel and annunciator are located at the doorman desk in a vestibule. There is also a communication station that allows communication with all the apartments in the building. These are located at the doorman's desk.

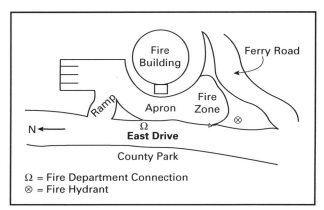

Ω = Fire Department Connection
⊗ = Fire Hydrant

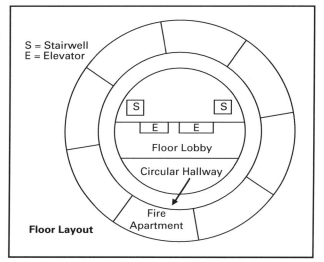

S = Stairwell
E = Elevator

Floor Layout

Time and weather

The time is 0530. The temperature is 68°F, with a slight wind condition swirling around the building.

Area and street conditions

East Drive is a two-way, two-lane street. The building, located at 820 East Drive, is set back from East Drive by approximately 100'. However, at the front of the building is a ramp that slopes up from East Drive. It is large enough to fit apparatus; however, there are usually cars and delivery trucks parked on the ramp and apron.

Fire conditions

Fire is seen issuing from a window about halfway up the height of the building. There are many people on the balconies.

Exposures

To the east of the fire building is a cliff. Along the perimeter of the building, running from the front to the south side is Ferry Road, which is a steep, winding road that leads down to the bottom of the cliff.

Water supply

The wet standpipe is located in the North stairwell. The fire department connection is located on East Drive at the west side of the building.

Response

Your response is three engines and a ladder company. An officer and two firefighters staff each engine. An officer and three firefighters staff the ladder truck.

MULTIPLE-CHOICE QUESTIONS

1. What is the most important size-up factor at this fire?

    A. Elevator control

    B. Auxiliary appliances

    C. Height

    D. Manpower and apparatus

    E. Location and extent

2. What action will you take regarding the elevator at this fire?

    A. Have maintenance control the elevator.

    B. Place the elevator in "bypass" or "independent service" mode to control the elevator.

    C. Place the elevator in fire service mode and operate it as an equipment shuttle.

    D. Have a firefighter take control of the elevator. As an added safety against unwanted door openings near the fire, operate the elevator only from the lobby to five floors below the reported fire. If any problems are encountered, abandon the elevator.

    E. Do not use the elevator under any circumstances. Have maintenance personnel shut it down under fire department supervision. Then lock out the power.

3. How would you control personnel flow to the upper floors?

   A. Order all responding personnel to assemble at the command post in the lobby. No one is to leave the command post until the order is given to proceed to the upper floor.

   B. Have personnel reporting to the upper floor stage at the operations post.

   C. Allow only essential personnel to the upper floors. Keep the tactical reserve at the command post.

   D. Have personnel reporting to the upper floor stage at the resource post.

   E. Establish a lobby control post. No one enters the stairwells until directed by the lobby control officer.

4. What is the best way for initial attack companies to access the fire area?

   A. Via the elevator. Get off five floors below the reported fire. Walk the rest of the way.

   B. Walk up via the stairwell to two floors below the fire.

   C. Take the elevator to the 21st floor. Descend via the enclosed stairwell and set up for the attack.

   D. Via an aerial platform to the highest floor possible. Then use the stairwell to reach the fire floor.

   E. One engine company take one stairwell. Another engine company take the other stairwell. Meet two floors below the fire. On the way up, direct occupants to areas of refuge and determine the best route for the attack. Relay this information to incident command.

5. The fire is in Apartment 16D. How will this fire be attacked?

   A. Connect to the standpipe in the enclosed stairwell on the 16th floor. Stretch two $2\frac{1}{2}$" fog lines to attack the fire.

   B. Connect to the standpipe on the 15th floor. Connect the second line to the standpipe on the 14th floor. Stretch a $1\frac{3}{4}$" solid bore line backed up by a $2\frac{1}{2}$" solid bore to locate, confine, and extinguish the fire.

   C. Connect to the standpipe on the 15th floor. Stretch a $1\frac{3}{4}$" solid bore line to attack the fire. Stretch a $2\frac{1}{2}$" solid bore line to back-up the first line from the same standpipe connection.

   D. Connect to the standpipe on the 15th floor. Stretch a $2\frac{1}{2}$" solid bore line to attack the fire. Stretch an additional $2\frac{1}{2}$" solid bore line to back-up the first line from the same standpipe connection via a large gated wye.

   E. Connect a solid bore $1\frac{3}{4}$" line to the standpipe connection of the 15th floor to attack the fire. Connect a second $1\frac{3}{4}$" solid bore line to the standpipe in the opposite stairwell to attack the fire from two directions in case one of the lines cannot advance due to adverse conditions.

6. How should the fire in the apartment be ventilated?

   A. From the floor above the fire, using a Halligan hook or similar tool, lean over the balcony to horizontally ventilate the fire.

   B. From the floor above the fire, using a Halligan tool and rope to swing the tool into the window to horizontally ventilate the fire.

   C. Using the stream to break out the windows from the doorway.

   D. The fire is already venting; the advance should not be impeded. Vent further only after the fire is under control.

   E. If possible, use a Halligan hook from an adjacent balcony to clear out any windows that may not be vented yet. Also, vent from the inside after the line has controlled the fire.

7. Suppose there were fire service elevators available. Where is the safest place to establish a Search and Evacuation post and what is the best way of getting there?

   A. The 21st floor lobby by way of the blind shaft elevator.

   B. The 21st floor lobby by way of the stairs.

   C. The top floor by way of the blind shaft elevator.

   D. On the roof by way of the blind shaft elevator.

   E. As this is a fire-resistive building, and therefore compartmentalized, no search and evacuation post need be established.

8. Companies are having a difficult time getting into the area of the apartment as the wind has shifted and is now blowing into the apartment. What alternatives can be used to control this fire?

   A. Cut a hole in the floor above the fire to vent the products of combustion.

   B. Cut a hole in the floor above the fire to vent the products of combustion. Place a PPV fan to appropriately blow the products of combustion out the window.

   C. Withdraw the line to a safe area. Breach the wall and use a fog stream to cool down the area, then attempt to attack the fire again.

   D. Use a Lorenzo ladder from the floor below to knock down the heavy fire. Keep the door to the apartment closed and the interior crew in a safe position.

   E. Attempt to reverse the flow of gases by using a solid stream in tandem with a fog stream. If possible, set up a PPV fan behind the attack crews to assist in pushing the products of combustion ahead of the streams.

9. The fire originated in the bedroom and spread to the rest of the apartment. There has been severe damage throughout. Where should overhaul operations be focused?

    A. The ceiling above the heaviest burning must be opened and a stream operated into the opening.

    B. Along the duct run of the HVAC unit.

    C. In the areas of outlets and light switches.

    D. In the area of the pipe chases in the bathrooms and kitchens.

    E. In the area of the compactor chute.

10. Suppose, as you arrived on the scene at this fire, you noticed that the female threads and swivels of the fire department connection are badly damaged. How would you get water to the fire?

    A. Use a spanner to connect the supply line to the FDC. Place a hose jacket between the supply line and the FDC to complete the connection.

    B. Stretch two 1³/₄" lines up the interior stairs. Use one as the attack line and one as the back-up line.

    C. Stretch LDH and a manifold up the stairwell to the floor below the fire. Utilize your attack lines from the manifold.

    D. Take a double female connection into the first floor stairwell and connect it to the first floor standpipe outlet. Attach a Siamese connector to the double female. Connect two 2¹/₂" lines to the Siamese. Pump into the first floor standpipe connection to supply water to the upper floors. This will allow you to bypass the out-of-service exterior standpipe connection.

    E. Utilize an aerial device as a portable standpipe. Raise the device to the highest floor possible. Connect a supply line and gated wye to the device. Stretch the attack lines from the gated wye to attack the fire

Scenario 7–3
High-Rise Under Construction Fire

Construction

The fire building is located in the waterfront area of the city. New construction has been going on for some time in this area as the waterfront undergoes a facelift. Among the buildings being constructed is the Ace Networks building, a 25-story fire-resistive high-rise office building. When completed, it will be the tallest building in the city. At present, it is only 10 stories high as the work of pouring the reinforced concrete floors continues. Floor 10 is being poured today. As a result, there is wood formwork in place where the 10th floor will stand. The wet concrete has not yet been poured. The floor below was poured two days ago, and is also a mass of formwork. The three floors below (floors 8, 7, and 6) are each being supported by a diminishing amount of formwork as the concrete has strengthened there in varying degrees. The floors below this have no formwork on them at all, but there is some debris stored there.

There are no sprinklers at this time. In addition, the metal stairwells, which are not yet totally enclosed, have only been completed to the fourth floor. Above this, the only access to the upper floors is by temporary ladders erected of 2" x 4" lumber. There is a construction elevator system located on the east side of the building. There is both a freight elevator and a passenger elevator.

Time and weather

The hour is 1325 on a cold, windy, Sunday afternoon.

Area and street conditions

Several fire-resistive high-rise office buildings are in the process of being erected and the area is bustling with construction apparatus and machinery, re-routed roads, and new water systems. Each day it seems the area changes.

There is a gate around the construction site that is locked after 1600 and all day Sunday. There is also a great deal of construction debris in and around the site, including many cylinders and tanks of flammable gases and liquids. The housekeeping procedures are shoddy at best.

Fire conditions

You arrive at this building to find smoke emanating from one of the upper floors. From your viewpoint, it looks like the smoke is issuing from the 7th floor.

Exposures

The most significant exposures are the lower floors of the fire building as well as the myriad of construction equipment and debris in the area. There is another building under construction directly to the leeward. It is separated from the fire building by 300'.

Water supply

The standpipe connection is at the front of the building. It is not known whether the standpipe is in service or if it has kept up with the pace of the construction.

Response

Your response is three engines and one ladder company. Each company is staffed by an officer and three firefighters.

MULTIPLE-CHOICE QUESTIONS

1. What is the most important safety-related action that can be taken to prevent firefighter casualties at this incident?

    A. Comprehensive pre-fire planning to stay abreast of changes to the building.

    B. Use of large diameter handlines in the fire area.

    C. Assignment of a safety officer.

    D. Institution of a proper incident management system.

    E. Notification of the construction company and a structural engineer.

2. What is the principal danger of fires involving formwork in buildings under construction?

    A. Spalling of the concrete.

    B. Ignition of the rapidly combustible formwork.

    C. Compete collapse of the reinforced concrete floor that is being supported by the formwork.

    D. Large, falling pieces of timber igniting nearby combustibles.

    E. Complete collapse of the formwork-supported concrete and successive collapse of the lower floors due to the impact load of the failing floor.

3. The amount of area around this building that should be considered a hot zone should be:

    A. The inside of the fenced-in area.

    B. The distance equal to the full height of the building plus one-half of that height.

    C. 100' in all directions.

    D. 200' in all directions.

    E. 200' on the side opposite the application of any hose stream.

4. What will be the best way to access the reported fire floor?

    A. Walk up the stairs and then up the construction ladders.

    B. Walk up the stairs to the 4th floor, use closet ladders to reach upper floors.

    C. Have an aerial platform operate as a manpower shuttle to the floor below the fire. Use a small fire department ladder to access the fire floor.

    D. Place the aerial to the 6th floor, carry up combination ladders to be used in place of the temporary stairs.

    E. Place the aerial to the fire floor on the windward side.

5. You are sizing up the perimeter of the building for the best access routes to the upper floor. One of the construction elevators has barrier tape over the opening with an out-of-service notice. The other car is available. It is supported by two steel cables and has its controls at a ground level station. What is your action regarding this elevator?

    A. This car can be used to shuttle manpower and equipment to the upper floor. It will hold up to 2000 lbs.

    B. This car can be used, but only two men can occupy it at any one time.

    C. The weight limit of this car is 250 lbs per support cable. Take caution not to overload the car.

    D. This car is for equipment only. No personnel should be allowed into the car.

    E. This car should not be used at all for any reason.

6. Recon has reported that the fire on the 7th floor is in a small pile of debris and plastic insulation materials. What are your actions regarding fire attack?

    A. Stretch an attack and a back-up line from the street up to the fire floor to extinguish the fire. Do not trust the standpipe.

    B. Supply the fire department connection. Operate an attack line and a back-up line from the standpipe on the floor below the fire. Extinguish the fire from the windward side.

    C. Use an aerial master stream to blow the burning debris off the building. Extinguish it at ground level.

    D. Use construction site extinguishers on the fire. Have personnel stand-by to stretch a line if required.

    E. Supply the fire department connection. Stretch an attack line and a back-up line from the standpipe on the 6th floor. In addition, stretch a 2½" line up the aerial to the 6th floor to be used in case the standpipe is out-of-service. Bring a gated wye to supply the attack lines.

7. Suppose the fire conditions have worsened to involve the formwork, extend to the floor above, and threaten the top floor. What action can be taken?

    A. Use a tower ladder to move from the lower floors to the upper floors, extinguishing fire as it goes.

    B. Place an aerial master stream to protect the 9th floor only. This will allow water to pour down the stairs to keep the fire below in check until companies can man positions to extinguish it.

    C. Establish a collapse zone and protect exposures as building collapse is imminent. Assign a brand control operation in the neighborhood.

    D. Use a tower ladder to move from the upper floors to the lower floors, extinguishing fire as it goes.

    E. Place master streams in service on all sides of the structure to extinguish the fire.

8. In regard to your action in the last question, what safety precautions will be addressed to safeguard personnel?

    A. Ensure no operations take place or personnel are present on the leeward side of the building, opposite the master stream nozzle. Do not operate stream until this is confirmed.

    B. Ensure the tower ladder buckets are in no way inside the aerial collapse zone at any time during the operation.

    C. Remove all personnel to lower floors to stand-by until the master streams knock down the heavy body of fire.

    D. Establish a safety division and rehab post.

    E. Ensure all power is shut off to the site.

9. The ability of the fire department to extinguish this blaze will primarily depend upon:

    A. The size of the fire.
    B. The location of the fire.
    C. The amount of manpower available.
    D. The weather.
    E. The service status of the auxiliary appliances.

ANSWER SECTION

SCENARIO 7–1
SHORT-ANSWER QUESTIONS

1. At any fire in a high-rise, command and control is a critical factor. Measures must be immediately taken to gain control of the situation. The command post must be established in the lobby, where the incident commander will set up his "field office" and develop the strategy to bring the incident under control. There may be times, when it is better, due to the potential communication problems of fire-resistant buildings, to set up the command post in the building's courtyard or other organization-friendly area outside the building. This information should be part of the preplan-generated directives. If however, which is usually the case in office and newer residential high-rises, building communication systems such as standpipe phone systems or PA system capability are available that make communication more effective, then the command post must be established in proximity to these areas. Many buildings now have designated fire command rooms. The fire department can set up operations in an area specifically designed for fire operations and must be instrumental in the planning of these areas

    Often, a chief officer is not the first to arrive, and a company officer may have established command prior to his arrival. In this case, a formal command transfer should take place with the company officer furnishing as much information as is known about the incident thus far. As is usually the case, initial-arriving companies will probably have begun the preliminary phases of the attack and the incident commander must be prepared to inherit the situation midstream with incomplete information. It is imperative that reports on conditions be made to incident command from companies operating in the fire area as rapidly as possible. This will allow the incident commander to begin to develop an action plan to confront the incident.

    The incident commander must evaluate the need for additional alarms based on current and projected conditions. He must consider the fatigue of personnel as well as the manpower demands of a fire in a high-rise structure. Fire operations in these buildings require a manpower commitment at least four to six times that of an "ordinary" fire.

    Utilization of a command company will assist the incident commander in making sense out of the chaos in the initial stages of the incident. It may be necessary to appoint the first-arriving company as the command company. The company commander can operate as the initial operations officer. This is because it is more important in regard to firefighter safety to control the upper floor operation early on than it is to develop a fire stream. Fires in this and other fire-resistive buildings will usually be sprinklered and/or compartmentalized to assist in confining the fire. Early organizational presence on the fire floor will be a major step in assuring firefighter safety and accountability. The remainder of the command company can be utilized as elevator control, aides to the incident commander, or any other job that is critical to a smooth and safe operation.

A lobby control post must be established prior to any companies other than the initial attack teams ascending to the upper floors. Here, companies can be directed to the proper stairwell or elevator and be given additional information about their assignment. This vital control point can also be the accountability system jump-off point. ID tags or accountability devices should be collected here and maintained until the company returns to the lobby. It is better to maintain them here than on upper floors, for companies may be assigned to several places during the course of the operation (i.e., SAE, rehab, staging, etc.) The officers in charge of these control points should have an aide to account for the assigned companies or keep track of them by using a marker on the wall (or other feasible area). Companies leaving any particular area are crossed off the wall in the previous area and then placed on the wall in the new area of assignment. This is easier than shuttling ID tags around.

No one should ascend to an upper floor without the knowledge of the lobby control post officer.

A FAST team or RIC should be immediately dispatched as soon as a working fire is reported. Once on scene, after checking in at the command post, they should immediately be sent to the operations post. It may be a good idea to assign a FAST Group comprised of several companies at these incidents due to the magnitude of the operation and the manpower requirements. Due to the size of the fire area in high-rise office buildings, the FAST Group should consider the use of one-hour SCBA cylinders.

Elevator control must be established if the elevator is to be used. This will usually be done by a member of one of the initial arriving companies such as the command company. This firefighter must remain with the elevator at all times unless properly relieved or the operation is de-escalated.

The operations post should be established by the next chief officer on the scene after a briefing at the command post. In this situation, the operations post would be on the 6th floor in proximity to the attack stairwell. The establishment of the attack, evacuation, and ventilation stairwells and the passing of this information to the incident commander is crucial for the safety of operating personnel both on the fire floor and being assigned at the command post.

In this case, as only two stairwells are available, the rear stairwell should be the attack stairwell by virtue of the fact that it pierces the roof, while the front stairwell should be utilized as the evacuation stairwell.

It is the progress reports furnished by the operations post that will be the criteria by which the incident commander evaluates the current action plan. This evaluation will lead to an escalation of the incident, if required, or a de-escalation as the incident stabilizes.

The 5th floor will be the location of the resource post, also staffed by a chief officer, if available. This resource post will be the gathering point for unassigned companies and their equipment. These may be additional alarm companies or companies from the initial alarm not yet assigned. It is crucial that close coordination between the command post, the operations post, the resource post, and the SAE post (when staffed) is maintained. The resource post is akin to the logistics section in the Incident Management System. The resource post is a "stuff-gatherer" for the needs of the incident. At large incidents, the resource post must ensure an adequate inventory of equipment is on hand to meet the requirements of the incident. This especially includes spare SCBA cylinders.

The resource post officer is also responsible for the establishment of the rehab post on the fourth floor. If possible, it is desirable to delegate the authority to operate this area to EMS personnel. If EMS officers are included in preplan sessions for these types of structures, they will already know their role. Taking care of this in advance will decentralize incident command and simplify the operation. This will lead to a safer, more manageable operation.

Finally, the Search and Evacuation (SAE) post will be established at least two floors above the fire. In large buildings, rapid establishment of this critical post will provide for a safer fireground. This is especially critical at this fire, as there are reports of people trapped on upper floors. A larger-than-usual manpower commitment may need to be committed to the SAE post at this fire. Often these upper floors are smoky and filled with panicking, frightened occupants. The decision to evacuate or protect-in-place, along with the size of the area to be covered by the search, will determine the manpower commitment required. More personnel are usually required to evacuate than to protect-in-place. The decision at this fire will depend on just how large the fire is. It is usually better, in most cases, to protect-in-place than to evacuate.

In this incident (as in all high-rise fires), the ability to quickly establish an effective command organization will often be the difference between uncoordinated, unsafe operations and safe, sane, effective mitigation of the incident.

2. Before any lines are stretched, the attack stairwell must be established. This will be in a strategic position that is close enough to the fire to keep the stretch short while maintaining the attack from the unburned side, preferably with any wind condition at the attack team's backs. At this incident, the rear stairwell on the west side of the building would be the best choice, as the front stairwell does not pierce the roof and should be used as an evacuation stairwell only. When only two stairwells are available, the attack stairwell may have to double as the vent stairwell. This will keep the other available stairwell free for recon and evacuation.

The first line should be stretched by the manpower of the first two engine companies. Time must not be wasted in getting this line into position. This may require an additional engine company if conditions are severe. The line should be stretched from the standpipe on the floor below the fire (in this case, the 6th floor).

As the floor area in the office section of these buildings is usually extensive, attack teams should anticipate a heavy fire condition, especially when the fire has had a head start due to the reflex time characteristic of high-rises.

A 2½" line with a solid bore nozzle, preferably equipped with a 1¼" tip should be used for the anticipated fire condition. This will give the attack reach and penetration to lob the stream over the maze of desks and cubicles that are likely to be found on the fire floor. (Fig. 7–1)

Fig. 7–1 This cubicle area is cluttered at the floor level, but wide open at the ceiling level. Use of a 2½" line is urged to provide reach and penetration. The stream can also be used to blast away ceiling tiles to provide access for reconnaissance and stream application above the ceiling.

The objective of this line is to quickly locate, confine, and extinguish the fire, while protecting the primary search being conducted by the ladder or rescue companies.

3. The second line should be stretched by the next two engine companies to be assigned. This line should also be a 2½" line with a 1¼" solid bore tip. If possible, the line should be stretched off the same standpipe where the first line was stretched. This would have required that the gated wye and pony length be placed on the standpipe connection by the initial attack team. If this is not the case, then the line will have to be stretched from

the standpipe connection on the fifth floor. The objective for the second stretched line is to reinforce the initial attack line, and if conditions permit, to check for extension in adjacent areas.

4. Another team of two companies should stretch the third line. This line should also be a 2½" with the same size tip as the initial attack line due to the floor areas and the potential area of involvement. This line can be connected to the standpipe on the fire floor, as it is going to the upper floors. It is imperative that the operations officer keeps abreast of conditions on the fire floor. If conditions are such that the initial attack teams are driven off the fire floor, the teams on the floor above may have to evacuate also or find an area of refuge. It might be easier and safer to connect the line for the upper floor in another stairwell, provided it is a safe stairwell. As there are only two stairwells in this building, and one is being used for the attack while the other is being used for the evacuation operation, then the choice is clear-cut.

Fig. 7–2 Photo by Ron Jeffers NJMFPA Autoexposure from the fire below caused these windows to fail. This will allow fire to enter the apartment above, especially if the wind is blowing into the building. Be cognizant of this potential and take early steps to prevent it.

The objective of the third line is to check for fire extension on the floor above and to protect the primary search. Areas of concern regarding upward fire spread include ducts, bathroom and kitchen pipe chases, utility and incinerator chutes, and auto-exposure via broken windows. (Fig. 7–2)

5. Both horizontal and vertical ventilation must be considered here. If you have the luxury of two ladder crews, one crew can work the fire floor and floor above while the other crew works on vertical ventilation and upper floor operations. Although these operations are typically referred to as ladder company work, they are support operations and can be accomplished by any company that is capable and properly equipped to accomplish the job.

Horizontal ventilation must be accomplished as thoroughly as possible opposite the attack line as well as on the floors above the fire. As these are double-pane windows, this may prove to be difficult to accomplish with tools from above. It will likely take twice as much work to clear the glass and, in

Fig. 7–3 Photo by Bob Scollan NJMFPA Horizontal ventilation of the fire area may be done from adjacent balconies using the reach of a pike pole of Halligan hook. Be sure to check the stability of the aluminum balcony railing before putting any weight against it.

addition, places personnel in a dangerous position. However, it may be accomplished with the use of an aerial or tower ladder. The fire is on the 7th floor; if the aerial can be positioned between the fountain and the building, the horizontal ventilation of the fire floor can be easily accomplished either with the tip of the ladder or by a man in the platform basket. (Fig. 7–3)

Vertical ventilation must also be addressed and accomplished. Through effective pre-fire planning, ladder company personnel should know which stairwell pierces the roof. Due to the two-stairwell profile of this building, the rear stairwell will serve as both the attack and ventilation stairwell. It is critical that attack and ventilation operations be closely coordinated. The roof team (roof division) must walk up the rear stairwell from the floor below the fire (operations post) to the roof. If both stairwells pierce the roof, operations will be more simplified as the roof team can walk up the stairs in the front stairwell, access the roof via that roof door, and open the door to the attack stairwell. This is not the case here, so the attack must be delayed while the roof is accessed via the only roof stairwell available.

On the way up, the team must ensure that the stairwell is clear of occupants. A ladder company member or part of the crew must remain at the fire floor and ensure the door between the stairwell and the hall is kept closed until the team reaches the roof. Opening the fire floor door prematurely can cause the roof team to be incinerated in the stairwell. For this reason, it might be a good idea to leave the biggest ladder company members to guard the door to discourage overzealous engine company members from jumping the gun and opening the fire floor door before the roof team has reached the roof. This is especially true in residential high-rises where apartment doors are often left or blocked open by fleeing occupants.

Opening the stairwell door in this case may pull the fire toward the stairwell like a flue, especially if the wind is blowing into the fire apartment from the exterior. Often, the gases behind the stairwell door are superheated but lack oxygen for ignition, especially at the upper regions of the hallway. Opening the stairwell door may cause ignition of these gases, causing a vent-point ignition and possibly a traveling flashover up the stairwell. Think a lot of fire here, folks. Attack crews, when properly notified to open the door and begin the advance, must be ready for substantial fire conditions. "Penciling" the ceiling with the stream on the way down the hall may allow the advance to make the fire apartment, abating rollover and pre-flashover conditions.

Let's get back to the roof team. Once the roof door is confirmed open, it is the duty of the ladder company members on the fire floor to check the fire's reaction to opening the stairwell door briefly. If the reaction is favorable, the roof door is removed. The stairwell door is closed again and remains closed until the roof team is confirmed out of the attack stairwell and in a safe location either on the top floor or back at the fire floor, depending on SOPs. If the fire's reaction to opening the roof door is unfavorable, then the roof door is kept closed while the attack is made. This may clog the stairwell with heat and smoke until the fire is under control, but may make the attack less punishing. The attack must, however, still wait until the roof team is confirmed out of the danger zone above the fire.

SCENARIO 7–2
MULTIPLE-CHOICE QUESTIONS

1.

A. *-2* While elevator control is an important factor in any high-rise operation, it is not one of the COAL WAS WEALTH size-up factors.

B. *+1* The serviceability of the auxiliary appliance at a high-rise fire is of the utmost importance. Not having the system available will cause extensive reflex times, resulting in an increase in the intensity and magnitude of the fire while lines are manually stretched up the stairs. This operation will also demand a much greater complement of manpower to accomplish, which may not be on scene when the problem of an out-of-service standpipe is discovered.

C. *+1* The 36-story height of this building is one of the key contributing factors to the fire problem at this incident. The difficulty of attacking a fire in this structure will be compounded exponentially by the height of the fire, which is out of the reach of aerial devices.

D. *+1* It has been stated that a fire in a high-rise building requires as much as six times the manpower of an "ordinary" building fire. This staggering commitment of manpower should be readily available if the fire forces stand a chance of successfully holding this fire to one apartment, no matter how compartmentalized the building is.

E. *+2* Just in case you forgot, the location and extent of the fire will directly affect all other operations at the incident. For those of you who wish to argue that the auxiliary appliance is the most important, consider that a fire on the lower floors, such as the first two, would be better attacked from the street, negating the need for the standpipe.

2. Recalling from the scenario that the building does not have a fire service elevator should lead to the best answer. This shows the value of critically reading the scenario.

A. *-1* This is a dangerous option. Maintenance personnel should be used in an advisory capacity only. They should be located and remain at the command post. To use a civilian as elevator control at a high-rise fire is incredibly irresponsible and dangerous.

B. *-2* The "bypass" or "independent service" mode of elevator control is to be used strictly by the maintenance personnel to hold and control the elevator while performing their routine duties or when people are moving in and out of the building. There are no guarantees that the car will respond properly and safely. Firefighters who attempt to take this car to the fire area are playing Russian roulette. This non-fire service method of elevator control is not to be trusted.

C. *0* Critical reading will reveal that, "there are no fire service elevators available in the building." Reading the scenario critically at least twice will eliminate this answer.

D. *-2* This is again dangerous and an unacceptable use of a non-fire service elevator at a high-rise fire. There are no guarantees that the car will not ignore the commands coming from inside the car and open on the fire floor due to the heat-sensing element in the car. Because the heat on the fire floor will be most severe, the elevator car may read this as the desired floor and go directly to this floor. This is exactly what happened at this fire. A firefighter, without SCBA, entered the elevator alone. The door opened on the fire floor. He was very lucky to find the exit stair in the blinding smoke.

E. *+2* This is the only safe and sane solution to the situation. Any elevator that is not fire-service capable should be avoided. It should be shut down to prevent any unauthorized use, and the power locked out by fire department personnel.

3.

A. *+1* This is safe and will eliminate freelancing. However, there are better ways of achieving this objective. Having all responding personnel assemble at the command post will cause unnecessary congestion and undoubtedly hinder the operation. The incident commander must delegate this control point to allow the command post to concentrate on development of a strategy to put the fire out.

B. *-1* Just as assembling the responding personnel at the command post will cause a logjam there, it will be more of a problem at the operations post, which will undoubtedly be in a smaller area. The control of personnel should be accomplished prior to their arrival at the operations post.

C. *-1* The tactical reserve does not belong at the command post, but at the resource post. There, they can be quickly deployed as the requirements of the incident dictate.

D. *0* This is ultimately where the tactical reserve (such as additional alarm companies) will wind up, but the method of getting to the resource post must be directed from the lobby. Lobby control will direct these companies to the proper stairwell or elevator. This will allow a safe and coordinated journey to the upper floors.

E. *+2* With the establishment of a lobby control post, freelancing is eliminated as far as fire area access is concerned. This critical control point must operate as the gate to the operational area and direct companies to these areas via the safest routes.

4. The key to this answer is in understanding the limitation of elevators. Even though the elevators are installed in separate fire-rated enclosures, it is no guarantee that they are safe to use without fire service control capability.

A. *-2* As stated in Question #2, the non-fire service elevator must not be trusted to safely transport manpower and equipment to any floor in the building. More bad can come of it than good. It is a chance you should not take under any circumstance when accessing a working fire on an upper floor in a high-rise.

B. *+1* Tiresome, but safe. This is where additional manpower and proper physical conditioning will pay dividends. It might be a good idea if the manpower is available to position a man in the stairwell on every other floor such as the 2nd, 4th, 6th, and so on. It may be easier to shuttle equipment from man to man until it reaches the resource post than it is for men to lug it up. However, this operation must be well coordinated for it to be successful.

C. *-1* There are no guarantees that the stairwell is not filled with superheated gases and possibly flame. It is always best to attack from below.

D. **0** This will take a very long time to get adequate manpower and equipment to the area of operations. It may not be a bad idea to get later-arriving equipment and men to the upper floors, but it is unorthodox and will not offer any advantages in the initial stages of the fire operation.

E. **+2** This is the best way to reach the upper floors, especially when there are two stairwells. A two-pronged reconnaissance operation can be launched using this method, clearing both stairwells on the way up. Recon teams should meet two floors below the fire and exchange information as well as issue a report to the command post. Once the attack stairwell is established and relayed to incident command, it is the responsibility of the lobby control officer to direct all personnel to the proper stairwell and possibly cordon off the unused stairwell.

5. Due to the unavailability of the elevator as access to the upper floors, the attack teams can expect heavier than usual fire and heat conditions. This may necessitate the use of a large diameter line.

A. **-1** The attack line should never, under any circumstances, be connected on the fire floor. This will cause line congestion in the area of the door leading to the hall. In addition, fog nozzles should be avoided if at all possible. Usually requiring 100 psi nozzle pressure, a stream of substandard quality will result due to the friction loss and elevation factors when fighting a fire on the upper floors. With inadequate water, a punishing advance for the attack team can be expected, while burns and withdrawal from the fire area are possible. Always think about the nozzle pressure when attacking fires in elevated areas and make every effort to minimize it.

B. **+2** Residential high-rises, with their inherent compartmentalization, may keep the fire to relatively manageable proportions. Especially if the apartment door is closed, the attack team may be able to stretch to the fire area with little problem, thus the use of the smaller, more maneuverable 1³/₄" line. (Fig. 7–4) With its 50 psi nozzle pressure, the chances of adequate water for the stretched line is good. If the team is lucky and the wind is on their side, they might be able to knock this fire down with an aggressive push into the fire apartment. In addition, the line is stretched from the standpipe from the floor below—a proper tactic.

Fig. 7–4 For residential high-rise fires, a 1³/₄" or 2¹/₂" attack line like this lightweight high-rise pack can be placed in service quickly. A smooth bore nozzle cuts down on the pressure required at the tip and is less likely to get clogged with debris.

If this were a commercial high-rise with its characteristic open floor areas and potential for larger fire loads, the only acceptable line stretched would be 2¹/₂". Depending on the building's ability to contain the fire, it is acceptable to use a 1³/₄" line in residential high-rises as many departments do. The key is to use a solid bore nozzle to cut down on nozzle pressure and, most importantly, to pump the line at the proper pressure. Even a 2¹/₂" line will be useless if not properly pressurized.

A gated wye should be placed on the standpipe connection to allow a second line to be stretched off the same standpipe, as long as it is the same diameter. If this fire can be controlled early, the back-up line can be the same diameter as the attack line and will be in service quickly. Connect a short 25' "pony" length of 2½" or 3" diameter directly to the standpipe connection and then connect the gated wye to the pony length. This will allow an additional attack line to be connected to the primary water source without hampering access to both the standpipe control wheel and the discharge gates of the gated wye. Attaching the gated wye directly to the standpipe connection may hamper the operation of the gate valve handles. (Fig. 7–5)

Fig. 7–5 Connecting a short "pony" length of 2½" or 3" hose directly to the standpipe connection will allow a gated wye to be added in an uncluttered area away from the main standpipe controls. With this set-up, two attack lines can be utilized from the main standpipe supply.

If a 2½" back-up line is warranted due to heavier-than-expected fire conditions, it is stretched from two floors below the fire. This line should be in place, even if it is not required. If the initial attack team is having difficulties, the larger line will be ready to go. When in doubt, go with the larger back-up line.

C. *0* The standpipe choice is correct, but stretching two lines of different diameter from the same standpipe is an invitation to disaster. While the nozzle pressures are the same, the pump pressures required to flow an equal amount of water is drastically different. To flow 150 gpm from a 1¾" line, the pump pressure required is about 20 psi per length of hose. To flow 250 gallons of water from a 2½" line would require about 10 psi per length of hose. This difference could cause dangerous pressures and accompanying nozzle reaction to occur if a line shuts down or bursts. Always avoid flowing different diameter lines from the same source.

D. *+2* The stretching of these lines takes into account the potential for a heavy fire and counters it with large flows, reach and penetration. If the line size and nozzle are the same, it is acceptable to flow two lines from the same standpipe provided that the proper volume can be delivered. This will be the most effective "one-two punch" that can be delivered here. A properly supplied standpipe is required to deliver 500 psi at the first outlet discharged.

E. *-2* It is always a mistake to oppose handlines. The potential for injury is great. Even though these lines are probably too small for this fire, advancing them together instead of opposite each other may provide the required fire flow.

6. Ventilation at a high-rise fire will always be a great challenge to the incident commander and his organizational staff. In high-rises, more than in any other structure, ventilating the fire correctly or incorrectly often makes or breaks the operation. In regard to vertical ventilation, the same considerations as were discussed in scenario 7–1 should be addressed. Strict control between fire floor and roof division (vent team) operations is critical. The question here, however, relates to the horizontal ventilation considerations of the fire apartment.

A. *-2* This is an unsafe action. First, the scenario states that fire is venting from two windows. To vent from the floor above will be extremely difficult and will most likely not be accomplished due to convection heat emitting from the parent body of fire. In addition, leaning over a balcony that has been exposed to the heat of a fire is akin to suicide. The fire may have spalled the concrete underneath the railing, causing it to be unstable.

Even more likely is that the aluminum balcony will have been greatly compromised by the heat of the fire below. Aluminum melts at 1200°F. The fire below may be issuing heat waves that are twice that. If the balcony is not already melted, it is a good bet that it is unstable. No operations should take place on the balcony directly above this fire.

B. **-1** This action is unsafe for the same reasons as the answer above. The only reason that is not a point less is that using the Halligan tool and rope are better than leaning over the balcony.

C. **+1** This will be an effective way to ventilate these windows if the stream can penetrate the window glass. Many times, this glass will be double- or triple-pane glass, making it difficult to break even with heavy tools. It is good thinking and does not put any personnel in harm's way opposite the line or directly above the fire.

D. **+1** The fire is already venting and this should take care of the bulk of the ventilation problem. However, the resourceful officer continuously seeks out ways to make the situation as tenable and safe as is possible. If more can be done on the part of the ventilation effort, these options should be explored and, if they can be safely done, accomplished.

E. **+2** This is also the best answer. This choice takes the venting operation a bit further and used a creative tactic to accomplish it. Functional fixity: don't be a victim. Answer choices that take an action one step further, while operating safely, will usually be best. In addition, as an officer, you should always be looking for ways to make the situation better. Additional ventilation of the fire area is one of those ways.

7. This question changes the building systems to include a fire service elevator. One of the key concerns of the SAE post officer is which stairwell, if any, is being used as the vent and/or attack stair. It is imperative that assigned personnel be made aware of the location of this artery and ordered to keep clear.

A. **+2** Since the scenario states that one elevator bank runs from the 1st to the 20th floor, while the other runs from the 21st floor to the penthouse, this is a safe use of the elevator. The fire is on the 16th floor. The upper bank elevator does not even stop until the 21st floor, completely bypassing the fire area. This first stop for this elevator is an ideal place to establish the SAE post. Personnel will arrive at the post and either be placed into service or held as tactical reserve. This area ensures the greatest level of control of personnel arriving at and operating above the fire. (Fig. 7–6)

Fig. 7–6 This blind shaft elevator serves only floors 21 through 36 and the penthouse. For a fire on a floor below the 21st, it is safe to use this elevator to reach upper floors SAE functions. The lobby on the 21st floor is an ideal place to establish the SAE Post

B. **+1** This, again, is the best place for the SAE post, but it is not necessary to walk up. Walking up to the 21st floor will take a long time and personnel will be fatigued before the operation even begins. Remember that the SAE post requires assigned personnel to check all floors above the floor above the fire. This will cover many floors. It is imperative that the proper amount of manpower be utilized and that they are kept as fresh as possible.

  C. *0* This is not a bad idea as deployment will only be in a downward fashion; however, it will not ensure the best control over personnel assigned to the SAE post. If personnel get off at floors other than the top floor, control is lost.

  D. *-1* The roof is not a good place for the reason that it is not an elevator termination point. Personnel will have to take the elevator to the top floor, get off, and then walk up to the roof. If smoke is venting in one of the stairwells, these men could be in danger.

  E. *-2* Unless the fire is on one of the top two floors or is very minor in nature, the establishment of an SAE post is mandatory. It is essential that all upper floors be checked (especially the stairwells) for victims.

8. The operations officer must be continually evaluating conditions to ascertain if the current attack plan is working. If it is not, there must be another plan already in his head, ready to be put into action. This may require additional manpower and equipment, so it is a good idea that they be available either at the operations post or as tactical reserve in the resource post. Whatever the case, as conditions change, operations must be prepared to change with them.

  A. *-2* First of all, a concrete floor is tough to breach and should not be the first alternative. In addition, using it as a vent opening will expose the apartment above, creating twice the amount of problems you started with.

  B. *-2* This will create the same exposure problem as the choice above, but the fan will now compound it. This is the worst decision you could make.

  C. *+1* Again, breaching the wall would be tough, but here it would not create the same problem of exposure as breaching the ceiling to vent the fire. If this choice is made, use a sledgehammer or maul. Ensure companies are withdrawn to a safe area while the fog attack is being attempted.

  D. *+1* A Lorenzo ladder is an improvised master stream utilizing a ladder pipe and a ground ladder. It is extended out the window from the floor below the fire and operated into the fire apartment to knock down the heavy fire. This is an alternative for sure, but one that will be a last resort as it will take a great deal of time to put it into operation.

  E. *+2* This is a viable alternative as the lines are already there. A fog nozzle, though not the first choice for nozzle at this type of fire, does have its applications and should be brought to the resource post with other equipment. If a fog nozzle is not available, you may be out of luck as it is sixteen floors to the apparatus. Many times, a high-rise kit will be equipped with a break-apart nozzle. There will be a fog tip, a small 15/16" solid bore tip called a St. Louis adapter, and the pistol grip and bale. (Fig. 7–7) The fog tip may be screwed right off the solid bore tip. The fog tip should be kept either in the officer's pocket or in the area of the standpipe operation, possibly near the control wheel operator. The fog nozzle pushes the products of combustion ahead of the stream while the solid stream penetrates to the seat of the fire. The PPV fan is used to reinforce the direction of attack.

Fig. 7–7 This break-apart high-rise nozzle is a versatile tool to have available for high-rise firefighting. The fog nozzle is kept with the extra equipment while the ¹⁵/₁₆" tip is used on the 1¹/₂" nozzle thread. This will give reach and penetration of the solid stream with the capability to switch to a fog nozzle if the need arises.

9.

A. **-1** The ceilings in this apartment are concrete and will be impervious to both overhaul tools and passage of fire. Spalling may occur, but it is not likely to be an avenue for upward fire spread.

B. **-1** The HVAC ducts in this building are contained units. This means that each apartment has its own unit. This is common in residential high-rises, especially older ones. The HVAC unit will not be a problem unless it needs to be opened because the fire started or extended there.

C. **+1** Manmade openings will always be an area to check, even in fire-resistive apartments. They may be traced back to the breaker or fuse box where they are then connected to the main electrical service in the building. If there is an electrical fire, especially one that involves the control panel, check around it and in adjacent areas without delay.

D. **+2** As in most apartments, these will be the paths of least resistance to other areas and the first place that fire may spread, especially if there is nowhere else to go (such as in a fire-resistive apartment).

E. **-2** The compactor chute is in the hallway and not in the immediate area of the fire. Overhaul efforts should not be focused in this area unless it is known that this area was involved.

10. This question tests your ability to adapt to a situation. Also gauged is your potential tendency to overreact to a problem. Tunnel vision often crops up in these situations and may hinder a rational approach to solving this problem.

A. **0** This is a nice try, but chances are you will not be successful. If the swivels and connections are damaged, you may be wasting your time. Don't waste time on something that is obviously doomed from the outset.

B. **-2** If you are going to waste time stretching hose up the stairs from the ground, it had better be of large diameter, for the friction loss alone will most likely cause nothing but dust to be discharged from the nozzle. If you calculate 20 psi per length for 150 gpm, then the needed pump pressure here, including nozzle pressure and elevation is more than 500 psi!

C. **0** This will take a mountain of manpower and an extended amount of time, but if you decide to stretch up the stairs to the 16th floor, then LDH is the only feasible way to overcome friction loss due to diameter and elevation. However, you are overlooking the obvious if you chose this.

D. **+2** Critical reading and a keen eye on problem-solving from the most simple to the most complex will lead you to zero in on this option. The statement said that just the fire department connection was damaged. It said nothing about the standpipe system as a whole. It is a mistake to read into this statement too deeply. Keep your options open. On a test, it might cause you to miss the question—99% of an answer often lies in the comprehension of the question. On the fireground, it may lead you to unnecessarily commit a great deal of manpower and time to a task that might not be necessary. One might argue that it may be likely that the standpipe itself may be suspect, but I say this: It is better to try this option before you try to stretch a mile of hose uphill (stairs). (Fig. 7–8)

Fig. 7–8 If the exterior fire department connection is out of service or there is a need to supplement the water supply, the first floor standpipe may have to be utilized. To connect to an interior standpipe for supply, use a double female and support the coupling with a rope tool.

E. **+1** If you decide not to use the standpipe, you at least get a point. This is the option that will require the least amount of time and manpower outside of answer choice "D." In a fire on the upper floor in a high-rise, the aerial is usually not used. Therefore, using it as a portable standpipe is acceptable in this case. The best aerial device to use would be the aerial platform for the large diameter waterway, with its accompanying low friction loss, can be put into service quickly with minimal manpower and can discharge 1000 gpm (2000 gpm if two pipes are on the platform). This can supply many handlines if the supply line layout is done efficiently, utilizing a water thief or manifold near the fire floor.

Passing Score for Scenario 7–2 = 14 Points

SCENARIO 7–3
MULTIPLE-CHOICE QUESTIONS

1. The construction process is a dynamic one. The almost daily change of the building and its features along with the construction perimeter will be non-stop. As a result, the area of operation at this fire will be precarious at best. Unfamiliar and challenging access, loose building materials, and dangerous processes are just a few of the features that will make the fireground a veritable minefield. For this reason, the safety of the operating personnel must be the primary consideration. This starts before the fire, during preplanning visits, but comes together under a strong incident management system that effectively incorporates gathered information into a solid action plan.

A. *+1* Ongoing pre-fire planning is the best way of staying abreast of changes to the building site. The old saying, "fail to plan, plan to fail" can have deadly consequences here if not adhered to. Fire personnel should make regular visits to the construction site. This will provide time to become familiar with the location of protective systems such as sprinkler and standpipe connections and discharges, building access and egress points, and to make note of the array of changes that takes place over the course of the construction period. Proactive steps in this area will be the first phase in regard to firefighter safety in this building, now and for years to come.

B. *-1* While the use of large diameter hose is necessary to operate safely, large diameter lines are not the most safety-related action that can be taken to prevent firefighter casualties.

C. *+1* Regardless of whether the building has been extensively preplanned or not, the assignment of a safety officer should be routine at this and all fires. Even in the best-known buildings, something can go wrong. However, the thorough and productive familiarization with the building before the fire and as structured and organized approach to the mitigation effort can make the job of the safety officer easier.

D. *+2* All the preplanning in the world will have little impact if the fireground is not organized. Safety is rooted in scene management. In some states, including New Jersey, the institution of an incident management system on the fireground is a law. In other states, the institution of an incident management system is a sound practice and is recommended by NFPA 1561 If scene management is a high priority on the command agenda, then pre-planning effort will not have been a wasted effort.

E. *+1* These are necessary notifications to make whenever a building under construction is involved in a fire of any significance. The expertise from these outside agencies will guide the incident commander in developing the plan of action. A strong incident management system will allow the capabilities of building experts to be fully realized. Ascertaining this knowledge beforehand through effective pre-planning will support the decisions that must be made. Ascertaining this information after the fire starts is reactive and may cause the incident commander to play catch-up or, worse yet, allow personnel to operate in hazardous areas that he may have found out at earlier time had he consulted these experts on an ongoing basis.

2. Formwork is constructed of wood (sometimes scrap) of varying dimensions, usually 2" x 4" (if you're lucky), 4" x 4", and plywood. The forms may be nailed together in a haphazard manner because they are used over and over again to hold the concrete. The formwork is also coated with a material to keep the concrete from sticking to it. This material is usually a combustible liquid. One construction worker told me that it is usually diesel fuel. This, of course, adds to the ignitability of the forms. While diesel fuel certainly does not ignite as readily as gasoline, once ignited, it can only complicate the problem.

Formwork is used to hold wet concrete. In reinforced concrete, the forms are built to the desired dimensions. Steel rods are strategically placed in the mold to give the concrete tensile strength when hardened. Concrete has great compressive strength, the strength that is applied by pushing two materials together. Steel has excellent tensile strength, the force caused when two materials are being pulled apart. Together, they join to make a strong product capable of holding up many floors, vehicles, and building equipment. (Fig. 7–9)

Fig. 7–9 The steel rods set in this floor will provide tensile strength. The concrete that will be poured around these rods will provide the compressive strength. It will take this mixture approximately 28 days to reach full strength.

Once the forms are erected, the concrete is poured. At this point, it is formless and has no strength. It will take approximately 28 days for the concrete to harden sufficiently where collapse is not a danger. The formwork will be mostly stripped off before then, however, with a relatively minimal skeleton of bracing left behind to support the "green" concrete for the rest of the curing process. A fire involving the formwork at any time during the construction process should be considered a severe collapse hazard and immediate steps should be taken to safeguard fire personnel.

A. **0** Spalling of concrete is not a danger in the early stages of concrete curing when the wet concrete is still in the formwork molds. Spalling may occur later in the life of the concrete as a result of direct flame contact. Moisture trapped inside the concrete vaporizes and expands due to the action of the heat of the fire. This expansion may cause small pieces or even large chunks to break off from the main body of concrete. The size of the spalling concrete will depend on the amount of moisture present, the heat applied, and the location of the weakest area of the concrete.

B. **+1** While it is true that the formwork will readily ignite, this has a cause and effect impact on the principal danger of fires involving formwork. Ignited formwork will rapidly lose its load-carrying capacity causing the collapse of the formwork and the mass of concrete it is supporting.

C. **+1** This is, as stated above, the result of the formwork igniting; however, the alert officer will always assume the worst. He will calculate how the initial collapse of the formwork-supported floor will affect the rest of the structure.

D. **+1** This is definitely a concern at these types of fires, as falling timbers can cause secondary fires if they land in areas where combustibles are located. However, this is not the principle danger of fires involving formwork.

E. **+2** The incident commander should always assume the worst when predicting a collapse. In this case, the weight of the collapsing formwork and concrete above it will likely cause a secondary collapse of the floors below as the impact of the failing floor pancakes the lower floors. Any fire involving formwork should be handled as the potential house of cards that it is. Prepare for the worst and hope for the best. At least, you will not be in the area of a complete and catastrophic collapse. (Fig. 7–10)

Fig. 7–10 Collapse of formwork can cause successive pancake collapses of the lower floors. If formwork becomes involved, withdraw, account for all personnel, establish collapse zones, and operate in a defensive manner.

3. Establishing a Hot Zone is not only limited to hazardous materials responses. A Hot Zone can be established at any incident and can be considered the area of operation inside which only essential personnel should operate. It is not to be confused with a collapse zone in this context.

A. **-1** The inside dimensions of the fenced-in area is unknown. Therefore, it cannot be assumed that it is a sufficient area for a Hot Zone to be established.

B. **-1** The distance equal to the full height of the building plus one half the height is a collapse zone. No one should be permitted inside the collapse zone. This question asks for the Hot Zone. Personnel are authorized to operate inside the Hot Zone, but only those personnel essential to the operation are allowed inside. For example, it is not an area where apparatus should be staged or medical operations posts established.

C. **0** For buildings of high-rise proportions, a 100' perimeter may not be sufficient an area to establish. Falling debris may land outside this area.

D. **+2** 200' in all directions is a safe area to be considered a Hot Zone. Areas beyond that distance can be considered safe havens for support agency locations and even the command post. Keeping the command post outside this perimeter gives the incident commander a wide view of operations and conditions.

E. **+1** While operations on the side opposite stream application should be completely prohibited, this is not as safe a response as answer choice "D." To limit your Hot Zone to this area invites trouble. Personnel free to wander around the immediate perimeter of the building in areas other than the established Hot Zone can easily be struck by falling debris caused by factors other than hose streams such as wind. Play it safe and establish a wrap-around perimeter at these incidents. (Fig. 7–11)

Fig. 7–11 Wind can cause a significant danger around the perimeter of a building under construction. Some lightweight steel trusses as well as a portion of the corrugated steel decking blew off this building during high winds. A Hot Zone was established until the winds died down.

4. Accessing the fire floor in buildings under construction will be treacherous at best, especially when the stairs are not yet completed or enclosed. Choose the safest route even if it takes more time. Site visits will play an invaluable role in this information.

A. *-1* If the construction ladders can be avoided, that is advisable. They are usually constructed of wood in a makeshift fashion, sometimes nothing more than nailed-together 2" x 4" lumber. If they can't be avoided, try to use them as minimally as possible. This fire can be better accessed than by walking this unnecessary amount of floors via the construction ladders.

B. *0* This is better than using the temporary construction stairs, but not by much. Closet ladders are usually only 10' and are extremely narrow. The floors between the buildings may be 12'. In addition, climbing a closet ladder is precarious at any height. Add full turnout gear and equipment and the task is even more hazardous.

C. *+1* The aerial platform is a safe way of accessing upper floors, but it can only carry limited manpower at one time. It must also leave to return to the ground, leaving the crews on the upper floors without a safe egress route. If a large amount of manpower has to evacuate the building quickly due to deteriorating structural conditions or heavier fire than they are capable of handling, the device will be unable to evacuate all men at one time, or worse yet, may be overloaded.

D. *+2* The aerial will provide a continuous stairway to the upper floors, allowing both access and/or egress of a large number of men at any given time. This is the most efficient way of accessing the fire floor, taking the least amount of time per man to get to the fire area. In addition, the combination ladder is safer than the closet ladder and more specific than the "small ladder" mentioned in answer choice "C."

E. *0* Placing the aerial to the fire floor is potentially dangerous, even if it is positioned at the windward side. If the wind shifts, the operating forces will be cut off from the safest point of egress. If the ladder is on the floor below, it will be in a safe area no matter which way the wind blows. If there are two or more aerial devices available, placing the second or third to the fire floor in a safe area is an acceptable practice.

5. This question tests your knowledge of construction elevators. In many scenarios, there will be straight knowledge questions directly tied into the scenario, such as in this question.

There are usually two types of elevators, also called hoists, at the construction site: the personnel elevator and the equipment and materials elevator. There are significant differences in each that should be known to all firefighters. (Fig. 7–12)

The personnel elevator will have the controls on the inside of the car. The personnel elevator will also have more cables supporting it than a materials elevator due to the safety factors of transporting men to upper floors. The materials elevator may only have one or two cables supporting it. There will also be a gate, usually of the pull-down type on the personnel elevator. However, some of these features, such as the gate, may also be found on the equipment elevator. The biggest difference in the two types of cars is the location of the operating controls. As mentioned before, the personnel elevator will have the controls

Fig. 7–12 The construction hoist is constructed of unprotected steel and may have few or inadequate connections to the building. Fire exposing this structure may cause it to fail early. Firefighters should never attempt to access upper floors via this hoist

on the inside of the car. (Fig. 7–13) This requires someone actually on the inside of the car to operate it. The equipment elevator will have the controls on the outside of the car, usually at the ground level. In older equipment elevator systems, the operator at the ground level used a bell to signal him when to move the car. The bell is similar to the SCBA bell, and some of these systems are still used. This is still another reason why personnel should not use the equipment elevator. An activated SCBA bell can cause the car to be inadvertently moved. More recently, however, walkie-talkies are used between the ground control station and the destination floor.

If the controls are not in the elevator or at ground level, the equipment elevator may be controlled via a system operated from a shanty at the ground level somewhere in the proximity of the elevators. If you see wires leading from the elevator shaft assembly to a shanty nearby, then this unit is a materials and equipment elevator and the elevator controls as well as the communication system is located there.

Fig. 7–13 The control panel will be located inside the personnel hoist. If it is not present, people, including firefighters were not intended to use it. Note the safety rules placard above the control panel.

The most important factor regarding these two types of elevators is that fire personnel should never, under any circumstances, use the materials and equipment elevator, to access upper floors of the building. This is an absolute rule that should never be broken.

A. *-2* Mistaking this car, obviously an equipment elevator, can have disastrous consequences. It is not designed to transport firefighters to the upper floors, regardless of how much weight you might think it may hold.

B. *-1* An equipment elevator should not be used for even one firefighter, let alone two.

C. *0* Regardless of how much weight the car may be capable of holding, it is not safe for personnel to use. If manpower is available and the situation is such that both types of car are available, a firefighter with the same duties as the lobby control officer can be assigned to ensure personnel are directed to the personnel elevator and that that no one use the equipment elevator.

Fig. 7–14 This makeshift shanty is actually used to control the materials and equipment hoist. It should never be operated by firefighters. Command should make use of a construction site foreman as a liaison at any fire involving a building under construction.

D. *+2* For the reasons stated above, no personnel should use this car to gain access to the upper floors. However, it is efficient to transport equipment to the upper floors via this elevator. It will save a tremendous amount of work and subsequent fatigue and possibly injury. It will also reduce reflex time. To most effectively accomplish this, two teams should be utilized to operate as an equipment shuttle post. One is stationed at the ground level while the other is at the resource post. Use of a separate frequency to coordinate this operation will also reduce the radio traffic on the fireground and command frequencies.

E. *+1* If you decide to play it safe and prohibit the use of the elevator, it is best to disconnect the power and utilize lock-out/tag-out procedures to prevent unauthorized operations.

6. As stated in the textbook high-rise chapter, the serviceability of the standpipe will be a major factor in the success or the failure of the fire operation. Pre-fire visits and inspections are the only way to ensure that standpipe systems are keeping pace with the construction process. In addition, these visits will ensure that access to the fire department connection is kept clear. Absent this prior knowledge, the incident commander can only hope that luck is on his side and the system is serviceable. For this reason, a contingency plan in case the system is not available for use must be addressed.

   A. *0* This choice automatically surmises that the standpipe will be out-of-service. Stretching hose from the street to the fire floor will be manpower-intensive and tiring. It is a good idea to at least hook to the fire department connection and supply it to ascertain if the system is serviceable.

   B. *+1* This surmises that the system is in service. Supplying the system and operating from the standpipe will be the quickest way of attacking the fire and ensuring it remains manageable; however, a contingency plan must be available in case the standpipe is unusable.

   C. *-2* This action is dangerous and will not guarantee where the debris will land when it is washed off the building. The incident commander who chooses this strategy may wind up with more than one fire as materials land in different places.

   D. *-2* The use of construction site extinguishers is a definite possibility for the recon companies arriving on the fire floor before any hose lines are stretched. This action can keep the fire from spreading and involving the formwork. However, the immediate stretching of a line is mandatory in case the fire is beyond the scope of the extinguishers. Waiting to stretch the line will cause a delay in its arrival at the proper place, thus allowing the fire to possibly spread even beyond the control of this line.

   E. *+2* This is the best strategy because it involves a contingency plan in case the primary plan of action is not effective. The thinking incident commander must address the "what if" factor and operate in a proactive manner. In this scenario, if the standpipe is unusable, the line from the aerial will be available almost immediately to attack the fire. A gated wye is a good piece of equipment to bring to the fire floor. If the fire is large, it can be attacked with the $2\frac{1}{2}$" line. If it is not, the $2\frac{1}{2}$" can be used as a supply line to supply two $1\frac{3}{4}$" attack lines. It can also be useful after the knockdown of the fire with the $2\frac{1}{2}$" to gate down to a $1\frac{3}{4}$" to overhaul the fire area. Plan ahead and you will have a better chance of not getting caught short.

   If the aerial is not available, another option, albeit slower and more manpower consuming, is to hoist a dry line up to the floor below the fire. If an aerial platform is available, but you don't want to tie it up as a portable standpipe, it can also be used to hoist the hose. Ensure that the line is tied off at the floor of operations and possibly on several floors below the fire before charging the line. The weight of the water may drag an unsecured line and the men assigned to operate it right off the side of the unenclosed building.

7. A fire that has involved the formwork in a building under construction as well as threatens to involve upper floors is an imminent collapse hazard. It will be immediately necessary to withdraw personnel from the building and establish a collapse safety zone.

   A. *0* Starting at the fire floor and directing the stream upward works well once the building is completed. This action would quench the main body of fire. This is imperative in a finished building because if the main source of heat is not cooled, fighting the fire from any other point will be a wasted effort. In an uncompleted building, the opposite procedure should be followed, as the openings in the construction will allow the master stream to flow down to the lower floors as well as hit the main body of fire. In this case, it is better to head off the fire.

B. *-1* Leaving the stream on the 9th floor in hope that it will find its way to the fire is a good plan, but the stream is better off moving from floor to floor to hit the main body of fire as well as the upper floor. An aerial master stream is not the tool of choice here. An aerial platform will be much more effective at this operation and, if available, should be used in lieu of the aerial ladder pipe. If the aerial is the only elevated master stream available, it will be best used to operate on the 8th and 7th floors. The latter is the fire area, while the former is the most exposed floor.

C. *0* This action completely gives up the fire building. There will be a point in the operation, if the incident escalates to such a juncture, that abandonment will be the only feasible and safe action. However, the white towel should not be thrown into the ring just yet. The incident commander has an array of options that may be used to stop the fire at this point. He must be willing to attempt to utilize his resources in the safest manner to stop the progress of the fire.

D. *+2* As stated above, a building under construction will not be equipped with many of the fire-restricting features found in a finished building. These will be holes in floors, stairways will not yet be enclosed, and the open floor areas will allow a stream to be directed in the most effective manner possible. Directing the stream from the upper floors to the lower floors will take advantage of these building openings and allow the discharged water to flow down to the fire floor, aiding in the extinguishment operation.

E. *-2* This is like opposing hose streams, but on a more magnanimous scale. The streams will push the debris all over the place, possibly into uninvolved areas. Coordination of master streams is the responsibility of the incident commander. He must forecast what impact, both positive and negative, the master stream will have on the fire, the building and exposures, the environment (think runoff), and the safety of personnel.

8. The safety of personnel is always the highest priority of the incident commander, especially when a situation escalates to where a defensive operation must be pursued.

   A. *+2* This will safeguard the most personnel. Master streams may wash a substantial amount of debris off the floor of operation onto the surrounding area below. This is especially applicable to buildings under construction where the floors may be wide open with no barriers to stop the debris from washing off the sides. Ensure all personnel are out of the danger area before the stream is put into operation.

   B. *+1* This is also a major concern whenever aerial master streams are in operation. Platforms and aerials often get too close to the building and are struck by falling structural materials above the device. The apparatus may be safely out of the collapse zone, but the position of the aerial device may not be. It is imperative that all personnel charged with operating these devices receive thorough training on not only collapse zones, but also appropriate safety zones when operating aerial devices. This choice gets the point only because answer "A" addressed the safety needs of a greater number of personnel, whereas this answer only addresses the safety requirements of a few members, namely the operators of the aerial devices.

   C. *-2* It may be acceptable in certain instances, to leave personnel inside buildings while master streams are operated, however, these personnel must be in protected areas where the destructive action of the master streams can do them no harm. This is not one of those instances. In this building, the possibility of a massive structural collapse precipitates the requirement that all personnel be completely withdrawn from the building and accounted for before any master stream device is put into operation.

D. *0* While these firefighter safety measures should be established at this and all significant fires, these are indirect methods of addressing the hazards of the situation. Direct action must be taken to safeguard personnel.

E. *0* This, again, will not prevent personnel from wandering too close to the building. Incidentally, once the site is shut down for the day, the power is usually shut down, too. It actually may be a good idea to have the power turned back on to operate the personnel and equipment hoists as well as temporary lighting during night operations.

9.

A. *+1* The size of the fire will certainly be a key factor as it will dictate the strategic mode. However, the ability of personnel to utilize the auxiliary systems directly impacts the size of the fire once companies are in place to attack it.

B. *+2* The location of the fire will dictate whether standpipes may be required or lines may be stretched to the fire floor manually. A fire on a lower floor will make life easier and render the question of the status of the auxiliary system a moot point.

C. *+1* This operation will, regardless of the magnitude, require a major commitment of apparatus and manpower. However, a fire on the upper floors will demand a greater commitment than a fire on a lower floor.

D. *0* Weather, while a significant factor at any fire, will not be one of the primary factors at this fire. Weather will also be influenced by the location. Fires on upper floors of tall buildings under construction may exhibit severe wind conditions that may whip fire into unmanageable proportions. In addition, a wide-open building in the winter may be subject to ice buildup. The dangers here include both slipping hazards for personnel as well as the added weight of ice to the structure.

E. *+1* The in-service status of the auxiliary appliances will make or break this operation, but usually only if the fire is on the upper floors. An out-of-service standpipe system at a fire on an upper floor will impact on required manpower as well as the size of the fire once lines are stretched. If the system is serviceable, the fire may be more rapidly extinguished. If it is not available, reflex time will be tremendously increased resulting in more severe fire conditions and a greater danger to operating forces.

Passing Score for Scenario 7–3 = 12 Points

CHAPTER EIGHT
SCENARIOS

SCENARIO 8–1
ROW HOUSE FIRE

Construction

The fire building is part of a two-story row house complex of wood frame construction.

Time and weather

Wind is from the North toward the B-side exposures, at 12mph. The fire was reported at 2200.

Area and street conditions

This is a run down section of town.

Fire conditions

There is heavy fire in the upper section of the house. Smoke is showing from the roof of the fire building and flames are showing at the top floor windows.

Exposures

The fire is located in the house in the center of the row. There are fifteen attached two-family houses in the row. The houses all share party walls, a common cockloft, and a flat roof. There is a scuttle hatch over the main stairwell on each building.

Water supply

Hydrants are located on each corner on the ends of the block. Each is served by a separate 12" main.

Response

The initial running assignment is three engine companies and one ladder company. All companies are staffed by one officer and three firefighters. Each additional alarm will bring two engine companies and one ladder company.

Multiple Choice Questions

1. What are the first actions for the first arriving engine company after a water supply has been secured?

   A. Stretch a 1¾" line into the fire building, coordinate your attack with ventilation operations.

   B. Immediately stretch handlines into Exposures B and D. Bring these lines to the top floor to stop any fire extension through the cockloft.

   C. Stretch one 2½" handline into the fire building to secure the stairway for occupant egress and search and rescue operations.

   D. Immediately enter the building and remove any occupants.

   E. Split the company into two crews. Have one crew stretch a handline into the fire building while the second crew goes to the rear to assist any victims that may be at the windows.

2. Considering the action you chose for the first engine company in the previous question, what actions should the second engine company take?

   A. Secure a secondary water supply. Take a 1¾" line into the exposure on the south side of the fire building.

   B. Secure a secondary water supply. Take a 1¾" line into the fire building and to the top floor. Back-up and support the initial attack.

   C. Immediately stretch handlines into Exposures B and D. Bring these lines to the top floor to stop any fire extension through the cockloft.

   D. Park the apparatus out of the way. Begin evacuation of the attached houses on both sides of the fire building.

   E. Enter the fire building with hand tools to initiate a primary search and rescue; begin horizontal ventilation on the upper floors.

3. What are the most appropriate initial actions for the first arriving ladder company at the fire?

   A. Raise a ground ladder to the rear windows of the 2nd floor of the fire building; begin vent, enter, and search operations on the top floors of the fire building.

   B. Raise the aerial to the roof of the fire building, immediately begin to cut a trench that will run parallel to the street directly behind the parapet, allowing for ventilation and inspection of conditions in the cockloft.

   C. Split the ladder company into two crews. Initiate a primary search and rescue in the fire building, ladder the attached building on the north side, begin vertical ventilation at the natural openings of the fire building.

   D. Due to heavy fire and dangerous conditions, keep the company together. Enter the fire building with hand tools to initiate a primary search and rescue, and perform horizontal ventilation on the upper floors.

E. Split the ladder company into two crews; send one crew into the fire building to search, rescue, and horizontally vent along with the engine company's attack. Send the second crew into the downwind exposure to evacuate occupants and check for any fire extension in the cockloft.

4. Would additional alarms or companies be required at this fire?

   A. Yes, a second alarm should be struck for this incident.

   B. No, the fire should be contained and all possible victims rescued with the manpower that is on hand.

   C. Yes, a second and third alarm should be struck based on the life hazard and fire spread potential.

   D. Yes, one extra engine and ladder company should be special called.

   E. Yes, a second alarm should be called due to the potential for fire spread to exposures.

5. Based on the information in the scenario, which size-up factors are of the most concern to the incident commander?

   A. Height, weather, and street conditions.

   B. Location and extent, life hazard, and exposures.

   C. Life hazard, construction, and exposures.

   D. Exposure protection, incident stabilization, and property conservation.

   E. Life hazard, incident stabilization, and property conservation.

6. Suppose this building was three stories tall. You have been ordered to stretch a third line to the top floor to protect the search operation being conducted on that floor. How would you deploy this line?

   A. Stretch the line around the rear of the building, through the back door, and up the stairs to the 2nd floor.

   B. Stretch the line through the front door and up the stairs to the 3rd floor.

   C. Raise a ground ladder to a 2nd floor window, stretch the line up the ladder, and then stretch via the interior stairs to the 3rd floor.

   D. Stretch the line up the ground ladder on the exposure, and bring it through the scuttle to the 2nd floor.

   E. Lower a rope out of the window of the 3rd floor, hoist the line up to the window, and stretch out on the 3rd floor.

7. You are the first-arriving company officer. As you arrive on the scene, you are met by a man in the street who tells you that his non-ambulatory handicapped mother lives in the house three doors south of the fire building. Light smoke is being reported at the eaves of Exposures B1 through B5. What is your response?

    A. "The battalion chief is parking his rig down the block and will be taking command in a couple of minutes. He will decide which actions to take from this point forward."

    B. "I will assign a member of the first due ladder company to remove your mother from the building."

    C. "All the men are assigned to making sure the fire doesn't spread. Additional companies are responding. It does not seem that your mother is in immediate danger. I will assign someone to your mother's building as soon as the other companies arrive."

    D. "I need to assess the fire situation first and I will send later arriving companies to evacuate your mother."

    E. "Everyone else is busy. I will go with you to help get your mother to safety."

8. Which exposure should receive the most attention?

    A. All exposures should receive the same degree of protection.

    B. Exposures A and B.

    C. Exposure B should be protected exclusively.

    D. Exposure B causes the most concern, but Exposure D must be entered.

    E. Exposure protection is not a problem. Put the fire out and all other problems disappear.

9. How would you order this exposure protected?

    A. Set up a portable deluge gun on the top floor. Pull the ceilings, and then withdraw to the hallway. Use a deck gun stream to extinguish any extending fire.

    B. Cut a trench between the fire building and the exposure. Stretch lines to the top floor and the roof. Pull the ceilings in the top floor apartment. Use streams to keep the fire from extending.

    C. Position two lines on the top floor. Horizontally vent the windows on the top floor. Pull the ceilings entirely. Direct streams into the cockloft to prevent the fire from extending into the exposure.

    D. Stretch two lines to the top floor. Open the walls in the apartment. Make exam holes in the ceiling on the exposed side. If any fire is found, pull the ceiling completely. Direct a stream into the cockloft. Make exam holes in the roof above the exposed apartment. Monitor the cockloft from above.

    E. Position two lines on the top floor. Open the apartment windows. Pull the ceilings entirely and direct the streams into the cockloft to keep the fire from extending into the exposure. Have the roof team cut exam holes in the roof over the exposed apartment and monitor for fire spread. If any fire is found, enlarge the opening.

10. Based on the information in the scenario, and the action taken in Question #3, what would be the most appropriate assignments for a second arriving ladder company?

    A. Assist Ladder 1 with the primary search.

    B. Split the company into two crews. One crew operates in Exposure B. Open the walls and ceilings and check for extension. The second crew goes to the roof of the fire building to assist in ventilation operations.

    C. Split the company into two crews. One crew goes to the rear of the fire building with ground ladders to conduct vent, enter, and search operations. The second crew goes to the roof of the fire building to assist in ventilation operations.

    D. Split the company into two crews. One crew goes into Exposure B to open the cockloft from the inside. The second crew goes into Exposure D to open the cockloft from the inside.

    E. Hold the company at the command post to use as a tactical reserve.

Scenario 8–2
Row House Fire #2

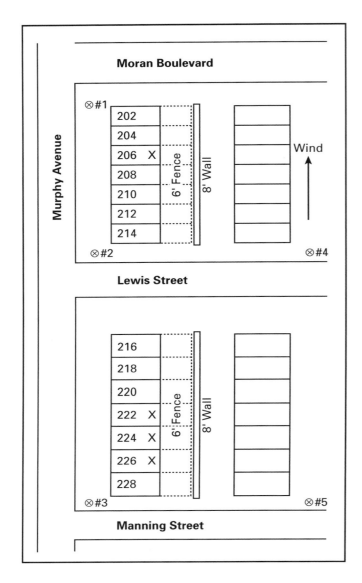

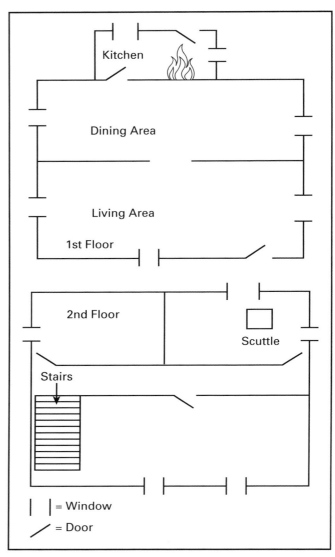

Construction

A fire has been reported in an area of town, which consists of two-story wood frame row houses. The fire has been reported at 222 Murphy Avenue. These particular row houses have had a one-story addition built at the rear that houses a kitchen.

Time and weather

The time is 1430. The wind is out of the south.

Area and street conditions

This area of town has been run down and is a known area for drug dealers. Some of the row houses are vacant. Vagrants have been reported in the area and have been suspected of accidentally starting several small fires via illegal cooking operations.

Fire conditions

As you arrive on the scene, you cannot tell if the fire is in 222, 224, or 226 Murphy Avenue. You just see heavy smoke. As you are sizing up this scene, a child runs up the block and tells you, "There is a fire next door to my grandmother's house down the block! She's 93, and doesn't walk so good." You can see light smoke coming from the top floor window of 206 Murphy Avenue, which is a vacant building.

Exposures

All the row houses are attached, with a common cockloft above the top floor. All the dwellings have a scuttle in the rear 2nd floor bedroom that leads to the roof. There is a 6' high fence in the rear (back yard) of all the row houses. Behind this fence is an 8' high block wall.

Water supply

The water supply is adequate.

Response

Your response is three engines and one ladder company, with four personnel on each, including the officer. You will receive the same assignment on a second alarm. You are the first company officer on the scene. The chief will be on the scene in two minutes.

Multiple-Choice Questions

1. Based on the information in the scenario, what is the most important size-up factor at this fire?

   A. Location and extent

   B. Construction

   C. Occupancy

   D. Height

   E. Command

2. Would additional alarms be needed at this incident?

   A. Yes, strike a second alarm. Direct all second alarm companies to 206 Murphy Avenue to handle the fire operations at that building.

   B. No, on-scene companies can handle all fire operations at this incident.

   C. Yes, due to fire spread potential caused by the inherent structural deficiencies of wood-frame row houses.

   D. No, due to the fact that the extent of the fire is not determined at this time. Proper reconnaissance reports from the first arriving ladder company will determine the need for additional alarms

   E. The chief will decide this upon arrival; he is only a few minutes away.

3. What are your orders for Engine 1?

   A. Establish a water supply at Hydrant #3; stretch a 1¾" line through front door of 222 Murphy Avenue; force entry; and attempt to locate, confine, and extinguish the fire.

   B. Establish a water supply at Hydrant #3; stretch a 1¾" line through front door of 224 Murphy Avenue; and attempt to locate, confine, and extinguish the fire.

   C. Establish a water supply at Hydrant #1; position the apparatus just past 206 Murphy Avenue; and stretch a 1¾" line through front door to the top floor to locate, confine, and extinguish the fire.

   D. Establish a water supply at Hydrant #2, position the apparatus on Lewis Street, and stretch a 1¾" line to the rear of 224 Murphy Avenue. Enter through the rear door to locate, confine, and extinguish the fire.

   E. Establish a water supply at Hydrant #3; stretch a 1¾" line through 228 Murphy Avenue to the rear of 224 Murphy Avenue to locate, confine, and extinguish the fire.

4. What are your orders for Engine 2?

   A. Establish a secondary water supply; go to 206 Murphy Avenue; and stretch a 1¾" line to the 2nd floor to locate, confine, and extinguish the fire.

B. Establish a secondary water supply, go to 224 Murphy Avenue, and stretch a 1¾" line through the front door to back up Engine 1's attack.

C. Establish a secondary water supply, go to 224 Murphy Avenue, stretch a 1¾" line through the front door to back up Engine 1's attack, and check for any fire extension.

D. Park the apparatus out of the way. Take a second attack line off of Engine 1. Stretch a 1¾" line through the front door of 224 Murphy Avenue to back up Engine 1's attack.

E. Establish a secondary water supply, go to 222 Murphy Avenue, and stretch a 1¾" line through the front door to provide reinforcement of the initial attack line and to check for any fire extension.

5. Ladder 1's officer informs you that the fire is located in the rear in the one-story kitchen addition at 224 Murphy. How should this fire be ventilated?

A. Access the 2nd floor roof via a ground ladder and cut a hole in the roof.

B. Break the windows in the rear of the kitchen, open the kitchen door, and, if necessary, cut a hole in the 1st floor roof.

C. Break the 2nd floor windows.

D. Use positive pressure ventilation at the front door.

E. Break the windows on the 1st floor in the kitchen and open the rear door. Cut a hole in the roof above the main body of fire.

6. You are ordered to perform post-control overhaul at the kitchen fire. How would you best carry out this order?

A. Open up the dining room walls.

B. Open up the walls and the ceiling in the kitchen.

C. Open up ceilings and walls on the 2nd floor.

D. Use a positive pressure fan at the front door.

E. Open up the exterior siding on the rear wall of the 2nd floor adjacent to the kitchen roof.

7. You have been ordered to supervise the ventilation operation. You also have to supply two men for a search of 224 Murphy Avenue. How would you direct your subordinates to accomplish this objective?

A. Keep the search team intact. Search starting on the 1st floor, then proceed to the 2nd floor.

B. Keep the search team intact. Search starting on the 2nd floor, then proceed to the 1st floor.

C. One firefighter searches the 1st floor and one firefighter searches the 2nd floor; both using a search rope.

D. Both firefighters search the kitchen, then split up to search the 1st and 2nd floor.

E. Both firefighters search the kitchen area only when a line has been brought into place.

8. Based on your initial action, what is the most important exposure at this fire scene?

    A. 222 and 226 Murphy Avenue.

    B. The row house in the rear.

    C. Treat all exposures equally.

    D. 216 Murphy, since it is at the end of the row.

    E. 204 and 208 Murphy Avenue.

9. The fire in 206 Murphy is on the 2nd floor. There is now fire showing at the front window. You are assigned to perform ventilation. Choose the best action:

    A. Ventilate horizontally at the 2nd floor windows.

    B. Access the roof of 208 Murphy Avenue via a 35' ground ladder and open all natural openings at the roof level of 206 Murphy Avenue. Cut the roof over the area of most involvement. Conduct reconnaissance and ventilation operations at the rear of the building. Report conditions.

    C. Use 24' ground ladder to open fire room windows, access the roof via the scuttle of 206 Murphy Avenue. Open all natural openings on 206 Murphy Avenue and cut the roof over the area of most involvement. Conduct reconnaissance and ventilation operations at the rear of the building and report conditions.

    D. Use the aerial to open or break the 2nd floor windows; ventilate the roof if necessary.

    E. Raise a 35' ladder to the rear of 206 Murphy Avenue and ventilate the top floor windows from the roof. Open any natural openings on the roof, cut the roof over the area of most involvement, and check the cockloft for any fire extension.

10. Where would the first line be stretched in the 206 Murphy Avenue fire?

    A. Through the front door to the 2nd floor via the stairway.

    B. To the common cockloft via 208 Murphy Avenue.

    C. Into 208 Murphy Avenue via the front door to support the rescue of the elderly woman.

    D. To the front of the building, use as an exterior line to knock down the heavy fire showing at the front windows.

    E. Up a ground ladder to the 2nd floor due to the possibility that the stairway in the vacant structure may be unsound.

SCENARIO 8–3
TOWNHOUSE DEVELOPMENT FIRE

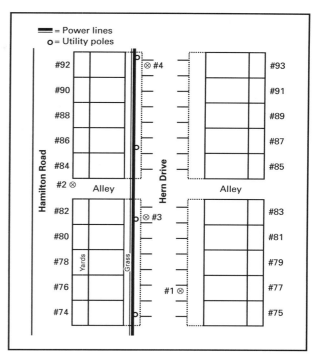

Construction

The townhouse development is ten years old. Each cluster consists of five connected units. The units are of lightweight wood frame construction with a plywood roof supported by a peaked lightweight wood truss roof system. Roofing materials consist of asphalt shingles.

Time and weather

The time is 0715 on a humid Friday morning, the temperature is 75°F, and the wind is from the south at 10mph. Heavy thunderstorms and winds have just passed through the area and provided needed relief from a recent drought.

Area and street conditions

Hern Drive is a one-way street running south to north. Electrical power has been lost in that part of the townhouses. There are designated parking spots perpendicular to the curb. The area between the curb and the townhouse is landscaped with concrete and grass. Hamilton Road runs parallel to Hern Drive.

Fire conditions

A call comes in from neighbors seeing heavy black smoke coming from the second floor of 78 Hern Drive. You notice a live power line down and moving from side to side in front of units 76 through 80. It has not been determined if there are any occupants in 78 Hern Drive.

Exposures

A firewall separates each unit from ground floor to roof. Each unit has a small backyard surrounded by a 6' privacy fence with gates usually locked at night. Townhouses across from Hern Drive mirror those on the opposite side in dimension and construction.

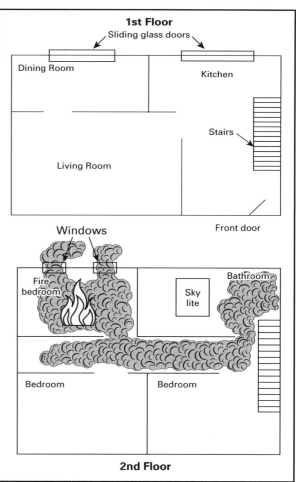

Water supply

There are four hydrants in the area, two on the same side of 78 Hern Drive. Also one is on the same grid across from 78 Hern Drive. The fourth is on Hamilton, on a different grid.

Response

You are first to arrive on the scene. Your response is three engine companies and one ladder company. Engine companies are staffed by an officer and two firefighters. The ladder company is staffed by an officer and three firefighters.

MULTIPLE-CHOICE QUESTIONS

1. Based on the information in the scenario, what would you do in regard to the downed power line?

    A. Have the electric company respond to the scene; advance a line to the rear sliding door.

    B. Radio the situation to dispatch, have them call the electric company; do not advance a line until the electric company arrives.

    C. Move the power line with rubber gloves and a hot stick. Notify the electric company.

    D. Instruct men to advance a 1¾" line through the front door and advise them to use caution. Notify the electric company.

    E. Contact the electric company by phone. Ask for advice on how to handle the situation.

2. What is the most important factor in your initial radio report?

    A. Building construction

    B. Establishment of the incident command system

    C. The presence of the downed wire

    D. Apparatus position

    E. Location of victim

3. Would an additional alarm be necessary at this fire?

    A. No, the fire should be contained and any possible victims rescued with the available manpower on-scene.

    B. Yes, due to the downed power line, contiguous structures, difficult forcible entry, and to keep people away from the power line.

    C. No, not until the situation is fully assessed.

D. Yes, as a possible life hazard exists; a second alarm is necessary to conduct primary search operations.

E. Yes, additional manpower is needed to protect the exposures.

4. What is the most important size-up factor to consider for interior attack?

A. Construction

B. Exposures

C. Occupancy

D. Auxiliary appliances

E. Location and extent

5. What are the first orders to give to the captain of the first-arriving engine company?

A. Drive around to Hamilton Street, establish a water supply at Hydrant #2 and proceed through the backyard to the fire building with a handline.

B. Establish a water supply at Hydrant #3; enter the front door with a 1¾" handline; attack the fire.

C. Establish a water supply at Hydrant #4; advance a line to the rear of the fire building to attack via the rear door.

D. Establish a water supply at Hydrant #1; use a deck gun to attack the fire until the power can be shut down.

E. Establish a water supply at Hydrant #3; enter via the front door with a booster line and attack the fire.

6. Power has been shut off; a second water supply has been established by Engine 2. What other action would you order Engine 2 to do?

A. Advance a second line to the front door; make a determination whether it is needed before advancing up the stairs.

B. Assist men from Engine 1 in advancing the initial hoseline and conduct a primary search.

C. Advance a second line into the fire building via the rear entrance. Back up Engine 1's attack and check for fire extension in the connecting rooms.

D. Stretch a second line to the exterior. Use this second line to wet down the roofs of the other townhouses.

E. Advance a second line through the front door; back up Engine 1's attack and check the connecting rooms for any fire extension.

7. The battalion chief orders ventilation operations to begin. As captain of the first arriving ladder company, what are your first actions?

   A. Cut the roof directly over the fire.

   B. Go to the 2nd floor, open a window, and use a fog stream to vent the fire area.

   C. Open the windows on the 2nd floor and use positive pressure at the front of the building

   D. Remove the skylight.

   E. Ladder the rear of the building and horizontally ventilate via the rear windows.

8. The battalion chief orders a primary search and rescue. As captain of the ladder company, what are your orders?

   A. Send the company to the 2nd floor and split the members into crews. Perform a systematic search and then move to the lower floors.

   B. Split the company into crews and conduct a systematic search, with the first crew upstairs and the second crew downstairs.

   C. Send the company to 1st floor and split the members into crews to conduct a systematic search. When the 1st floor is completed, search the 2nd floor.

   D. Keep the search crew intact and search off of the hoseline.

   E. Conduct a systematic search of the 2nd floor and use chalk to mark searched rooms.

9. A number of factors may affect the conditions of building that are being checked for hazards. Overhaul has started. What factor should be given the most consideration in overhauling the scene?

   A. The degree of fire intensity and overall building dimension.

   B. The overall building dimension and weather conditions.

   C. The amount of water applied and overall building dimension.

   D. The amount of water applied and degree of fire intensity.

   E. The weather conditions and degree of fire intensity.

10. What is the best set of actions to take during post-control overhaul?

    A. Open the ceiling above the fire area, check the cockloft for any fire extension, and remove the electrical outlets and check behind them for hot spots.

    B. Cut a hole in the roof (if not already done), check the cockloft for any fire extension, and cut a small hole in the wall to check for hot spots.

    C. Preserve the scene for fire investigators, remove any smoldering furniture, and cover any remaining furniture with salvage covers.

    D. Check for any smoke issuing from around the baseboards, listen for any crackling sounds, and feel the walls for any signs of heat.

    E. Keep a charged line ready and make a small hole in the wall in the fire area.

SCENARIO 8–4
GARDEN APARTMENT FIRE

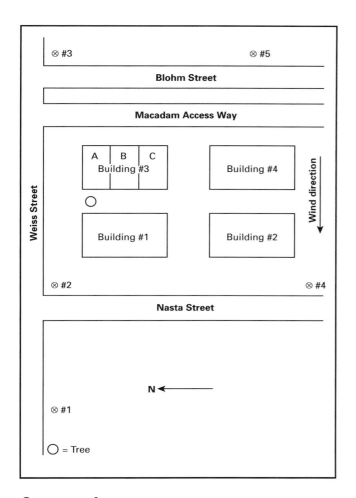

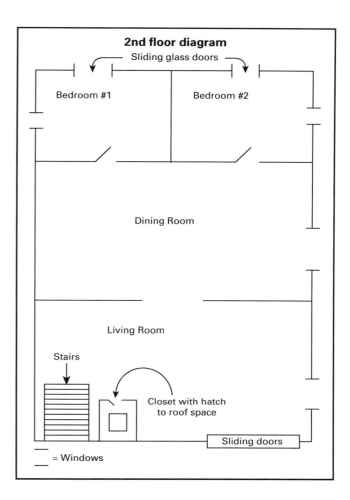

Construction

A fire has been reported in a development consisting of rows of two-story, wood-frame garden apartments. All interior walls are covered in gypsum board terminating at the 2nd floor ceiling. On the 2nd floor of each dwelling, there is a 3' x 4' hatchway for access into the common roof space. The roofs are severely sloped. The living room, dining room, and kitchen are located in the front of each dwelling, with the bedrooms in the rear. There is a sliding glass door at the rear of each bedroom.

Time and weather

The time is 0100. The temperature is 77°F, with a 5mph wind blowing out of the west. There hasn't been any rain for nineteen days, and all vegetation is very dry.

Area and street conditions

The development faces two streets; Building #1 and Building #2 face Nasta Street. Building #3 and Building #4 face Weiss Street. The distance between the buildings is 30'. The attached dwellings of each building are lettered A, B, and C, from north to south. Weiss Street is to the north of the development, and is a two-way street. There is a macadam access way between Blohm Street and the front of Building #3 and Building #4. It is large enough to fit apparatus into.

Fire conditions

The fire was called in by a neighbor, stating that there is heavy, black smoke coming from the 2nd floor of Building #3. Upon arrival, you see heavy smoke and fire blowing out of the rear of Building #3, dwelling A, on the 2nd floor (Apartment A2). A pine tree in the rear has a few branches burning next to the window.

Exposures

In each row of apartments, there are three attached dwellings. Each attached dwelling is a two-family occupancy. Above this ceiling, there is a common roof space above all three attached dwellings.

Water supply

Water supply is adequate for the fire load presented by the area.

Response

Your response is two engine companies and one ladder company. An officer and three firefighters staff each piece of apparatus.

MULTIPLE-CHOICE QUESTIONS

1. What is the most significant fire spread factor at this fire incident?

    A. The proximity of exterior exposures

    B. The common roof space above the apartments

    C. The combustibility of the exterior

    D. The open stairwell

    E. The combustible roof

2. What would be your orders for Engine 1? You will be supplied by Engine 2.

    A. Position the apparatus on Weiss Street (adjacent to access way) and stretch a 1¾" line through the front door to attack the fire on the 2nd floor of the fire apartment.

    B. Position the apparatus in the access way just past the front door. Stretch a 1¾" line through the front door to attack the fire on the 2nd floor of the fire apartment.

    C. Position the apparatus in front of Building #3 on Blohm Street. Stretch a 2½" line through the front door to attack the fire on the 2nd floor.

    D. Position the apparatus in front of Building #3 on Blohm Street and stretch a 1¾" line to the rear to extinguish the tree fire and to protect Building #1.

    E. Position the apparatus in the access way and have two firefighters stretch a 1¾" line through the front door to attack the fire on the 2nd floor. The officer will stretch a 1¾" line to the rear to protect exposures and to extinguish the tree fire.

3. Would additional alarms be necessary in this situation?

    A. Yes, some possible life hazard exists; a second alarm should be requested.

    B. No, the fire should be contained. Any possible victims should be rescued with on-scene manpower.

    C. Yes, due to the potential spread of fire to exterior exposures.

    D. Yes, due to the inherent deficiencies in modern wood-frame dwellings and the threat of fire spread to uninvolved interior exposures.

    E. Yes, due to the lack of moisture content in the surrounding vegetation and its ability to spread fire to exposures.

4. The fire is in Bedroom #1. The carpet, mattress, and wall near the window are burning. The fire is being extinguished by Engine 1. What are your orders for Engine 2?

    A. Establish a water supply for attack at Hydrant #3. Position the apparatus on Blohm Street. Stretch a 1¾" line through the front door to back up Engine 1's attack.

    B. Establish a water supply for attack at Hydrant #3. Position the apparatus on Weiss Street adjacent to the access way. Stretch a 1¾" line through the front door to back up Engine 1's attack and proceed to Bedroom #2 to check for extension.

    C. Establish a water supply for attack at Hydrant #1. Position the apparatus on Nasta Street. Stretch a 1¾" line between Building #1 and Building #3 to protect exposures and extinguish the tree fire.

    D. Position the apparatus on Weiss Street adjacent to the access way. Stretch a 2½" line from Engine 1 through the front door to back up Engine 1's attack.

    E. Establish a water supply for attack at Hydrant #2. Position the apparatus on Blohm Street. Stretch a 1¾" line through the front door of Building #3, Apartment B2 (upstairs). Check the hatchway for any fire extension into the common roof space.

5. The battalion chief has arrived on the scene and has ordered a search and rescue operation. How should this operation be accomplished?

   A. Keep the company intact. Search the 1st floor apartment and then 2nd floor apartment.

   B. Keep the company intact. Search Bedroom #1 and then Bedroom #2.

   C. Split the company into two crews. Crew 1 searches Bedroom #1 and Bedroom #2 and Crew 2 performs a secondary search.

   D. Split the company into two crews. Crew 1 searches Bedroom #1 and Crew 2 searches Bedroom #2.

   E. Split the company into two crews. Crew 1 searches the fire apartment (A2) and Crew 2 searches the 2nd floor adjacent apartment (B2).

6. The fire has been confined to Bedroom #1. You are ordered to perform post-control overhaul. What are your actions?

   A. Pull the ceilings and the walls in Bedroom #1; have a charged line standing by.

   B. Check the roof space via the hatchway.

   C. Check for smoke issuing from cracks around the baseboards, listen for crackling sounds, and feel the walls for signs of heat.

   D. Open inspection holes in the dining room and living room.

   E. Open the ceiling in the apartment below, place salvage covers on the room's contents, and remove excess water with a wet/dry vacuum.

7. The fire has been confined and extinguished in the bedroom. The battalion chief has ordered you, as captain of Ladder 1, to ventilate the structure. How would you carry out this order?

   A. Cut a hole in the roof directly over the fire; knock out the 2nd floor windows.

   B. Open the windows and the sliding glass door on the 2nd floor.

   C. Use a 24' extension ladder to ladder the building; break the 2nd floor windows.

   D. Close all of the doors except for the fire room and use PPV.

   E. There is no need to ventilate as the fire is venting itself.

8. What is the most important size-up factor in considering and assessing this fire?

   A. Life hazard

   B. Exposures

   C. Height

   D. Construction

   E. Weather

9. Regarding fire extension, which exposure should receive the most attention?

    A. Building #1

    B. Building #3, Apartment B2

    C. Building #3, Apartment A1

    D. Building #4

    E. Building #3, Apartment C2

10. The fire has been extinguished and the overhaul is nearly complete. When is it safe for the firefighters to take off their SCBA?

    A. Once the smoke has cleared to the point where objects can be clearly seen.

    B. Once the smoke has cleared completely and it is determined that there are no hot spots.

    C. Only after you have removed your mask and made a determination that masks are no longer needed.

    D. Only after they have left the building.

    E. Only firefighters who are checking for extension in other rooms may remove masks.

ANSWER SECTION

SCENARIO 8–1
MULTIPLE-CHOICE QUESTIONS

1. This scenario offers no diagram, so a picture must be formulated in the incident commander's mind. This will be similar to listening to the size-up report from another officer who has arrived on the scene and begun to take action. His report becomes the beginning of your own mental process of strategy creation prior to arrival. An effective size-up report by the first-arriving officer will best aid in the accomplishment of this objective as it will allow responders not yet on the scene to mentally visualize what is happening before they arrive and prepare themselves to go to work. From this pre-arrival evaluation of already known facts, the arriving incident commander can formulate a mental picture of what needs to be done, which can then be compared to what has already been done once he or she arrives on the scene. From this information and your own scene size-up, you can continue with the current action, modify it, or change it as necessary. (Helpful hint: If this were a real test, it may be helpful to draw a picture based on the information given in the narrative.)

Fig. 8–1 Old row frames present a significant fire problem. The wood is dried out, the integrity of the cockloft is likely to be compromised, and the entire row is completely combustible. Treat these buildings as a single structure and be prepared to fight fire in the attached exposures. Manpower will be the name of the game.

A. **+2** This action is short, sweet, effective, and will lead to the accomplishment of the most important objectives—that of a coordinated attack with attendant support operations. A line stretched anywhere else would not follow the Point of Entry Rule of Thumb and allow the fire to extend. In old row frames, unless there is cellar-to-roof flame, it is critical that the initial attack line be stretched to the seat of the fire. (Fig. 8–1)

B. **0** Stretching to the top floor of the adjacent exposures to stop extension via the cockloft is a proper tactic at any contiguous structure. It should not be the position of the first line, however. Allowing the parent body of the fire to spread unchecked will lead to surrender of the fire building and probably the exposures as well.

C. **+1** This tactic is acceptable, but it is more efficient and quicker to stretch the lighter, more mobile 1¾" line. In addition, stretching the line to the seat of the fire will not only quickly make progress on extinguishment, but also more than likely protect the stairs and occupant escape at the same time.

D. **-2** The job of the first-arriving engine company is to get water on the fire, period. The response here is three engine companies and a ladder company. The ladder company will handle the search and removal of occupants. The engine company may make a search in the area of the attack line during the stretch and, if possible, during the attack, but search and removal is not their primary mission, getting water on the fire so that the search can be conducted more safely is. Even if there is someone showing at a window, the best action for the engine company to take at this fire is to get a line on the fire and leave the rescue to the ladder company. The only time the engine company should deviate from this mission is when they are the only ones on the scene and do not expect the arrival of any more apparatus for quite some time. At that point, if there is a threatened victim showing at a window, then all engine company personnel should involve themselves in the rescue. This scenario makes no such statement, and with the response of a ladder company on the first alarm, any rescue and evacuation should be left to them.

E. *-1* Splitting the engine company to accomplish tasks other than stretching the initial attack line is almost a guarantee that nothing will be accomplished. The entire company, with the exception of the pump operator, should be involved in the task of stretching the initial attack line. As a matter of fact, if the apparatus is placed just past the front of the building in an attack pumper configuration, the pump operator may even help the stretch to the front door. He should never, under any circumstances, go into the building, but should return to the pump panel as soon as the line is properly flaked out in the street and is making its way to the seat of the fire. This is an option for an extra set of hands when an insufficiently staffed engine company is on the response. In large buildings and where warranted, the first line should be stretched by the first two engine companies. This will ensure that it will get there in the least amount of time.

2. This question will assume that the first line was effectively stretched to the proper place by the first engine company.

A. *0* The wind is from the north, so the southern B-side exposures should be the first exposures protected. However, this should not be assigned until a back-up line is stretched to the fire building. This is both a safety and reinforcement action. If the first engine company has trouble with the volume of fire or has to back out due to a burst line or injury, the fire building may be lost if a back-up line is not in place to continue the attack. The incident commander must pursue an offensive/defensive strategic mode at this fire. In this fire, and just about all fires of this type, take care of the fire first and then worry about exposures. If you can do both simultaneously due to a sufficient amount of manpower, great; if you can't, establish priorities based on current conditions. Hit the fire hard and the problems will diminish.

B. *+2* The second line must reinforce and support the initial attack. At the same time, a secondary water supply is established so that future hydraulic needs can be met, if required. Also, if the first water supply proves to be inadequate due to a defective hydrant, the second supply will be able to quickly take up the slack and continue the attack. Don't allow yourself to get caught short on your water supply if your initial attack source fails. Be prepared so that the later need to scramble is eliminated.

C. *-1* Here, not only is a secondary water supply not established, but a back-up line is not positioned to reinforce and support the initial attack. Get the second line to the same place as the first line, and then take care of the exposures.

D. *-2* No secondary water supply is established. No line of any type is stretched by this company. This is not a one-line job. In fact, it will be at least a three-line job. Initial arriving engine companies should not be assigned to building evacuation, especially for the exposures. This will be the task of later-arriving companies.

E. *-2* The tactics mentioned in this answer choice is also not the job of an initial arriving engine company. This is the job of the ladder company. If there were only two engine companies on the first alarm, then these tasks may be proper, but this is not the case here. Let the ladder companies do the ladder work. If you have a ladder company on the response, engine companies must stretch hose, establish water supplies, and put water on the fire. The use of manpower in this fashion will give rise to conditions getting better sooner than later.

3. The first-arriving ladder company should also focus its efforts on the main problem, the fire building.

A. *0* Unless the way is blocked by fire, ladder company personnel must enter the building from the front entrance to initiate a primary search. Most conscious victims will attempt to flee via habitual exits, such as the front entrance. If they are overcome while attempting to do this, they will often be found behind doors or in close proximity to an exit. VES (vent, enter, and search) operations are also proper tactics, but, in this case, it may be quicker and safer to perform these tactics from the front, as fences and poor access to the rear of these types of buildings may impede laddering efforts. If it is necessary to get to the rear, bringing the ladder through the ground floor of an adjoining occupancy may be the answer. Take a moment to size up the apartment to ensure the ladder can be brought through.

B. *-1* The aerial or ground ladder should be raised to a windward exposure, if possible. Raising to the roof of the fire building does not allow a sufficient path of escape if the roof conditions deteriorate. Raising to the windward side will provide a point of egress in an uninvolved area. In addition, it is more advantageous to cut a large hole in the roof and to open and inspect all natural openings at this type of building than it is to cut a trench. First of all, the amount of cutting required will allow a fast spreading fire to blow right past the trenching operation. If a trench is necessary, it may be necessary to give up several roofs to provide space between the trench operation and the fire so that it may be properly finished. Second, the trench tactics here are improper. The trench should not be cut parallel to the street as only one or two joist bays will be vented. It should be cut perpendicular to the street, so all the joist bays, which run parallel to the street, are exposed.

C. *+2* Splitting the first ladder company to accomplish both primary search and roof ventilation is proper and will provide for both the safety and preservation of the lives of both the occupants and the firefighters operating inside the fire building. The ladder is properly raised to the northern, windward exposure, and roof operations are begun to clear the building of the products of combustion, allowing the search and the hose advance to be properly conducted in the safest manner possible.

D. *-1* If conditions are marginal, it may be proper to keep the ladder company together. This would be the decision of the company officer or the incident commander. However, in this instance, the tactics chosen do not account for the vertical ventilation of the structure. Heavy fire exists in the upper region of the building. If the roof is not opened, at least the natural openings such as scuttles and skylights, the fire will spread with the wind throughout the cockloft. Steps must be taken immediately to localize the fire through vertical ventilation.

E. *-1* Here, again, no vertical ventilation is performed. This illustrates an old fireground proverb: the incident commander who places vertical ventilation low on the list of priorities in contiguous structures creates many parking lots. (Fig. 8–2)

Fig. 8–2 Contiguous structures require manpower to both confine the fire (venting directly above it) and to get ahead of the fire (entering, monitoring, and protecting attached exposures.) If this is the usual result of your fireground strategy, it may be time to re-evaluate your approach or change careers.

You may have noticed that there is no mention of cutting the roof over the fire. This is because the question addresses initial ladder company operations. To attempt to quickly confine the fire, the scuttle must first be opened, which is quicker and more effective than cutting the roof (at least initially). To be sure, the roof will be cut at this fire, but not initially. Do the things first that make the interior operation more tenable, and then work on the more time-consuming roof-cutting operation.

4. Additional alarms should be struck based on current and forecasted conditions. Contiguous structures must be considered as one whole structure when predicting the potential fire load and spread.

A. *+1* Recognizing the potential for fire spread demands the striking of additional alarms, possibly upon arrival. The scenario states that heavy fire is showing in the upper regions of the building. This will require a major commitment of manpower to confine, control, and extinguish.

B. *-1* Three engine companies and a ladder company are on the initial response. The total complement of members is sixteen. For this fire, this is inadequate. It will take approximately this many men to cover the fire building. That leaves no one to address the exposures. Unless the fire is of a very minor nature, a second and possibly a third alarm are almost always required for combustible, contiguous structures.

C. *+2* The response on all additional alarms is two engine companies and a ladder company. This response can be utilized as a task force, providing coverage for the reinforcement of the initial response as well as coverage of the most threatened exposure. A second and third alarm response will bring an additional four engine companies and two ladder companies. In addition, chief officers should be summoned to decentralize command. Assigning a chief officer to the fire building as the interior division supervisor as well as to each threatened exposure as Exposures Bravo and Delta supervisors, respectively, creates a safer fireground through reduced span of control. This will allow for the formation of two task forces; one to operate in the leeward exposure and one to check on the windward exposure, which cannot be ignored. If the windward exposure is not threatened or if the fire is in a corner building, the third alarm task force can be held as tactical reserve at the command post to provide relief of the initial companies. If they are committed, it will be necessary to strike a fourth alarm to provide a tactical reserve. Remember that any time you are at the command post and have no one standing by in reserve—you are out of resources to assign. The heavy fire condition in this building will require operations to be conducted on both the leeward and the windward side of the fire. It is probable that a fourth alarm will be required to ensure a tactical reserve is on scene.

D. **0** It is unproductive to special call apparatus to the scene of a serious fire. Dribbling in companies is an inefficient method of providing reinforcement for the operating forces and will undoubtedly lead to catch-up firefighting. If you think they may be required, request a standard alarm with a greater complement of resources. Companies not needed can be held in staging or at the command post as a tactical reserve or released from the scene.

E. **+1** This is an acceptable answer, but due to the fact that answer choice "C" requests the second and third alarm based on the life hazard (our highest priority), it is a better answer and one that works in a proactive manner in properly staffing the fireground.

5.

A. **0** The height is only two stories, reachable by ground ladders and/or an aerial. This factor does not pose a problem. There is no mention of the weather, other than the wind. While the 12 mph wind is a factor, it is not the most important. Also, the street conditions are not mentioned; therefore, nothing out of the ordinary in this regard should be encountered.

B. **+2** This is the best answer. The location and extent of the fire will determine both the life hazard and the degree to which the exposures are threatened. (Fig. 8–3)

C. **+1** The difference between this and the previous answer choice is the substitution of the size-up factor "construction" for the "location and extent" factor. While construction will greatly influence the combustibility and spread of the fire, the location and extent of the fire will determine how much and to what extent the construction of the building and the surrounding exposures will be threatened and/or damaged.

Fig. 8–3 Photo by Bob Scollan NJMFPA
The location and extent of the fire will always have a major impact on the fire problem. While a top floor fire will take more time to reach and may spread to adjacent structures via the open cockloft, a lower floor or cellar fire has the potential to trap occupants above the fire as well as expose the remainder of the building.

D. **-1** Exposure protection must be a concern to the incident commander during the size-up process, but incident stabilization and property conservation are not size-up factors. They, along with life safety (rescue), are fireground priorities.

E. **-1** Again, although life hazard as a size-up factor will play a significant role in the development of the action plan, as above, the other choices are fireground priorities and not size-up factors.

6.

A. **0** Access to the rear yards of contiguous row frames will often be impeded by fences from one end of the row to the other. As the line is stretched, the fences will have to be jumped, which may cause injuries; bridged, which requires a ladder; or breached, which causes unnecessary damage. Whichever way this is accomplished, it will take a considerable amount of time requiring that an alternative route should be sought. (Fig. 8–4)

Fig. 8–4 The rear of these contiguous structures is a maze of fences, sheds, and vegetation. It may be easiest to use the rear yard of the exposure directly behind the fire building.

B. *-1* There are already two lines in the stairway. Adding another line would lead to line congestion, impeding the stretch to the point where the line will probably not reach the desired point. This may also impede the movement of the first two lines. Three lines should never be stretched via the same entry point.

C. *+2* Using a ground ladder serves two purposes. It gets the line to the floor of operation without further congestion of the stairwell. Second, the ladder provides an alternate means of egress from the 2nd floor. As soon as enough hose is on the floor and flaked out, ensure the hose is secured so that when it is charged the weight of the water does not drag it out of the window. To safely operate on the 3rd floor, it is most desirable to bring the line to the floor below, then to stretch to the floor of operation via the interior stairs.

D. *-2* The scuttle of the fire building is used to exhaust the products of combustion. Stretching a line via this opening is tantamount to stretching down into a chimney. It is neither safe nor necessary.

E. *+1* Using a rope to hoist the line is a great idea, and is used by many engine companies to avoid stairwell congestion. However, it is safer to hoist the line to the 2nd floor, pull in enough hose to get to and operate on the 3rd floor, and then advance up the stairs from the 2nd to the 3rd floor to protect the search operation as ordered.

7. This is a question that requires a decision to be made based on current conditions and the perception of the incident commander as to the urgency of removing the woman.

A. *0* Passing the buck and having someone make your decision shows poor leadership qualities. It is death on an exam and will eventually cause subordinates to lose faith in you as a leader. Make a decision and take some action to address the situation.

B. *-1* Assigning a member of the first-arriving ladder company to a task in Exposure B2 is a waste of manpower that can be clearly used more effectively elsewhere, namely for search and support operations in the fire building. The initial tasks regarding this fire are numerous. The firefighter should not be spared. On the fireground, especially in the early stages of the operation, there will be many people and conditions that will pull at the sleeve of the incident commander. Decisions based on priorities must be made quickly and decisively. This question and situation is a distracter. How well you handle this non-priority says plenty about your readiness to command a fire.

C. *+2* This reply shows both leadership and an understanding of the priorities on this fireground. The priority here is the fire building and confinement of the fire to the building of origin. This is where manpower should be committed at this time. The woman is in no immediate danger. Her removal can wait until later-arriving companies are on the scene. If the fire is controlled by properly assigned first alarm companies, she may not require removal at all.

D. *+1* This is also a good reply, but not as thorough or decisive as the answer given in answer choice "C."

E. *-2* You are the incident commander. Leaving the command post to perform a task that is not a high priority, especially as soon as you arrive on the scene is incomprehensible. Your job is to coordinate the initial action and develop an action plan. This will be impossible while you are removing a handicapped grandmother.

8.

A. *-1* Giving all exposures the same amount of attention is a non-answer. Playing it safe here cost you a point. With the wind direction clearly established at this fire, the most severe exposure is clear-cut.

B. *-2* Exposure A is a street. Streets usually do not require protection.

C. *+1* Exposure B is the downwind exposure. It is the path of least resistance for fire travel and must receive the highest priority. However, it is not to be protected exclusively.

D. *+2* This is the most complete answer. Not only does the answer address the most severe exposure as the highest priority, but it also recognizes the fact that the windward exposure, Exposure D, is also a concern. Attached buildings must be checked on both sides of the fire, with a lesser priority and possibly even a smaller commitment of manpower assigned to the upwind exposure.

E. *0* The incident commander who chooses to ignore attached exposures is either a victim of tunnel vision or incredible irresponsibility. The incident commander who ignores and/or is not prepared for fire spread, especially in an attached building, is playing a dangerous game. In an attached building where a heavy body of fire exists, companies should be sent into both exposures to "pinch off" the fire should it involve the original building to the point that it must be surrendered to the fire.

9. It is essential that fire officers know how to apply both manpower and equipment along with the proper tactics to protect an interior exposure. Knowledge and aggression are the keys to safely and effectively operating in this environment. It is also essential, in order to maintain control over the operation, to assign a division commander to each exposure.

A. *-1* It is not necessary to place a deluge gun on the top floor. A handline is a much more mobile stream and will get the job accomplished more effectively. Interior exposure protection in contiguous structures is down and dirty firefighting. It is not a job for the timid. Get in there, rip the ceilings down, and aggressively apply streams as necessary to keep the fire confined to the building of origin.

B. *-1* By the time the trench is cut, it is likely that the fire will have passed the intended "stop" point. Trench cuts must be placed when there is time and where there is space to complete it before the fire gets there. The position of this trench will not be successful in stopping this fire.

C. *+1* These are good tactics, but there is one thing missing. It will be necessary to at least provide some examination holes from the roof to monitor the cockloft. The exam holes will allow the roof team to monitor the cockloft while the ceilings are being pulled. Exam holes are much more quickly accomplished than pulling the entire ceiling, especially if tin ceilings are present. If the fire has extended, the exposure roof should also be cut to attempt to localize the fire. The horizontal ventilation will provide some relief for the interior teams, especially where the lines are operating. Operating hose streams cause smoke to lose its buoyancy and bank down as well as produce steam. The open windows will help in alleviating this discomfort.

D. *0* The entire ceiling must be pulled. If the fire extends into the cockloft of the exposure, it may be too late to get the ceiling down and the fire may extend past the team. In addition, opening the walls may invite fire spread into the apartment to the extent that tenability may be affected to the point of withdrawal. When possible, keep the walls intact and open the ceilings completely. If the walls must be opened, this should be done after the ceilings are down.

E. **+2** These are the best tactics. Lines are in place and windows are vented. The ceilings are completely pulled and streams applied to beat the fire back. In addition, the roof team is busy cutting exam holes and monitoring the cockloft. If fire is present, they are ready to enlarge the openings to attempt to confine it. This is aggressive interior exposure protection at its best.

10. The fact that there is a question asking about the duties assigned to a second ladder company is a hint that at least a second alarm is required. It is important that, after reading the scenario, you read all the questions. Many times, there is additional information in the questions that will assist in making the best decisions.

A. **-1** The apartments are usually limited in size in row frames. Ladder 1 is already assigned the primary search of the fire building. Assigning the second ladder company to this duty violates the principle of economy of forces and is redundant. There are other tasks that need to be addressed.

B. **+2** Pre-control overhaul duties in the most severe exposure will be required as soon as possible. This is an ideal place to assign the second ladder company. It will also be imperative to reinforce this operation with hose line support. In addition, the roof duties required at this fire, both on the roof of the fire building and on the exposure roof will be manpower-intensive. Confinement operations must take place from above as well as from below.

C. **+1** As mentioned earlier, access to the rear will be difficult at best. With ground ladders, it may be next to impossible. Still, the reinforcement of the search from the rear may be required. Answer choice "A" assigned the entire second ladder crew to this duty. This is not necessary. Two men to the rear will suffice. Sending the second team to the roof is an excellent tactic. This fire will require relatively extensive roof operations on the roof of the fire building as well as on the roof of exposures.

D. **+1** These are also acceptable tactics and will check the spread of the fire on two fronts. It may be better, in this case, to leave the windward exposure to the companies arriving on the third alarm. When the leeward exposure is threatened to the magnitude that it is in this fire, committing an entire task force (in this case, the second alarm companies) to that exposure will be the most efficient use of manpower. Spreading the troops too thin may result in the accomplishment of nothing. Reinforce all operational positions with proper manpower as well as command supervision to give the assigned crews the best chance for success.

E. **-2** A tactical reserve is nice to have, but it is not the place for the second due ladder company; not now, not at this fire situation. The time to utilize this company is now, as there are many tasks to be accomplished and probably still not enough manpower to do them. This choice is essentially inaction, never a desirable action on a still-escalating fireground.

Passing Score for Scenario 8–1 = 14 Points

Scenario 8–2
Multiple Choice Questions

The incident commander's ability to effectively fight a fire is directly proportional to the information he has about that fire. In this scenario, there are many distractions. Not only are you met with a heavy smoke condition in the area that obscures the true seat of the fire, but you are also confronted with a report of a second fire. Your ability to ask for and receive information regarding the true location of the fire will impact heavily on your ability to successfully extinguish it. (Fig. 8–5)

Fig. 8–5 Photo by Ron Jeffers NJMFPA
Heavy smoke will often mask the true seat of the fire, especially on days where humidity and extreme cold prevent smoke from lifting. Proper recon and timely reports will help guide the incident commander in creating and adjusting the action plan.

On the fireground, this information can be ascertained via reports transmitted from the various unseen areas, namely the interior, rear, sides, roof, and any unseen shafts. Without proper information, attack lines can easily be stretched to the wrong location, requiring repositioning. Repositioning lines will take time and allow the fire to spread to uninvolved areas, further jeopardizing any occupants and the structural integrity of the building. In this case, it is better to prepare the lines and stand by at the ready until the recon team, usually the ladder crew, can locate the seat of the fire. It may seem like eternity, but will probably only take a minute or two. In fact, while the engine crews are preparing lines, the ladder crews should be actively seeking the location of the fire from as many vantage points as is safely possible. An effective initial scene assignment for ladder operations will best assure this.

In this particular scenario, it is mandatory to round out information not furnished in the scenario by carefully examining the diagram and looking at the questions before you make any definite decisions regarding strategy and accompanying tactics such as apparatus and line placement, ventilation operations, and the like. The diagram actually shows the seat of the fire, but it does not show which building is involved. That information is found in Question #5.

1.

A. **+2** While not immediately obvious from the initial information in the scenario, the location and extent of the fire will be the guiding factor in the placement of attack teams and support crews. You cannot begin to stretch lines until this information is available. You are dispatched to 222 Murphy Avenue, but are met with a heavy smoke condition which obscures where the fire is actually located. There will be times when the fire is reported in one place and it is actually in another. In addition, the presence of this heavy smoke condition also suggests the possibility that more than one building may be involved upon arrival. Attack operations may have to wait until the teams conducting recon of the area find the fire location.

In the scenario, if you examine the questions first, the location of the fire is not given until Question #5. That information will have an impact on the first five questions. This is why it is important to look at the questions for extra information before making decisions with inadequate information.

B. **+1** Wood-frame row houses not only offer a fire extension threat via interior voids such as pipe chases and the open interior stairwell, but will also cause fire to spread beyond the building of origin due to inherent construction deficiencies. Extension to adjoining buildings via the combustible cockloft will be a major hazard. It is critical that the roof be opened in an attempt to confine the fire to the building of origin. Failure to open the cockloft will result in lateral fire spread to adjacent cocklofts. In addition, the combustible exterior will cause fire venting from windows to ignite exterior walls and spread fire from building to building via the exterior. A well-placed line operated by a knowledgeable and disciplined crew will extinguish a combustible wall fire without jeopardizing interior crews with opposing exterior streams directed into occupied, interior working spaces.

C. **0** The occupancy—residential—presents no great hazard from the firefighting point of view. While the life hazard found in occupied residential structures will be a major concern, perils associated with other type occupancies will not be present.

D. **-1** The buildings are two stories in height. There are no wires to contend with according to the diagram. Ground ladders as well as aerial devices will easily reach all accessible areas of the building and exposures.

E. **-2** Command is not one of the thirteen size-up factors. Therefore, it cannot be considered as a valid answer to this question.

2.

A. **+2** You must take some sort of action step in regard to the report of fire in the row of houses on the next block. It is better to fight the fire you have been dispatched to with the first alarm assignment than it is to split the first alarm companies to fight both fires. Odds are that you will not be successful in either endeavor. Sending a second alarm assignment to fight the other fire is the best answer among these choices.

B. **-2** Three engine companies and a ladder company are totally inadequate to handle this fire, the coverage of the exposures, and the fire reported on the next block. This fire requires manpower. On most departments, the only way to secure it is through additional alarms and/or mutual aid. As always, help should be requested early to stay ahead of the fire.

C. **+1** The structural deficiencies of wood frame contiguous structures have already been discussed in depth. Sufficient manpower will allow the incident commander to address the exposed areas with both ventilation and line support. Insufficient manpower will allow the fire to march down the block, causing the total loss of the fire building and the top floors and roof areas of all the leeward (and possibly windward) exposures.

D. **0** Failure to request an additional alarm for the reasons stated is both a display of inaction and a gamble. The amount of smoke upon arrival coupled with the report of an additional fire points to a major incident. Striking at least a second alarm to bolster the current manpower assignment is both a safety-oriented and a needs-forecasting action.

E. **-1** You are the incident commander and are acting in the same capacity of the chief at this time. It is your responsibility to evaluate current conditions and request assistance based on that evaluation. Passing the buck to wait for the decision of a higher-ranking officer is indicative of poor leadership.

3. If you did not read all of the questions, the placement of the first and second lines as well as the location of and tactical decisions regarding primary search and ventilation would be mere guesses. On the fireground, use all the resources available to you to allow you to make the best decisions on the fireground. Take a few extra seconds to evaluate the situation and let the reports from the unseen areas assist in guiding your strategic and tactical decisions.

    A. *-1* Stretching the initial attack line into 222 Murphy Avenue will accomplish nothing. This is not the location of the fire. From information gathered from company reports, especially the ladder officer's report to lead off Question #5, the astute strategist would be stretching lines to the proper area with confidence gained from reliable reports.

    B. *+2* Once the true location of the fire is revealed, these bread-and-butter tactics will produce the best results. If you stretched into 222 Murphy Avenue based on the report of the dispatcher, you began operating in the wrong place. Having to reroute your lines may take time, allowing the fire to intensify to where offensive operations may not be safe. Remember that dispatch reports will not always be accurate. Only with proper size-up and building coverage can the most accurate decisions be made. If you stretched lines based on the report of the ladder officer, your decision was correct because it was based on the best and most up-to-date information available.

    C. *-2* This is classic tunnel vision. The report from the child on the street led you to fight the fire down the block. The conditions in 222/224 Murphy Avenue are much worse and this was your dispatch order. There is definitely a fire there. If there wasn't any fire at 222 or 224, you could assume the address was wrong and reroute to 206 Murphy Avenue. This is not the case here. It is also stated that 206 Murphy Avenue is vacant and is only showing light smoke. In addition, the reported occupant is next door. The same cannot be said about 222/224. The life hazard profile is as yet undetermined; therefore it must be assumed to be severe. You must fight this fire.

    D. *-1* Using Hydrant #2 will cut off the access from Lewis Street. It is better to use Hydrant #3. In addition, the method of stretch is from the burned side and will likely push the products of combustion throughout the rest of the structure. Also, the path of the stretch, over a 6' fence, is not the safest, most accessible path of least resistance to accomplish the objective of placing the line between the fire and potential victims.

    E. *-1* These are basically the same tactics as far as line placement and direction of attack as the previous choice. In almost all cases, it is better to attack from the unburned side, usually the inside, and push the fire out the path of least resistance, in this case, being the back door. Attacking the fire from the rear will create more problems than it solves, both to attackers, to occupants, and to searching firefighters.

4. More times than not, if your tactics regarding the positioning and objectives of the initial attack line are improper, the second and all other subsequent line placements will be improper as well.

    A. *-2* If you chose to fight the fire at 206 Murphy Avenue, you are already two points in the hole. Now you are four points down. If you chose to take the second line to the same place, whether as a back-up or as the initial line while the first line attacked the fire at 224 Murphy Avenue, you come up a loser as well. Do not dilute your forces to fight two fires at once. Put all your effort into one or the other, based on priorities and conditions. The heavy smoke and fire condition as well as the life potential at 224 Murphy Avenue is a higher priority than the light smoke condition at the vacant building on the next block. Breaking up your forces in this manner or completely neglecting to address 224 Murphy Avenue will likely result in two parking lots and possible injuries and death.

B. *+1* These are proper tactics regarding the placement of the second line. In almost all cases, the second line must back up the first. Firefighter safety is the most appropriate reason for this action.

C. *+2* These tactics take the proper actions in answer choice "B" one step further. Not only is the initial attack line properly reinforced, but a check for fire extension is also accomplished. If it is proper and safe, accomplishing as much as possible in the assigned area of operation is the mark of an effective and resourceful officer.

Sometimes the test taker will choose what looks like the correct answer without looking at the other choices. In this case, if you chose answer choice "B" because it looked like the best answer, it cost you a point. Make sure you check out all the answers before deciding which is best.

D. *0* The tactics here are acceptable, but it is imperative to establish a second water supply, especially in attached wood-frame buildings. The fire extension potential is severe. The entire row must be considered as the present fire load. Take this into account and order at least two water supplies to meet this fire load potential.

E. *-1* This line, like the initial line stretched to 222 Murphy Avenue, will waste time and manpower. It will also cause an escalation of fire conditions in the fire building next door. Proper building coverage and timely reports along with a disciplined approach on the part of both the incident commander and the attack team will ensure both attack and back-up lines are stretched to the proper location in a timely fashion. It may be more effective in the first few minutes for the attack teams to drop lines, flake them out, and stand by until the fire's location is discovered. This act may save both occupants and property when compared with a head-down "damn the torpedoes, full speed ahead" hose stretch to the wrong location due to a lack of information.

5. The information in the first sentence of this question is the key to the entire scenario. On the fireground, this information is best gathered by companies on reconnaissance missions. This information, revealing the true seat of the fire, will guide the incident commander in the development of a proper action plan and direction of attack.

A. *-2* Cutting a hole in the second floor roof would not be the most efficient way to ventilate this fire. In fact, this opening may spread the fire by pulling it up through the spaces in the exterior walls. For sure, the cockloft, the exterior walls, the baseboard, and possibly the walls on the interior of the exposed wall will have to be examined, but the primary way to localize this fire would not be cutting the 2nd floor roof.

B. *+1* The fire must be localized. Breaking the windows in the kitchen as well as opening the kitchen door can quickly do this. These actions, along with a quick and aggressive hose stretch should confine this fire. However, this fire, which upon arrival has all the indications of being somewhat substantial, should be further localized by immediately cutting the 1st floor roof. This fire must be slowed down before it spreads to the main portion of the dwelling. From there, it can spread to the 2nd floor via the open stairwell. It will have access to the cockloft, multiplying the problem. Take all steps necessary to confine this fire to the 1st floor kitchen.

C. *-1* Breaking windows on the 2nd floor is inefficient and will do nothing to localize the fire in the kitchen, which is on the 1st floor. Worse yet, this venting action may pull fire toward the 2nd floor, where the greatest life hazard is likely to be.

D. *0* While this may seem an efficient ventilation strategy, it may have disastrous consequences. There is no guarantee that the fire has not found its way into adjacent walls and other concealed spaces. The draft created by the fan can push fire into uninvolved areas, threatening both the rest of the fire building and the exposures. If there is any chance that construction voids may be present, the vent fan is best left on the apparatus until the fire is definitely under control.

E. *+2* If immediate steps are not taken to confine this fire to the area of origin, the result could be a drawn-out multiple alarm affair. The spread of fire must be localized if the initial companies stand a chance of making the stop. In addition to opening doors and windows in the fire area opposite the line advance, the roof must be cut. Don't overlook any vent pipes on the roof, which may be used to vent the cooking area. These areas should be examined; in fact, it is probably a good bet that the cut will be made in the area of these natural arteries as the fire and heat take the path of least resistance.

6. An extensive knowledge of building construction will be the best ally in determining where to conduct overhaul.

A. *+1* The dining room is directly adjacent to the kitchen. It should be checked for fire spread. However, it is not the most likely path of least resistance, especially if the attack was launched from the proper direction. If the attack was initiated from the rear, it would be likely that the fire would have extended into the dining room. A proper attack would have pushed the fire right out the kitchen door, providing protection for the dining room and rest of the dwelling.

B. *+2* The ceilings and walls of the kitchen would be the most likely path of least resistance for fire extension. If these areas are not quickly opened, the fire may extend via roof joist bays to involve adjacent areas such as the adjoining exterior and interior wall. Much like salvage, where you should think like water, in both pre- and post-control overhaul, you must think like fire and you will be led to the most likely paths of fire extension.

C. *-1* Opening up the ceilings and the walls on the second floor without first addressing the fire area causes unnecessary secondary damage. If evidence gathered by opening the immediate fire area warrants further opening in adjacent and remote areas, openings are then justified. Indiscriminately damaging a structure is both very unprofessional and may also pull a smoldering fire into an uninvolved area. Start in the area of most fire involvement and work outward from there.

D. *-2* Using a positive pressure fan as your main post-control overhaul strategy is inefficient and improper. Overhaul involves opening structural areas to both get ahead of and confine (pre-control) or expose concealed fire (post-control). Simply operating a fan into the structure doesn't accomplish this objective. While it is true that once the fire is definitely under control, a fan blowing fresh air into the structure may benefit firefighters working inside; this fan operation must be carefully controlled and shut down if hidden fire is discovered. Carefully coordinated and monitored PPV may be used to enhance the post-control overhaul operation, but in no way is it a substitute for it.

E. **+1** The adjoining exterior wall will be one of the paths of least resistance for fire travel to the cockloft. This is why this area must be opened and evaluated. A fire may quickly traverse wall voids and, if you are not examining this area, may be through the roof before you realize it has extended. In addition, any roof holes cut in the kitchen roof must be made in such a way that they do not expose this combustible exterior wall. It may be necessary to use a stream to keep the wall wet if heavy fire is venting out of the 1st floor roof vent opening.

7.

A. **+2** Since there are only two men searching the structure, it is best to keep them together. The search should concentrate on the fire area first (the kitchen and 1st floor). Then, other areas of concern should be addressed, namely the 2nd floor. It may be acceptable to split these members if they are to be operating with a hose team on each floor. This would get the search done in half the time and provide a safety margin by having search members operate in close proximity to other personnel. However, that choice is not among these options. Therefore, safety in numbers must be the rule.

B. **+1** If the sleeping areas are located on the 2nd floor, the point will be given if that is the first area searched. However, as a rule of thumb, any occupant in the fire area will usually be in more danger than anyone on the floor above.

C. **0** It is best to keep the team together. The smoke condition is severe in the fire building. While the search rope will provide a path to the exit, it is best to operate with a partner for the sake of firefighter safety in case one member gets in trouble or finds a victim.

D. **-1** This method will quickly provide a search of the fire area, but will not provide sufficiently for the safety of the men in the rest of the dwelling. At least answer choice "C" gave them a rope.

E. **-1** Firefighters conducting primary search must quickly cover all tenable areas as quickly as possible. If the search team waits until a line is stretched to the fire area, the window of survival for victims may have closed. In addition, once the line is opened, the fire area may not only become untenable due to the creation of steam, but any visibility due to the thermal balance of the heat will be destroyed. The best time to get the victims out is as soon as possible. As the attack is launched, conditions for victims without SCBA and protective clothing will get worse before they get better.

8. The scenario states that the wind is out of the south. The diagram gives the direction of the wind. There is no mention of velocity.

A. **+2** The heaviest smoke and probably fire is located in 224 Murphy Avenue. Therefore, the attached exposures should receive the most attention. 222 is on the leeward side, so initial exposure protection and opening up efforts should be concentrated on this exposure. When manpower is available, the windward exposure (226) should be entered and examined as well.

B. **-1** The row house complex on Side C is not in the path of the wind. Even if the fire building becomes heavily involved, it will become a priority only after the attached exposures are protected. The 8' block wall will be effective in blocking a great deal of the radiant heat being emitted.

C. *-1* All exposures are not created equally. The threat of fire travel will almost always be more severe on the downwind, leeward side of the fire. (Fig. 8–6) All things being equal, the priority must be based on this fact. This rule of thumb can be modified based on the layout of exposures. If the leeward side is unattached and the windward side is attached, then the priority may shift to the closer exposure regardless of the wind. The same may be true if the fire building is a three-story building, the leeward exposure is one story, and the windward exposure is four stories with windows facing on the fire building roof. In this case, if the fire breaks through the roof, the windward exposure may be more severely threatened. The incident commander must take all these things into consideration when planning his strategy.

Fig. 8–6 Photo by Bob Scollan NJMFPA
The direction of the smoke drift will cue the incident commander as to which exposure should receive priority. In this case, the D exposure is leeward; however the presence of smoke at the roof level of the B exposure suggests an open cockloft and the need to check this area as well.

D. *-2* There is a lot of real estate between the fire building and 216 at the end of the row. If this is your most important exposure and you commit companies there, you will have an awful lot of explaining to do to the owners of 222, 220, and 218 Murphy Avenue.

E. *0* The fire is apparently much heavier and more of an extension threat in the 224 block.

9. The fire in 206 Murphy should be under someone else's command as it is a separate incident.

That doesn't mean questions cannot be asked about it. Although the fire building is vacant, the attached exposures are not. In fact, the report of the elderly occupant in the leeward exposure demands an offensive attack be attempted to confine the fire to the building of origin.

A. *0* The fire is already venting from the 2nd floor windows. It will not be a priority to open these windows. However, ventilation from the roof and at the rear will be necessary to allow for the safer and more expedient advance of an attack line. A top floor fire will always benefit from topside venting.

B. *+2* These are proper tactics for an attached fire in a row house, especially on the top floor. It is safest to access the roof of the windward exposure than it is to ladder the roof of the fire building. Laddering the windward exposure leaves a clear retreat path if conditions on the fire building's roof deteriorate. (Fig. 8–7) The scuttle must be opened to clear the stairwell. In addition, as at all top floor fires in a flat-roofed building, the roof should be cut over the area of most fire involvement to localize fire spread in the cockloft. Further ventilation should also be conducted. The roof is the best place from which to horizontally vent rear windows. A tool on a rope or a Halligan hook can easily accomplish this task. Finally, it is imperative that a report of conditions be made to the command post from the roof team. The roof is an area out of the line of sight of the command post that is critical to the success of the fire operation. If the roof must be abandoned, the result is usually forfeiture of the fire building. For this reason, the command post must constantly know what is going on at the roof level. Regular progress reports must be issued.

Fig. 8–7 Photo by Bob Scollan NJMFPA
It is best to ladder the windward exposure to access the roof. This provides an egress route from a relatively safe area. Fire is evident in the cockloft as shown by the cornice fire. This roof must be opened without delay to keep the fire from spreading to the attached building.

C. **-1** Finishing the horizontal vent operation with a ground ladder is not a bad idea, but the fire is already venting there. The crew can be better utilized elsewhere. In addition, although the roof operation is correct, the roof access is not via the safest, most accessible path of least resistance. It has already been established that the scuttle is less than desirable as a path to the roof and even more treacherous as a retreat route.

D. **-1** This is a waste of the aerial. The fire, as stated, is already venting from the front windows. In addition, no roof operation is being conducted. The scuttle must be opened and the roof must be cut. At top floor fires, roof venting is necessary. Get it done early and the chances of having to chase the fire down the block will be minimized.

E. **-1** First, it will be a monumental task to ladder the rear of the fire building. It is better to ladder the exposure to reach the roof. Otherwise, these tactics are not that bad. The point is lost because the location of the ladder will be solely responsible for none of the well-intentioned roof operations to be completed.

10.

A. **+2** Stretching to the seat of the fire via the front door is the safest, most effective path of least resistance to attack the seat of the fire. The victim reported next door (208) and any other victims, both in the fire building and in adjacent exposures, will benefit most by a direct and aggressive attack on the main body of fire.

B. **-1** The first line should not go to the common cockloft of the exposure. This is basically a surrender of the fire area. While lines must be stretched to this area (top floor of adjacent building) to cut off any extending fire, the initial lines should be directed toward attacking the main body of fire.

C. **0** It is not likely that a line will be required at this time to make the rescue (actually, it is a removal) of the elderly woman. The fire is in the adjacent building and will only become a threat if tactics such as these are used. Put water on the fire as quickly as possible and both the fire and the rescue problem will diminish.

D. **-1** Even though the scenario states that this is a vacant building, the potential for squatters is a reality in all so-called vacant buildings. For this reason, an initial defensive attack is improper and will result in the writing-off of any potential occupants in the fire area. If the scenario had stated that 206 was in a deteriorated state and unsafe to enter, a defensive posture may be the best strategy. Unsound structures, due to lack of maintenance or previous and repeated fires are candidates for this strategy. This was not stated in the scenario. It cannot be assumed. A cautious offensive attack and primary search are mandated here. If interior crews report that the structure is unsafe to operate in, then the strategy must be changed. The incident commander must be cognizant of the dangers of vacant buildings and carefully evaluate reports issued by crews operating in the different areas of the building.

E. **-1** The stairway is not stated as being unsound. If reports or previous knowledge reveal this, then this tactic, or better yet, a defensive strategy, may be the only option. A line up the ladder will not only take time, but there is fire showing at the front windows. For this tactic to work, the line may have to first knock down the venting fire before climbing and entering the window. This would jeopardize any lives inside the building. As conditions allow in regard to firefighter safety, try to launch the attack via the front door and main interior stairs as much as possible.

Passing Score for Scenario 8–2 = 14 Points

Scenario 8–3
Multiple-Choice Questions

1. Many firefighters have been killed due to improper operations in the area of power lines. At times, this is due to tunnel vision. It is of the utmost importance that all firefighters be aware of their surroundings. Electrical hazards in and around fire buildings must be addressed. All personnel must be made aware of their presence and the accompanying danger area. Control zones must be established and all personnel must be kept clear until the power can be shut down.

   A. **+2** The first and best action to take is to notify the utility company to respond as soon as possible. At the same time, recon teams should be searching for anther possible way into the building. The potential life hazard as well as rapid fire spread due to the combustibility of lightweight wood frame construction is great. A line must be stretched and a primary search must be conducted. Although the front door may be the most effective path of least resistance, it is not the safest route. Therefore, the rear sliding door is both the safest and most effective path of least resistance to accomplish the objective of both stretching the attack line and entering the fire building for primary search.

   B. **0** While the immediate request for the utility company is proper action, to stand by and let the building burn is unacceptable. The power line is down between the two poles at the front of the building. It is safe to go through the alley to Hamilton Road to launch an attack.

   C. **-1** Moving any power line, even if it is assumed to be uncharged, is inadvisable. Having equipment such as rubber gloves and a hot stick does not make it permissible to perform this task. This is only asking for trouble; leave it to the experts and work around it if possible. (Fig. 8–8)

   D. **-2** Operating anywhere near the live power line at the front of the fire building is suicide.

   E. **0** Contacting the power company is a necessity here. Asking for their advice on how to handle the situation is inappropriate; they deal in electricity. We deal in fire strategies and tactics. The best action is to get them on the scene as quickly as possible and to conduct our operations in as safe a manner as possible while steering clear of the hazard. Establishing a Hot Zone and announcing the hazard and accompanying control orders over the radio as an emergency transmission will reach the most operating personnel in the shortest amount of time.

Fig. 8–8 Photo by Ron Jeffers NJMFPA
Electrical hazards on the fireground must be identified and isolated. Power lines should only be handled by the experts. Never place yourself or your crew in a situation they are not equipped or trained to handle.

2. The initial radio report to dispatch should include the most germane information regarding the incident. Upon arrival, this information may not be totally complete due to unseen hazards and conditions. However, the purpose is to begin to develop an organization and initiate the formulation of an action plan as well as relay to incoming companies the current situation as seen by the first-arriving company officer. You will note that most of the answer choices are both acceptable and applicable to the initial radio report.

A. *+1* The presence of lightweight wood-frame construction denotes the presence of the truss. In most cases with this type of construction, the trusses will be of two major types: peaked trusses in the roof area and parallel trusses used for the flooring. Both may fail in as little as five minutes of fire exposure and should raise the flag of caution. Recalling the "B" in C-BAR as well as the C-HOLES acronym, construction is a large part of the initial radio report. Awareness of this will cue arriving firefighters and the chief officer of the hazards to be found in this type building.

B. *+2* The establishment of incident command by the first-arriving company sets the stage for all other actions taken on the fireground. The effective development of the organization to combat the fire is the single most important action taken on the fireground, not only on behalf of the future existence of the building, but more importantly, on behalf of firefighter safety. Firefighter safety is best achieved through a structured and orderly approach to the problem.

C. *+1* This piece of information should be included in the building description portion of the initial radio report. It will notify responding companies and establish an awareness of an existing fireground hazard even before they step off the apparatus. The reason that the establishment of incident command is more important than the downed wire is that it is the incident commander who will begin to develop a safe and effective plan to confront the situation while addressing the dangers of the uncontrolled hazard.

D. *0* Effective apparatus positioning will come about as a result of incident command planning, disciplined officers, and proper knowledge of how to best accomplish fireground objectives while keeping the operating forces from becoming exposed to the downed power lines. (Fig. 8–9)

Fig. 8–9 Photo by Bob Scollan NJMFPA
In contiguous structures, position apparatus with the fire's anticipated progress in mind. Here, a Telesquirt was positioned next to an aerial ladder to knock down fire on the building's exterior as well as provide for protection of the leeward side of the fire if necessary

E. *-1* There is no mention of a victim at this time. Because of the hour of the morning, just as at any hour of the day, the building must be assumed occupied until proven otherwise. The only way to confirm the presence of any victims is by conducting a rapid primary search.

3.

A. **-1** If the downed power line was not part of the situation, this would be a routine fire if the attack teams could quickly access the seat of the fire. However, the presence of the power line will not only delay what is an otherwise routine operation, but it will cause hoselines and manpower to be diverted away from the most effective path of least resistance into the building.

B. **+2** There is always potential for extensive fire spread in a contiguous structure. Couple this with the fact that the power line is causing a problem: the fire attack, primary search, and support operations will have to be carried out from an entry point other than the front door, delaying operations, and necessitating tight scene control. Any one of these factors could cause the proverbial fly in the ointment that quickly turns the fire operation from a potential winner into a loser. Take the necessary steps as soon as possible to address the possibilities. On the fireground, it is always better to be safe than sorry. (Fig. 8–10)

Fig. 8–10 Photo by Bob Scollan NJMFPA
Especially at contiguous structure fires, ensure adequate manpower is requested as soon as possible. Incident command should have no less than two companies standing by at the command post to assign those tasks that arise as conditions change. With no tactical reserve, any unexpected move by the fire can cost an incident commander the war.

C. **0** In this and almost all situations, waiting until the situation has been fully assessed can cause the incident commander and the entire operation to fall well behind. This is especially true in the critical initial stages of a fire operation. It is likely in this case that all responding companies will be assigned. The potential problems make a tactical reserve a definite necessity, if not to perform other tasks, but to relieve working personnel. As stated many times in this workbook, it is always better to be proactive in your approach to manpower requirements than to place your personnel in jeopardy by being caught short. Nowhere is this need more evident than in a contiguous structure.

D. **+1** Recognizing the need for manpower to both conduct a primary search of the fire building and also recon and evacuate other threatened, adjacent areas, is sound reasoning for summoning additional alarms. These extra personnel will come in handy if the fire extends beyond the building of origin.

E. **+1** Summoning additional personnel for exposure protection is a way of staying one step ahead of the game. There is no worse feeling than standing alone at the command post with tasks to be accomplished and no one to assign them to. Don't put yourself in this position. It is both lonely and counter-productive. Share the command post with some friends. In the long run, you'll be glad you did.

4.

A. **+1** The lightweight construction of the building is a major factor in the offensive/defensive decision. Lightweight wood trusses can fail in as little as five minutes when exposed to fire. Heat exposure can also cause the sheet metal surface fastener to curl away from structural members, causing collapse. Generally, if the fire has not reached the flashover stage, the chances of collapse are not as great as when flashover conditions have been reached. The fire then begins to attack the structural members. The decision to withdraw the troops will be based on conditions visible at the command post as well as reports from the interior of the building.

B. **+1** Attached exposures, especially wood-frame, will always be a significant factor in size-up. It is one of the factors that prompts the incident commander to request additional alarms to provide coverage of the as-yet uninvolved exposures. In contiguous structures, it is best to ensure the complement of on-scene personnel is consistent with the potential involvement.

C. **0** The occupancy is residential and does not pose an extraordinary threat to emergency responders.

D. **-1** Auxiliary appliances are not present. In this type of structure, one of the best defenses against fire extension and collapse of the light-weight wood trusses is the installation of a wet sprinkler system to keep fires in check until the fire department arrives. The fire department should urge that all contiguous structures, especially ones constructed of light-weight materials, be completely protected by automatic wet sprinkler systems. (Fig. 8–11)

Fig. 8–11 Photo by Ron Jeffers NJMFPA
This condo complex under construction nearly caused a conflagration and was completely destroyed. The auxiliary appliances had not yet been installed when the fire broke out.

E. **+2** The location and extent of the fire has the most impact on the operation. The location of the fire in relation to the downed power lines is a limiting factor that cannot be overcome until the utility company responds. Therefore, the incident commander must adjust his strategy and action plan, taking into account the dangers present. The primary responsibility of the incident commander is the protection of the responders.

5. If you follow the Point of Entry Rule of Thumb, this answer should not shock you.

A. **+2** The best action is to avoid the downed power line entirely until power can be shut down. This would entail positioning the apparatus at Hydrant #2 on Hamilton Road and stretching the line via the rear yard. Technically, this is not the path of least resistance, but it is safer than using the front door. Safety of operation is always of paramount importance and often the deciding factor in determining where and how to attack the fire.

B. **-2** Utilizing Hydrant #3 and stretching to the fire via the front door is extremely unsafe and an almost criminal display of tunnel vision. Personnel should be kept away from the downed power line until power shutdown has been accomplished. It is safer to enter the building from the rear.

C. **+1** While Hydrant #4 is out of the immediate danger zone, it is an extremely long and time-consuming stretch to the rear of the building. Operating completely at the rear is a better option.

D. **-1** Even though the power line is down across the front of the building, it is no reason to deluge the fire area with a master stream; to do so writes off any occupants. This is an offensive fire. Offensive fires are fought from the inside of the structure. Find a safe way into the building and take care of business.

E. **-2** The hydrant is unsafe, the entry point is unsafe, and a booster line should never, ever, be stretched to the interior of a building to attack a working structural fire.

6.

A. **0** Since the power has been shut down and is no longer a threat, the front door now becomes the safest, most effective path of least resistance. Therefore, the second line can be positioned in the quickest manner by stretching through the front door. The line should, however, be stretched to the 2nd floor to provide reinforcement of the initial attack line.

B. **0** The potential problems created by having to stretch the initial line from a position other then the front entrance may require the assistance of the second engine company to expeditiously and effectively attack the fire. Often, townhomes and newer construction will have large and confusing layouts, further complicating the stretch. If assistance is required and the second engine company is not assigned to the initial line as per SOP, proper communications requesting this assistance should be made without delay. However, with the power now shut down, if the first line is having difficulty, the second line stretched from the front of the building may make the fire floor before the initial line. Coordination through timely and effective communications is required here. In addition, while all engine company members should be searching off the line as the advance is made, the primary search responsibility is usually the responsibility of the ladder company.

C. **+1** Stretching the second line via the rear is not efficient now that the power lines have been de-energized. In addition, the only hydrants available for the establishment of a second water supply are at the front of the building, making the stretch from this point less time-consuming and fatiguing. The problem of opposing lines shouldn't be a problem here. If the fire was on the 1st floor, lines attacking from both the front and the rear would be dangerous, but as the fire is located on the 2nd floor, both the lines should meet at the stairwell where they are then advanced up the stairs to the 2nd floor, establishing a same-direction attack.

D. **-1** Unless there is involvement and the threat of fire spread across the combustible exterior walls, the second line is generally advanced to the interior of the structure in proximity to the initial attack line as reinforcement and adjacent area extinguishment. It may be necessary, if the combustible exterior is involved, for the second line to sweep the ignited wall before entering the structure. For the safety of the interior attack team, it is imperative that the second line be stretched to the same area as the initial attack line. If this is the case, a third line must be rapidly stretched to address the problem of exterior spread. If no line is available, a deck gun may do the trick. Aim the stream effectively and keep the stream from entering the interior. The objective is to extinguish the exterior fire. Once that is accomplished, the exterior line should be shut down.

E. *+2* Stretch into the building via the safest, most effective path of least resistance and not only will less barriers be encountered, but the time it will take to get the line to the desired area will be minimal. Once the power line is de-energized, the front door is the best route for the line.

7. Remember that venting operations should also follow the Point of Entry Rule of Thumb. When addressing ventilation, it is probably more accurate to call it the Point of Exit Rule of Thumb.

A. *0* Cutting the roof is usually warranted only when the fire is directly below it and is generally not an initial action regarding ventilation of a fire building. This is a peaked, truss roof. If the ceiling is of open construction without an intervening barrier between the roof boards and the living area, such as in a cathedral ceiling or atrium ceiling, then cutting the roof may be required. It is critical that this operation be conducted with extreme caution and usually only from an aerial device. (Fig. 8–12) A roof ladder will be ineffective and dangerous because the hooks, which are meant to be secured over the ridgepole, will have nothing to grab on to should the roof fail. This is because there is no ridgepole in lightweight truss roof construction.

Fig. 8–12 Photo courtesy of Capt. Mike Oriente NHRFR Buildings constructed of lightweight materials must raise the flag of extreme caution for the incident commander. Consider any fire involving the roof area to be destroying the integrity of the trusses. Withdrawal from the main fire area is required—there will be a collapse.

If the roof trusses are involved in fire, then the roof is not a safe place to be. As a matter of fact, the interior companies should be withdrawn and a defensive strategy pursued due to the likelihood of truss collapse. No evidence in the scenario suggests truss involvement. Therefore, unless a scuttle is available in the area of the fire, it is best to provide horizontal ventilation to the area before attempting to open any area of the roof.

B. *-1* Fog ventilation is not a tactic that should be used until the fire is definitely under control. It is certainly not the first ventilation tactic used at this or any other fire. It is also not a ladder company duty.

C. *0* Positive pressure may work if it can be controlled. It should also not be the first ventilation strategy pursued. The problem is that there is too much that can go wrong in this type of structure. Hidden fire can be pushed into uninvolved areas and spread to adjacent exposures via unprotected openings such as pipe chases and wire penetrations. In any buildings with the possibility of void spaces, ventilation is best controlled by natural means until the fire is definitely under control.

D. *+1* Removing the skylight is an acceptable tactic, but not as the first action. It is not in the area of most fire involvement, but is above the adjacent room. It is best to wait until the fire is under control to open this skylight for fear of pulling the fire into this uninvolved area. (Fig. 8–13)

Fig. 8–13 Skylights can be a great asset to vertical ventilation operations of lightweight truss roofs. Take into consideration the effect this venting will have on the fire Opening a vertical artery remote from the main body of fire may pull it into uninvolved areas.

E. *+2* The ladder at the rear of the building will accomplish two objectives. First, it will be a means to vent the windows of the fire area. These windows will be opposite the attack and are the quickest and most effective means of removing smoke and heat from the fire area. Second, it can serve as a second means of egress to crews operating on the 2nd floor. Keep your ventilation simple, consider heat flow in regard to paths of least resistance, and chances are that your tactics will be correct.

8.

A. *+2* The floor area can be quite large in these buildings. Split the team into two members to most effectively cover the area. Concentrate on the fire area and fire floor before covering other areas.

B. *+1* As there is no fire downstairs, it should receive a lower priority in the primary search operation. It is better that both teams initially operate on the fire floor.

C. *-1* It makes no sense to initially conduct a search on a floor that has no fire on it. Remember that the search priorities in multi-story buildings are the fire floor (2nd floor in this case), floor above, top floor, and finally all other floors. The floors below the fire receive the lowest priority unless circumstances dictate otherwise.

D. *-2* It is not necessary to keep the search team intact in this structure. Unless the area is untenable, it is also unacceptable to wait until the hoseline is stretched to initiate a primary search. The concept behind the primary search is to get in and get out, searching as much tenable area as possible before conditions deteriorate and chances of survival are unlikely.

E. *+1* It does not state whether the team is split or not, but using a systematic search in addition to providing a means to communicate which areas have been searched is both efficient and safe. Just the fact that the doors are being marked to avoid duplication of effort suggests that more than one team is involved in the search.

9. Picture this: it is time for post-control overhaul. A safety survey is being conducted as to the fire's effect on the integrity of the building. This question addresses the reader's ability to apply building construction and other factors to the incident and make a decision about the structural integrity based on an analysis of these factors. Let's take a look at these factors one at a time.

Degree of fire intensity. Lightweight wood construction is generally made up of 2" x 4" components. Wood burns at a rate of 1" every 40 minutes. If the wood were being attacked on both sides, it would take only 40 minutes to burn completely through. It must be assumed that the structural integrity of the wood will have been compromised to the point of failure before the 40 minutes are up. This is the most important of the factors presented. (Fig. 8–14)

Fig. 8–14 Welding operations caused these wood studs to burn completely through without ever extending to any nearby combustibles. The seat of this smoldering fire, which must have burned for well over an hour, was found with a thermal-imaging camera.

Overall Building Dimension. Large buildings create large fire loads. The amount of time it takes to bring the fire under control will likely be in direct proportion to the amount of area offered by the building to burn. Remember, too, that an intense fire occurring over a relatively confined area can still create a severe structural hazard.

Weather conditions. Whether extremely hot and humid or cold and wet, weather conditions will have a substantial impact on fire department operations. They may also have an effect on the building. Extreme cold will cause ice to form, creating a much greater load on the building. Heat causes things to dry out, lowering ignition temperatures. While weather and especially wind conditions will have a major impact on exposures, firefighter safety, and additional alarm decisions, it usually does not apply to structural stability. This scenario, which offers high humidity along with heat, will have a great impact on fire operations, but not so much on the building.

Amount of water applied. Water weighs eight pounds a gallon and has the ability to cause both excessive secondary damage and collapse. Water accumulation will have a great effect on the building, especially if a defensive strategy is pursued. The incident commander must keep a keen eye on what effect water is having on the building. If there is little runoff, think water accumulation and collapse. At the Roc Harbour fire discussed in the text book, one of the major influencing decisions regarding the abandonment of the fire area was the report of a bowed floor, indicating a structural weakness in the floor truss. Companies were withdrawn to the adjacent areas to fight a defensive/offensive battle. The floor later collapsed.

The scoring for the answers to this question are as follows:

A. +1

B. −1

C. 0

D. +2

E. 0

10. Secondary damage during overhaul should be consistent with the fire damage already suffered by the structure. On one hand, it is dangerous and counterproductive to overhaul a pile of rubble. On the other hand, excessive damage after a small fire is overkill and unprofessional.

    A. *+2* These are thorough tactics and are not only consistent with the path of least resistance that the fire will take, but are proactive in minimizing property damage. Open the ceiling over the immediate fire area because this represents the area of the most prolonged heat exposure. Check the cockloft via the least damaging means necessary. This may be possible by opening an attic scuttle sometimes found in closets or in top floor hallways. Checking the cockloft space in this way will eliminate the need to excessively rip open ceilings. Also, check behind manmade openings such as electrical switch plates and light fixtures. Fire will penetrate the wall spaces much quicker in these areas than it will where there are no openings.

    B. *-2* It isn't necessary to cut a hole in the roof during this fire operation. It is out of the question to cut the roof during overhaul operations. There are easier ways of checking the cockloft. It is the responsibility of the officer to find them.

C. **+1** Preserving the scene for the fire investigator is the proper action, especially when there is evidence of arson. All fires must be investigated. However, proper and safe post-control overhaul must be conducted to ensure that the fire is extinguished. Removal of smoldering furniture, if it can be done safely, is a good example of fuel removal; fire can burrow in upholstered furniture for hours. Use extreme caution when carrying out this task. Covering furniture with salvage covers is a prudent action to take when overhaul operations are underway. These are all proper actions to take at the fire scene. The only problem here is that none of the actions stated are specific ways to properly open up and examine the area for hidden fire. These are actions that should take place while the building is being opened up. This is the reason the point is given.

D. **+1** Two of the three things stated in this choice are correct; one is not due to the location of the fire. While it is proper action to listen for sounds of fire and feel walls for the presence of heat, a fire on the top floor will not allow crews searching for fire extension to open baseboards on the floor above simply because there is no floor above. Instead, the cockloft or attic must be checked.

E. **0** While it will be necessary to open walls in the fire area, it is more likely that fire extension will follow a vertical path before it follows a horizontal path. For this reason, the area directly above the area of most involvement must be opened as well.

Passing Score for Scenario 8–3 = 14 Points

Scenario 8–4

Multiple-Choice Questions

This is another scenario where you are confronted with several problems—open, attached construction; the ignited tree at the rear; and fire showing from the rear of an assumed-occupied building. In these cases, the best action to take is an aggressive interior attack to locate, confine, and extinguish the fire. If this can be successfully done, all other problems will diminish.

1. Your knowledge of the fire spread characteristics in this building is directly related to your understanding of building construction, allowing you to make the best judgments given the circumstances of each incident.

   A. *+1* The combustible exposures are relatively close to the fire building. The ignited tree can act like a fuse to Building #1. However, the ignition potential the fire building offers to itself by way of large unprotected openings is more significant and must be addressed first.

   B. *+2* The common roof space above the adjacent apartments in the fire building will offer an unobstructed route for fire spread in the fire building; your first priority. Fire spreading to this area can burn the roof off the building as well as the top floor of the adjacent apartments. It will also create a more extensive flying brand and radiant heat problem, exposing the adjacent buildings. It is best to address this fire spread potential before all others. Letting this go unchecked will likely cause other exposure protection measures to fail. In almost all cases, the best way to protect exposures is to extinguish the main body of fire.

   C. *+1* The combustible exterior walls of the fire building can allow fire to spread along the exterior even as interior fire containment and extinguishment tactics are being conducted. If the incident commander is not careful and cognizant through either direct observation or reports from the companies, this exterior spread can drive fire past attack positions and into upper windows and roof openings. In these garden apartments, there are often vent openings just below the eaves, which act to circulate air in the roof space. Fire licking up an exterior wall can easily enter this space. Be aware of this potential and take aggressive steps to avoid it. (Fig. 8–15)

   Fig. 8–15 The walls that extend through the roof of this building are designed to act as fire stops at the roof level. Consider them compromised until proven otherwise. Note how they don't even extend to the ends of the roof. The combustible exterior of this structure can spread fire past these walls to adjacent units, negating their so-called benefit.

   D. *+1* Stairwells in garden apartments will usually be one of three designs. There may be one entrance on the ground floor with a separate entrance and stairwell leading to the 2nd floor. Entrance doors will be side by side. There may be an open, exterior stairwell that serves the individual apartments. Finally, the apartments may be located in a common hallway and stairwell. In this last case, a fire blowing out of an apartment door on the 1st floor can expose the stairwell and apartments on the 2nd floor. Doors should be solid core and self-closing to minimize this potential.

E. *+1* The ignition of the combustible roof will spread flying brands to the roofs and grounds around the complex. For this reason, if the fire breaks through the roof of the fire building, it may be necessary to use master streams to both attack the main body of fire and keep the adjacent roofs wet to prevent ignition. This may require more than one stream. It may be best to allow the fire to burn away a major portion of the roof, localizing itself, before using a master stream into that area. Do not get lulled into concentrating only on the fire blowing through the roof of the fire building. Take steps to prevent the ignition of nearby exposure roofs and interiors.

2.

A. *+1* This fire should best be attacked at its seat via the safest, most effective path of least resistance. The fire is blowing out of the rear of the building. Stretching the initial line from the front will establish the proper attack direction and push the fire from the structure. However, as you will be supplied by the second-arriving engine company and are basically attacking using the booster tank, the attack can be started from a position closer than Blohm Street. Critically read the scenario and a better apparatus position can be established.

B. *+2* The scenario states that the macadam access way is large enough to fit the apparatus for a reason. It places the attack pumper closer to the building, making for a shorter and therefore more rapid stretch and attack, while at the same time minimizing friction loss. Positioning the apparatus here ensures the most direct path to the fire.

C. *0* Again, positioning on the access way is more effective and will result in a quicker stretch. In addition, the large diameter line is not required for the fire attack in this structure. The rooms will be small and stretching the line up the stairs and through the apartment to the rear where the fire is located will be more easily accomplished with the smaller, more mobile 1¾" line.

D. *-1* The position of the first line in all offensive fires is to the seat of the fire while at the same time protecting occupants, that is, placing the line between the fire and the occupants. The fire at the rear can wait. There is no life hazard in the tree. If the fire in the apartment is not attacked, victims may not be rescued and the fire may grow to involve adjacent apartments and the cockloft.

E. *-2* Assigning two firefighters to stretch an attack line into a structure while the officer stretches a line alone to the rear is a gross failure regarding company supervision. The job of the company officer is to protect his company. This can be most effectively done through proper supervision. The actions displayed in this answer choice are not characteristic of the effective supervisor and leader.

3.

A. *+2* The hour of the day suggests that occupants are sleeping. This will always complicate the rescue problem. I have heard civilians say that the smell of smoke will wake them up. This is just not true. Sleeping occupants who inhale carbon monoxide will actually fall into a deeper sleep. This is why people sometimes do not wake up during a fire and die of smoke inhalation in their sleep. The best defense against this is properly located and operating smoke detectors. An activated smoke alarm has a better chance of waking occupants well before lethal concentrations of fire gases permeate the area. In this situation, six apartments have to be searched and/or evacuated. Request the proper manpower to get it done quickly and efficiently.

B. **-1** The presence of attached buildings along with the hour of the day requires that an additional alarm be requested. Twelve men and a chief officer is not enough men to address all the tasks that will be required to control this fire, search and evacuate the buildings, check for fire extension, and provide relief for the operating companies. Give the men a chance by reinforcing the operation early.

C. **+1** Exterior exposure spread could be a real problem, especially if the original fire proves beyond the capability of the initial assignment. Having the manpower on scene to address this potential is proactive and proper. Remember, though, that life takes precedence over exposure protection.

D. **+1** The inherent deficiencies in wood-frame dwellings is a genuine concern at this incident and a definite reason to request additional alarms. However, the life hazard potential is the most urgent and important justification for the striking of additional alarms.

E. **0** The lack of precipitation will leave the area ripe for ignition, especially surrounding vegetation which could spread fire from building to building. However, this reason does not hold as much water (no pun intended) as the other, more point-worthy answer choices.

4. This question gives some specific information about the location and extent of the fire as well as the extinguishment progress of Engine 1.

A. **+1** Hydrant #3 will provide the quickest supply to Engine 1's attack. The supply lines can be quickly stretched across the property from Blohm Street. The line is stretched to properly back-up the initial attack line.

B. **+2** The attack is properly supplied here as well. The difference in the apparatus positioning between answer choice "A" and this choice is not a factor. Either way, the supply line can be quickly stretched to the intake of the attack apparatus. The difference here is that not only is Engine 2's line reinforcing the initial attack line, but is also operating to prevent fire extension. These actions offer correct tactics and a little more initiative on the part of the company officer.

C. **0** The fact that "the fire is being extinguished by Engine 1" is not sufficient reason to neglect to reinforce the initial attack line. Remember that in virtually all interior, offensive attack operations, the second line must back up the initial attack line. The exposure and the tree can wait. The safety of the firefighters on the initial line must receive priority in line placement decisions. Hydrant #1 is also not the best choice for hydrant selection. The protection of exterior exposures should not be necessary, as the first line is extinguishing the fire. The extinguishment of the tree fire may be accomplished from the sliding glass doors by either the first or second line. If this is not possible, then a third line may have to address this elevated fire.

D. **-2** In this scenario, it is the duty of the second arriving engine to operate as a water supply for the initial attack engine. Failure to accomplish this jeopardizes both the safety of the attack team and the chances of success for the operation. Stretching a 2½" line from Engine 1, while a proper tactic if the water is available, will only deplete the already limited water supply more quickly. Most engines carry 500 gallons in the booster tank. A 1¾" attack line flowing 150 gpm will run out of water in a little over three minutes. Attempting to supply a larger diameter back-up line without a continuous water supply will guarantee a waterless attack and probably crew withdrawal in a very short time.

E. *0* The most significantly exposed apartment should be the place that the third line is stretched. Extension checks are mandatory at all fires. Contiguous structures such as garden apartments with common roof spaces are no exception. However, do not address this problem in lieu of properly stretching a back-up line.

5.

A. *-1* The size of both the apartments and the rooms in them will be limited. For this reason, it is most efficient to split the company to conduct a rapid and efficient primary search of the fire area and apartment. In addition, searching the 1st floor before searching the 2nd floor, where the fire is located, may cost victims their lives as search teams may not reach them in time after dilly-dallying around on the uninvolved 1st floor. Get to the fire area to give victims the best chance of survival.

B. *0* The search effort is better focused in this answer choice, but it is still counterproductive to keep a four-member company intact to search such small rooms. The search can be done in half the time if the company is split into two-member search crews.

C. *-1* This is even more detrimental to the search operation than keeping the company intact. Keeping one crew "on the bench" until the primary search is complete is a waste of manpower. It is better to split the company to cover the area. If the same company must conduct the secondary search, let them switch areas of responsibility.

D. *+2* This is the best course of action to take. The company is split into two crews. Each crew takes a bedroom in the fire apartment, giving any potential victims the best chance of survival.

E. *+1* This is less efficient than the two-crew fire apartment search, but due to the inherent small room sizes in these structures, this search strategy is acceptable.

6. The fire has now been confined. Post-control overhaul operations should focus on the areas most susceptible to fire extension in these buildings.

A. *+2* The ceilings and the walls in the area of most fire involvement will be the most likely path of least resistance into the open roof space. A charged line should always be available when all overhaul operations are being conducted.

B. *+1* The crawl space should also be checked for presence of fire. Although it is in the other side of the apartment and is not a likely area where fire will spread into the common roof space, it is an area that must be checked and accessing it via the safest, most effective path of least resistance will more efficiently accomplish this objective.

C. *-1* This shows a lack of awareness regarding the location of the fire. The fire is on the top floor. Therefore, any baseboards will be at floor level and need not be checked. If the fire were on the 1st floor, then the opening of the baseboards on the 2nd floor would be an appropriate action.

D. *0* This area will not be a path of least resistance. The fire has been confined. Opening above the area of most involvement must be accomplished before openings in adjacent and further rooms are made and inspected. In fact, the only time these remote areas should be opened is if the overhaul of the immediate fire area shows signs of fire travel in that direction.

E. *-1* These are salvage operations, not overhaul tactics. Damaging the ceiling of the apartment below is unnecessary and shows a lack of professionalism. If water must be removed, there are better ways.

7. The key phrase to this question is "confined and extinguished." This signifies that the fireground priority of incident stabilization has been accomplished and the priority of property conservation should now be the main focus of both the incident commander and the operating companies.

   A. *-2* This is unnecessary damage. This ventilation strategy is proper if the fire was not yet under control and threatening to extend via the open roof space to the rest of the building. Opening the roof over the fire and knocking out the 2nd floor windows would then be the best method of localizing the fire. Let the damage to the structure be in direct proportion to the situation at hand. At this time, these actions are contradictory to property conservation and are more along the lines of property destruction.

   B. *+1* Opening windows and the sliding glass doors in the fire area is the best way of preventing damage and letting the natural air currents allow the smoke and heat to dissipate.

   C. *-1* Window breakage after fire extinguishment is unprofessional. It is likely that windows in this area were broken during the fire attack and rightfully so. If they were not, you missed your chance and probably made the attack effort more punishing than necessary. Post-extinguishment operations are not the time for this.

   D. *+2* Once the fire is extinguished, properly-controlled PPV will be the most effective means of removing residual and lazy smoke from the structure. While I am not an advocate of PPV-assisted fire attacks, it can be invaluable in clearing the area for post-operations, providing fresh air and visibility for crews still operating in the area. Selective clearing of the area is best. Closing and opening doors to channel the airflow as desired best accomplishes this. Strict monitoring for flare-ups of any hot spots is critical.

   E. *0* Upon arrival, it is true that the fire was venting itself. This was a plus for the attack operation, making advancement and other support operations less punishing. However, the fire is stated as being extinguished. Officers are responsible for developing solutions to best address problems and situations on the fireground. This answer choice deserves a zero because the decision was worthless.

8.

   A. *+2* The time of this fire is 0100. It must be assumed that the life hazard will be severe. This will increase the problem of manpower available versus tasks required and will be the biggest reason to call for additional alarms.

   B. *+1* Exposures on the interior include all the adjacent apartments in the fire building, especially those on the 2nd floor. This exposure will be threatened through the open roof space. In addition, the bone-dry vegetation and the combustible exterior of the adjacent, leeward exposure will become a major concern should the fire escalate beyond the apartment of origin and especially if the roof becomes involved.

   C. *-1* The height is two stories. There are no overhead power lines in either the diagram or the scenario. Access to the structure with ground ladders should not be problematic.

   D. *+1* The construction of the fire building and the exposures can create some major headaches for the incident commander. The presence of the open roof space and combustible exterior can spread fire not only on the interior of the building, but also on the exterior, potentially igniting vegetation and the combustible exposures.

E. **0** While weather is not the most important size-up factor, it should not be ignored. The temperature is 77°F. This is not the best condition in which to operate, but it is also not the worst. It is also nighttime; the sun will not be blazing down on the operating forces making it feel hotter than it is, as would happen during the day. The breeze should also aid in firefighter comfort if it doesn't spread the fire to the exposures. However, the heat will cause fatigue. The incident commander should take this into account and have additional companies on the scene to provide relief to operating companies.

9. Consideration must be given to both interior and exterior exposures. Based on the location of the fire and the path of least resistance, priorities for exposure protection can be established.

A. **+1** Building #1 is directly behind the fire building as well as on the leeward side. With the ignition of the adjacent tree and the potential for the dried vegetation to act as a fuse, the protection of this exposure should be considered. However, in offensive operations, interior exposures will usually receive a higher priority in regard to fire spread potential.

B. **+2** The roof is open over the entire building. Therefore, the likelihood that the fire will penetrate into the adjacent apartment earlier than any other exposure must be considered. Failure to address this area and the area above can cause the loss of the entire building as well as ignition of adjacent exterior exposures. (Fig. 8–16)

C. **0** Apartment A1 is directly below the fire. If this question asked which exposure should receive the most attention in regard to salvage operations, this would be a **+2** answer choice. A check must be made in areas of the pipe chases and other voids for any drop-down fire, but the fire extension threat will be greater in other areas.

Fig. 8-16 Photo by Lt. Joe Berchtold, Teaneck (NJ) Fire
If the fire involves or threatens to involve the roof area, it will be necessary to cut the roof in order to slow the lateral spread of fire in the cockloft. Keep the roof personnel to an absolute effective minimum. There are too many people on this roof.

D. **-2** For Building #4 to be imminently threatened, the apartments adjacent to the fire apartment as well as the roof must become involved. In this situation, it is best to concentrate exposure protection, both interior and exterior in the fire building in the former case, and on the leeward exposure in the latter. If Building #4 were to become an imminent ignition threat, a lot has to go wrong in the fire building, indicating a lack of aggressiveness, the lack of a proper fire containment strategy, and a probable failure to place additional companies in the likely paths of fire travel. Get personnel ahead of the fire and open the building up well before the need is apparent.

E. **0** Due to the presence of the open roof area, Apartment C2 may become a major manpower-commitment area, but not before the adjacent Apartment B2. Get forces into Apartment B2 early and check for extension using tactics that are consistent with the magnitude of the threat before performing the same task in Apartment C2. If the fire in Apartment A2 is small, a few examination holes as well as a check of the cockloft via the hatchway at the front of the apartment should suffice. If, on the other hand, the fire next door is severe, get the ceilings down in the adjacent apartment and prepare for the worst. In both cases, charged lines should be standing by to extinguish any fire that shows in the exposed apartment. It's always easier to be ready for fire extension than to be sorry later that you weren't.

10. Safety must be the overriding concern for the incident commander at all times. During the latter stages of the operation, firefighters often get injured because they let their guard down. It is at this time that control by the incident commander must remain strong. This control also translates very strongly to company officer discipline.

A. *-1* Fires that are under control even after the smoke has cleared can be deadly to firefighters. Remember that carbon monoxide is colorless and odorless, as are some of the other poisons that are present as a result of incomplete combustion. At times, the telltale aroma of the byproducts of combustion may be masked by other odors. Sometimes, the effects of chemical poisoning due to inhalation of the products of combustion may not be apparent for several hours. Don't take the chance. Firefighters should wear SCBA until the operation is complete and they have exited the building.

B. *-1* When was the last time a firefighter smelled or saw carbon monoxide or any other colorless gas? That's one special nose. Unless the atmosphere can be confirmed safe by reliable testing methods, firefighters should operate with SCBA.

C. *-2* Using this method of testing the atmosphere is gross negligence and sets the poorest example of leadership. Officers should set the tone for safety at all times. An officer who cannot be trusted to comply with an obvious safety procedure such as a mandatory mask rule cannot be trusted to make proper decisions on the fireground.

D. *+2* This is the safest and best rule to live by. No building, especially after the fire has been knocked down, is worth the lungs or the life of a firefighter. Officers must take proactive steps to safeguard their companies. This starts and ends with safety. Remember, also, if this question appeared on a test like this, how could any candidate in their right mind answer this question any differently?

E. *-1* Firefighters checking for extension may also be overcome by the products of combustion. Voids can contain concentrated products of combustion long after the fire is knocked down in the main fire area. A lungful of bad stuff may end both a career and a life.

Passing Score for Scenario 8–4 = 14 points

CHAPTER NINE
SCENARIOS

SCENARIO 9–1
CHOPPER'S TAVERN FIRE

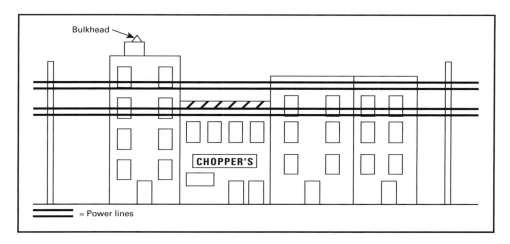

Construction

The fire building is a two-story, Class 3 mixed-use occupancy. The ground floor is occupied by Chopper's Tavern. The top floor is split into two apartments. There are tin ceilings throughout the building. There is a one-story roofed extension at the rear of the building, which serves as the kitchen and storage for the bar. There is no fire escape, but the 2nd floor windows can be accessed via the extension's roof. There is a cornice at the front of the building and a scuttle at the rear of the roof.

Time and weather

It is 0420 on a warm, clear night. The wind is blowing at 15mph toward Exposure B.

Area and street conditions

The streets are wet from an earlier storm. The traffic is particularly busy for this time of night.

Fire conditions

You are responding to a fire reported by a passerby from a cell phone. Upon arrival, you observe heavy smoke emanating from the roof area. Dark smoke is issuing from around the cornice. You can't see any fire on the top floor.

Exposures

Exposure B is a four-story wood-frame tenement, and Exposure D is a three-story wood-frame residential occupancy.

Water supply

The water supply is adequate for the fire load profile of the area.

Response

On the response are three engine companies and two ladder companies. Each engine is staffed with one officer and two firefighters. The ladder is staffed by an officer and three firefighters.

SHORT-ANSWER QUESTIONS

1. The fire is in the cockloft. Where do you position the first line? What are the objectives?

2. Where do you position the second line?

3. Discuss ladder company operations required at this fire.

4. Where would additional lines be placed?

_____

_____

_____

_____

_____

_____

5. Suppose this became a defensive operation. Discuss what fire control tactics you would use.

_____

_____

_____

_____

_____

6. Many times, when the cockloft is involved in fire, the top floor is free of smoke and men are able to operate in a smoke-free environment. Why is it important to be cognizant of the dangers that exist, but are unseen?

_____

_____

_____

_____

_____

Scenario 9–2
GB's Hair Salon Fire

Construction

The fire building is of ordinary construction, and is a two-story taxpayer with a flat roof. The 1st floor is occupied by GB's Hair Salon. The 2nd floor is occupied by a dance studio, an assembly occupancy. The rear half of the studio has been partitioned into what appears to be an office or an illegal apartment. There is a wood door separating these two adjacent areas. Access into the 2nd floor is on Side B. There is a fire escape on Side B of the fire building (no roof

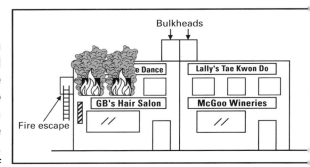

access) and at the rear of Exposure D (roof access). There is a stairway leading to the roof terminating at a roof bulkhead. The bulkhead on the roof is shared with Exposure D. There is also an interior stairway in Exposure D leading to the roof, separated by a fire rated partition from the stairway in the fire building. Both buildings have a cockloft with a firewall separating them. It is not known if the firewall is unpierced for its entire length. There is a 3' parapet running around the perimeter of both buildings.

Time and weather

The time of the fire is 0130. There is a 20mph wind blowing from the rear of the building toward the front.

Area and street conditions

Streets in the area are narrow, but traffic is light due to the lateness of the hour.

Fire conditions

Upon arrival, you observe that lights on the 2nd floor are still on and that there is fire showing out of two front windows on Side A.

Exposures

Exposure A is a street. Exposure B is a 25' wide parking lot serving exposure B1, a two-story noncombustible bank. Exposure C is a one-story garage of ordinary construction separated by a 5' alley. Exposure D is a street.

Water supply

The water supply is adequate and is fed by a looped 12" grid.

Response

The initial running assignment is two engine companies and one ladder company. The engine companies are staffed by three personnel, while the ladder company is staffed by four. There is an officer assigned to each company. Each additional alarm will bring two engine companies and one ladder company.

Multiple-Choice Questions

1. After securing a water supply, what would be the first actions of the first arriving engine company at this fire?

   A. Stretch a 1³/₄" line into fire building up to the 2nd floor. Remain at the top of the stairs between the fire occupancy and the stairway until all occupants have been evacuated. Then move in and attack the fire.

   B. Use the deck gun to knock down any visible fire at windows, while the 1³/₄"line is being stretched into position. Shut down the deck gun when the attack line is in position. Attack the fire.

   C. Stretch a 1³/₄" line into the building to the top floor. Attack the fire, coordinating operations with the vent and search teams.

   D. Split the company into two crews. One crew stretch 1³/₄" line to the top floor to protect egress. Second crew takes a 1³/₄" line up the Side B fire escape to attack the fire in coordination with the vent team.

   E. Bring a line into the hair salon. Ask for a haircut and a shave while waiting for a drop down fire.

2. What action would you order the second arriving engine company to perform?

   A. Secure a secondary water supply. Stretch a second 1³/₄" attack line to the 2nd floor landing. Back-up the first attack line and protect the egress via the interior stair.

   B. Secure a secondary water supply. Stretch a second attack line up the fire escape on Side B of the fire building. Provide additional attack power to the first line by hitting the fire from the fire escape window position.

   C. Secure a secondary water supply. Stretch a 2¹/₂" line into the building to the top of the stairs. After ensuring the progress of the initial attack line, stretch into the adjacent area and check for any fire extension.

   D. Split the company into two crews. After ensuring a secondary water supply, the first crew assists on the initial attack line. The second crew executes VES operations in the adjacent area, checking for any fire extension.

   E. Split the company into two crews. First crew to assist people down the fire escape on Side B of the fire building. The second crew performs a recon mission in Exposure D; evacuate and report on conditions.

3. Depending on the action you chose for the initial attack line, explain the strategic concept or reason for the tactical action?

   A. Knock down any heavy fire from the outside to allow safer and easier advancement of the initial attack line.

   B. Use a pincer attack to surround the fire and keep it in one side of the building.

   C. Use the initial attack line in a holding action to protect the main area of occupant egress.

D. Use the line to drive the products of combustion out of the building, continuing the original venting direction of the fire.

E. Bring the line to the eventual path that the fire will take by evaluating fire spread possibilities and getting resources there early.

4. What initial actions would you order for the first arriving ladder company?

A. Split the company into two crews. Crew 1 operates in the dance studio and conducts a primary search and horizontal ventilation. Crew 2 operates in the adjacent area with the same assignment.

B. Split the company into two crews. Crew 1 ladders the front of the building with a ground ladder and performs horizontal ventilation of the fire occupancy. Crew 2 uses an aerial to access the roof, and then opens all natural openings and reports conditions.

C. Split the company into two crews. Crew 1 goes to the fire floor and conducts primary search and horizontal ventilation operations of the fire and adjacent area. Crew 2 goes to the roof via the adjoining bulkhead and opens all natural openings, horizontally vent from the roof, and report conditions.

D. Split the company into two crews. Crew 1 goes to the fire floor and conducts primary search and horizontal ventilation operations, pulling ceilings to expose the cockloft for recon of any fire extension. Crew 2 goes to the top floor of Exposure B with the same orders.

E. Split the company into two crews. Crew 1 goes to the roof via the aerial. Open all natural openings, horizontally vent the fire apartment from the roof. Crew 2 goes to the roof via the Exposure B bulkhead, clearing Exposure B on the way up, and finally assists Crew 1 with venting operations.

5. Should a second alarm be sounded for this fire?

A. No, this is a routine fire. Companies on the scene can handle it.

B. Yes, a second and third alarm should be sounded, due to the exposure potential.

C. Yes, due to the existence of a life hazard.

D. Not until better information on the location and extent of the fire are available.

E. Yes, if reports of cockloft involvement are received from the interior teams or the roof.

6. The engine company attacking the fire reports a severe heat condition in the fire area. From these reports, you suspect superheated gases are present in the cockloft. What is the best way, in the proper order, of alleviating this condition and easing the engine company's advance?

A. 1. Provide more horizontal ventilation at the top floor windows.

2. Pull the ceilings in the fire area.

3. Use a fog stream into the cockloft space to cool gases.

B. 1. Make examination holes in the roof to locate the highest concentration of gases in the cockloft.

    2. Provide more horizontal ventilation at the top floor windows.

    3. Open a hole as directly over the fire as possible to vent the cockloft.

C. 1. Provide more horizontal ventilation at the top floor windows.

    2. Open a hole as directly over the fire as possible to vent the cockloft.

    3. Direct hose stream from above into the opening after the gases have vented to cool the cockloft.

D. 1. Provide more horizontal ventilation at the top floor windows.

    2. Open a hole as directly over the fire as possible to vent the cockloft.

    3. Push down the ceiling with a hook from the roof to vent the apartment.

E. 1. Provide more horizontal ventilation at the top floor windows.

    2. Make examination holes in the roof to locate the highest concentration of gases in the cockloft.

    3. Open a hole as directly over the fire as possible to vent the cockloft.

7. In considering the construction of the building and the location and extent of the fire, what would be your greatest concern regarding building collapse at this fire?

    A. Floor collapse due to the accumulation of runoff water on the 2nd floor.

    B. Roof collapse due to fire weakening the main roof joists.

    C. Collapse of the fire escape on Side A due to heat-related failure of the anchoring points.

    D. Parapet wall collapse due to flame exposure on the free-standing parapet wall (unequal heat expansion of the bricks).

    E. Bulkhead stair collapse due to the weight of firefighters and firefighting equipment along with fire damage and water destruction.

8. If this fire had occurred on the 1st floor in the hair salon, where would be the most likely path fire would take to reach the 2nd floor of the structure?

    A. Via pipe chases in the bathroom and sink areas.

    B. Via autoexposure from the front windows.

    C. Via light fixtures in the ceiling.

    D. Via interior wall studs beneath the windows.

    E. Via the interior stair adjacent to the hair salon.

9. Based on the information in the scenario and the action taken in Question #4, what would be the most appropriate assignment for a second arriving ladder company?

    A. Split the company into two crews. Crew 1 goes to the top floor of Exposure D to check for fire extension and to evacuate the building. Crew 2 goes to the fire building to perform a primary search and to check for drop-down fire extension into the hair salon.

B. Operate in Exposure D to check for fire extension on the top floor and to evacuate any occupants.

C. Split the company into two crews. Crew 1 goes to the rear and side fire escapes to assist any fleeing occupants. Crew 2 goes to the top floor of Exposure D to check for fire extension and to evacuate any occupants.

D. Split the company into two crews. Crew 1 performs a utility shutdown in the fire building and in Exposure D. Crew 2 to assist in roof operations on the fire building.

E. Split the company into two crews to operate in Exposure D. Crew 1 stretches a line to the top floor to control any fire extension. Crew 2 opens the ceilings on the top floor to expose any extending fire in the cockloft.

10. You are assigned to Ladder 1. While searching the fire apartment, you have become separated from your partner. You have become disoriented by the smoke and are lost. Your air supply is also running low. What is the best action to take to ensure your safety?

A. Turn on your PASS Unit, calmly wait for rescue.

B. Find your way to a wall, follow it to a window, break the window, and then straddle the window to attract attention.

C. Find a hoseline, follow the male coupling out of the building.

D. Find a hoseline, follow the female coupling out of the building.

E. Work your way up to the roof and escape by exiting through the scuttle, crossing over to the adjoining building and going down the fire escape, through the scuttle, or down the aerial.

SCENARIO 9–3
BERGEN AVENUE TAXPAYER FIRE

Construction

8225 Bergen Avenue is an old two-story, Class 3 mixed-use occupancy occupied by a Beauty Salon. It is one of a row of attached identical structures. There is an apartment on the 2nd floor. The access to the 2nd floor is through a door adjacent to the Beauty Salon door.

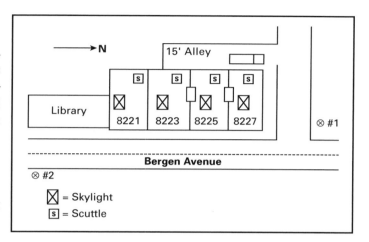

Time and weather

The temperature is 62°F with the wind out of the north at 10mph.

Area and street conditions

Bergen Avenue is a two-way, two-lane street that is fed by smaller streets that run through adjacent residential areas. All of the side streets are one-way streets.

Fire conditions

Upon arrival, the Beauty Salon is closed, as are all the stores in the row. How long they have been closed is not certain. There is no visible flame from the front of the building. This is because the show windows are covered by steel roll-down gates secured by at least half a dozen padlocks. There is heavy smoke emanating from behind the gates. The glass on the front door of the Beauty Salon, which is not covered by the gates, is stained black. There is evidence of smoke on the 2nd floor.

Exposures

The occupancy on the B side is a real estate office, the occupancy on the D side is a roofing company. Attached to the building on the B side (Exposure B1) is a one-story library that has been recently renovated. It has a bowstring truss roof. On the C side is a 15' alley that runs behind the store and sometimes has delivery trucks in it. Street addresses 8221 through 8227 Bergen Avenue are housed in the same building, each occupied by a store on the ground floor and an apartment on the 2nd floor. There are firewalls extending through the roof for a height of 12" between each occupancy in the building. On the roof of each occupancy is a large skylight and a scuttle that leads to the top floor of the dwellings. In between 8223, 8225, and 8227 is an open shaft. Windows face on this shaft on all floors in the fire building and on the 2nd floor in the exposures.

Water supply

Water supply is adequate and hydrants are well-spaced.

Response

Three engine companies and a ladder company respond. A second alarm will bring an additional engine, ladder, and rescue company. Three firefighters, including the officer, staff the engine companies. Four firefighters, including the officer, staff the ladder companies and the rescue company.

MULTIPLE-CHOICE QUESTIONS

1. What would be the quickest way of forcing the roll-down gates covering the show windows?

   A. Use a duckbill lock breaker along with a maul to remove the locks.

   B. Use a miner's pick to provide a shearing force against the shackle of the padlock.

   C. Avoid cutting the locks by employing an inverted "V" or teepee cut to remove the gates.

   D. Use a circular saw with an aluminum oxide blade to remove the locks.

   E. Do not cut the locks until topside vertical ventilation has been accomplished. A back-draft could occur. Then after the ventilation is complete, use the three-cut method to remove the gates, thus avoiding the locks.

2. What would be your orders for the crew of Engine 1? The report from Side C is that there is fire issuing from two small windows at the rear.

   A. Establish a water supply at Hydrant #2. Position in front of 8223 Bergen Avenue. Place a 2½" line in a flanking position at the front of the store. After the roof venting has been completed, advance into the store through the front show windows. Extinguish fire as you advance.

   B. Establish a water supply at Hydrant #1. Position in front of 8223 Bergen Avenue. Place a 2½" line in a flanking position at the front of the store. After the roof venting has been completed, advance into the store through the front door. Extinguish any fire as you advance.

   C. Establish a water supply at Hydrant #1. Position in front of 8227 Bergen Avenue. Stretch a preconnected 1¾" line through the residential door up to the 2nd floor to cut off the fire spread and to protect the primary search.

   D. Establish a water supply at Hydrant #1. Position in front of 8223 Bergen Avenue. Stretch a preconnected 1¾" line into store through front door to locate, confine, and extinguish fire while protecting primary search.

   E. Establish a water supply at Hydrant #2. Position in front of 8227 Bergen Avenue. Place a deck gun on sidewalk in flanking position at front of store. Open a deck gun to knock down heavy fire in store once roof ventilation has been completed and roll-down gates and store windows have been taken out. Stretch a preconnected 1¾" line through the residential door up to the 2nd floor to cut off the fire spread and to protect the primary search.

3. What are your orders for Engine 2?
   A. Stretch a second 2½" line to flank the front of the store. After ventilation is complete and the gates have been removed, attack in coordination with Engine 1's line.

B. Secure secondary water supply. Stretch preconnected 1¾" line into ground floor of 8223 Bergen Avenue to ensure fire does not extend horizontally. Work in coordination with the ladder company pulling ceilings in store.

C. Secure secondary water supply. Stretch second 2½" line to flank the front of the store. After ventilation is complete and the gates have been removed, attack in coordination with engine 1's line. Then, after the fire is knocked down, stretch to the 2nd floor to cover any extension on the floor above the fire.

D. Secure secondary water supply. Stretch a 2½" line into store to back-up Engine 1's attack line.

E. Secure secondary water supply. Stretch a preconnected 1¾" line to the 2nd floor through the residence door to cover any fire extension of the floor above the fire.

4. Would additional alarms be required at this incident?
   A. Not unless the library was directly exposed.
   B. Yes, a suspected life hazard requires the striking of a second alarm.
   C Yes, due to the potential fire extension problems of the adjacent occupancies and the library.
   D. No, place the lines in service quickly and mitigate the problem.
   E. Yes, due to the fact that a backdraft condition is possible, additional manpower must be summoned due to the possible damage profile of the incident.

5. What are your orders for Ladder 1?
   A. Split the company into two crews. Crew 1 will operate in the store after removing the roll-down gates and taking out the windows. Crew 1 conducts a primary search and further horizontal ventilation of the store, attempting to locate the seat of the fire for the advancing engine companies. Crew 1 makes the floor above for primary search as quickly as possible. Crew 2 will ladder the roof of the upwind exposure, opening all natural openings to vertically ventilate the building. Crew 2 recons the shaft areas and adjacent roof openings for cockloft involvement.
   B. Keep the company intact. Ladder the roof of the upwind exposure. Open all natural openings on the roof to alleviate the backdraft condition. Conduct a secondary search of the store and the floor above when conditions permit. Horizontally vent as required.
   C. Split the company into two crews. Crew 1 ladders the roof of the upwind exposure. Open all natural openings on the roof to alleviate the backdraft condition. Check all other roof openings for signs of fire extension. Request engine company support as necessary. Crew 2 takes out the windows and roll-down gates after vertical ventilation has been completed. Support hoseline advancement into the store by removing obstacles and checking for fire extension on the fire floor and the floor above. Horizontally vent as required.
   D. Split the company into two crews. Crew 1 conducts a primary search of the fire floor after forcing the roll-down gates. Horizontally ventilate as the search progresses. Coordinate with the engine company that's advancing the line. Crew 2 operates on the 2nd floor conducting a primary search, horizontally venting, and pulling ceilings to expose any fire traveling in the cockloft.

    E. Keep the company intact. Force the doors and gates of all the stores. Allow engine companies to break glass using a master stream, bringing the fire back to a free burn. Split the company into two crews. Crew 1 moves in with a line advancement on the 1st floor, searching as the line advances. Crew 2 works on the floor above with the second engine company.

6. What are your orders for Rescue 1 and Ladder 2?

    A. Conduct a primary search of fire floor and the floor above 8223 Bergen Avenue.

    B. Split Ladder 2 into two crews. Crew 1 assist in roof operations of fire building and exposures. Crew 2 operates in 8223 Bergen Avenue, conducting a primary search and opening ceilings on the 2nd floor to check for any extension. Rescue 1 assists in the fire building in the primary search and horizontal ventilation of the fire floor and the floor above.

    C. Split Ladder 2 into two crews. Crew 1 assists in the vertical ventilation of the roof along with Ladder 1. Crew 2 operates in 8223 Bergen Avenue, conducting a primary search and opening the ceilings on 2nd floor to check for any extension. Rescue 1 will stand by at the command post as the FAST team.

    D. Keep both companies at the command post as a tactical reserve.

    E. Rescue 1 will check conditions in the library and report back to the command post. Ladder 2 will operate in 8223 Bergen Avenue, conducting a primary search, pulling ceilings to check for fire extension, and horizontally ventilating as required.

7. How would you address the problem of the open shaft between the fire building and the exposures?

    A. Stretch a line to the roof with a Bresnan Distributor. Lower the line into the shaft adjacent to the windows to create a water curtain.

    B. Close the windows on the shaft. Remove all combustibles from the area.

    C. Stretch a line into each exposure to the 2nd floor to protect against fire spread into the exposure. Operate the line only if necessary.

    D. Stretch a line to the downwind exposure. Operate the line across the shaft into the fire building to knock down the heavy fire and aid the companies advancing into the fire building.

    E. Since there are firewalls between the buildings, the shafts are not a concern at this time.

8. There is an elderly, bedridden woman on the 2nd floor of 8221 Bergen Avenue. She is on oxygen 24 hours a day. How would you address this situation?

    A. Protect her in place. There is no need to evacuate her as the fire is two buildings away.

    B. Send a crew to remove her from the building. Turn off the oxygen cylinders.

    C. Send in two policemen to remove her.

    D. Have Rescue 1 go into her apartment to calm her down and monitor both her and the fire's conditions.

    E. Send a crew to remove both her and the oxygen bottles and equipment immediately.

9. There is heavy fire in the rear storage area partially obstructed by a partition wall and some showcases. The first lines find that even with vertical and horizontal ventilation complete, they still cannot advance into the rear of the 1st floor. What action will you take?

    A. Evacuate the building and conduct a roll call. Set up collapse zones and master streams. The building is doomed. Take defensive positions and protect exposures.

    B. Back them out and reposition them to the rear of the building to attack the fire from there.

    C. Stretch a line around to the rear of the building. Pull the interior crews out. Knock down the heavy fire from the rear. Shut down the rear line. Attempt to advance from the front again.

    D. Stretch a line around to the rear of the building. Use this line to knock down the heavy fire at the rear. This action should alleviate the interior conditions and advancement should be easier.

    E. Set up a master stream at the front of the building. Pull the interior crews out. Knock down the heavy fire with the deck gun. After the fire has been knocked down, attempt to advance from the front again.

10. During the post-control overhaul, a safe has been found in the closet of a room on the 2nd floor. There is still a lot of deep-seated fire in this room below the ceiling materials and debris. The floor of the room has been heavily damaged and has burned through in spots. What is the best course of action to take regarding firefighter safety?

    A. Have a crew remove it to a safer area so overhaul of the room can continue.

    B. Keep crews from operating beneath this area. Use the reach of the stream to overhaul the area.

    C. Use a heavy tool to push the safe through the hole in the floor. This will send it to the 1st floor and eliminate the problem. Alert the 1st floor crews and ensure all personnel are out of the area on the 1st floor before accomplishing this tactic.

    D. Have the safety officer recon the area and report back to you with recommendations for how to operate safely in the area

    E. Have dispatch make an announcement over the air as an emergency transmission to all operating personnel of the location of the object. Keep all personnel from operating beneath the area. Use the reach of the stream to complete overhauling in the area of the object.

Scenario 9–4
Isabel's Embroidery Fire

Construction

You have arrived at a three story, mixed-use occupancy of wood frame construction. Isabel's Sewing and Embroidery occupy the ground floor store. The ceiling is tin. There is a 30" x 30" trapdoor leading down to the cellar at the rear of the store. There, the cellar can be accessed via a set of narrow wooden stairs. A heavy-duty, 1 3/8" case-hardened padlock secures the sidewalk cellar door. The cellar is piled high with tied-up cardboard boxes and old unused material. Four apartments, two on

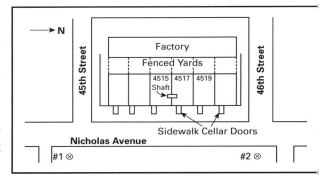

each side, occupy the top two floors. There is a fire escape on both the front and back of the building. Access to the upstairs apartments is via a door adjacent to the store entrance at the front of the building. There is a scuttle hatch on the roof that leads to the top floor hallway.

Time and weather

The time is 1548 on a Friday afternoon.

Area and street conditions

The location of the fire is 4517 Nicholas Avenue. Nicholas Avenue is a one-way, one lane street running south. The cross streets, 45th Street to the south, and 46th Street to the north, both run west. Nicholas Avenue is a main shopping thoroughfare as well as a heavily-used artery in the city. There are many people in the area and several apparatus are reporting that they are stuck in traffic and will be delayed.

Fire conditions

There is heavy smoke evident on the first floor and pushing from around the sidewalk cellar door. Flame is issuing from the windows in the cellar grates at the front of the building. The owner of Isabel's tells you that the store is empty.

Exposures

The fire building is attached at both sides to similar structures. The fire building, as well as the attached exposures is 25' wide and 75' deep.

Water supply

Hydrant spacing is good with a hydrant on its own main on each corner.

Response

You have three engine companies and one ladder company responding. All companies are staffed by an officer and three firefighters.

Multiple-Choice Questions

1. What would be the best way of opening the sidewalk cellar opening?
   A. Attack the hinges at each side of the door.
   B. Use a saw with an aluminum oxide blade to attack the padlock.
   C. Ask the owner for the key.
   D. Use a heavy duty pair of bolt cutters to cut the lock shackles.
   E. Use two pry bars to snap the hasp to which the padlock is attached.

2. Where should the first line be stretched?
   A. Stretch a 1³/₄" line to exterior cellar door entrance. Advance line into cellar to confine and extinguish the fire.
   B. Stretch a preconnected 1³/₄" line into door leading to upstairs dwellings. Use line to protect evacuation and prevent fire extension into the dwelling areas.
   C. As a preconnected 1³/₄" line is being stretched into store, use deck gun to knock down heavy fire at cellar door entrance. Shut down deck gun when line is at the trapdoor. Advance down the interior cellar stairs. Locate, confine, and extinguish fire via the interior cellar stairs.
   D. Stretch a 2¹/₂" line into cellar via the exterior cellar door. Advance a line into the cellar to locate, confine, and extinguish the fire.
   E. Stretch a 2¹/₂" line into store to trapdoor opening. Advance a line into the cellar via the interior stairs to confine and extinguish the fire.

3. Where would you stretch the second line?
   A. Stretch 2¹/₂" line to interior trap door. Remain at the top of the opening to protect the artery against fire extension into the store and to protect egress for initial attack team.
   B. Stretch a 2¹/₂" line to the exterior cellar entrance to reinforce the initial fire attack operation. Advance into the cellar as a back-up line.
   C. Stretch a 1³/₄" line to the exterior cellar entrance. Advance into the cellar to reinforce the initial fire attack operation.
   D. Stretch a preconnected 1³/₄" line into Exposure D to protect the light shaft between the buildings. Report to the command post on the fire extension status.
   E. Stretch a preconnected 1³/₄" line to the 2nd floor to check for any fire extension in the apartments.

4. Where would you stretch a third line?
   A. Stretch a 1³/₄" line to top floor of Exposure D to protect against extension via the shaft.
   B. Stretch a 1³/₄" line through the dwelling door to the 2nd floor to check for fire extension.
   C. Stretch a 2¹/₂" line into the store to protect against vertical fire spread into the store.
   D. Stretch a 2¹/₂" line into the adjacent store to protect against lateral fire spread via the tin ceiling.
   E. Stretch a line into the adjacent cellar to check for extension via openings in the cellar walls.

5. The owner informs you that there are two cellar windows at the rear of the building. What is the best way of accessing the rear of this building to effect ventilation?

   A. Go around to the side street. Access the rear via the rear yards.

   B. Go through the cellar of the fire building.

   C. Go to the 2nd floor of the fire building. Go out the window onto the rear fire escape. Climb down the fire escape to access the rear yard.

   D. Go through Exposure B and out the window onto the rear fire escape. Climb down the fire escape and over the fence to access the rear yard.

   E. Access the rear via the roof of the fire building by descending the fire escape.

6. The main body of fire is located near the rear of the cellar. There is an extremely heavy smoke condition in the store. How would you order this fire ventilated?

   A. Open all natural openings on the roof. Break out the cellar windows at the front of the building and the windows of the store.

   B. Open all natural openings on the roof. Open the scuttle to check and monitor the cockloft. Cut the roof over the hot spot. Break out the windows in the store.

   C. Open the scuttle to alleviate the heat conditions on the top floor. Break out the cellar windows in the rear to horizontally ventilate the cellar. Break out the store windows.

   D. Open all natural openings on the roof. Open the scuttle to alleviate the heat conditions on the top floor. Break out the cellar windows at the rear to horizontally vent the cellar. Cut a hole in the 1st floor above the main body of fire.

   E. Open all natural openings on the roof. Open the scuttle to alleviate the heat conditions on the top floor. Check and monitor the cockloft. Break out the rear cellar windows to horizontally vent the cellar. Also, take out the store windows. Cut a hole in the 1st floor near the front show windows.

7. How would you direct the ladder company to conduct a primary search at this fire?
   A. Split the company into two crews. Crew 1 searches the store. Crew 2 searches the apartments on the 2nd and 3rd floors.

   B. Split the company into two crews. Crew 1 searches the cellar using a lifeline. Crew 2 searches the remainder of the building.

   C. Keep the company intact. Conduct a primary search of the apartments first and then the store.

   D. Split the company into two crews. Since the building has been confirmed evacuated, Crew 1 evacuates Exposure D. Crew 2 evacuates Exposure B.

   E. Split the company into two crews. Crew 1 searches the cellar using a lifeline. Crew 2 searches the upper two floors, then ensure that Exposure D is evacuated.

8. What would be the most significant safety concern for crews operating in the cellar?
   A. Floor collapse above them due to heavy fire conditions.

   B. Collapse of water-soaked stock.

   C. Cellar stair collapse.

   D. Running short of hose.

   E. High heat conditions due to inadequate ventilation.

9. Would additional alarms be required at this incident?
   A. No, the fire should be contained to the cellar.

   B. Yes, the entire building is exposed to fire extension.

   C. Yes, as the building is attached and there is a shaft between the two buildings, inviting the possibility of fire spread into the exposure.

   D. No, as there is no life hazard.

   E. Just request a second ladder company to handle the evacuation of Exposure D.

10. What operations would be conducted in Exposure D?
    A. Evacuate the building. Check the cockloft for fire spread. Stretch a line to the 1st floor to protect against extension via the shaft.

    B. Stretch a line into the cellar. Evacuate the building.

    C. No operations should take place in Exposure D. Instead, operate in Exposure B to prevent fire extension. Also, evacuate the building.

    D. Evacuate the building. Stretch lines to the cellar and the 1st floor to prevent against lateral extension. Open unexposed windows to relieve heat conditions. Also, monitor the cockloft for fire spread.

    E. Evacuate the building. Open unexposed windows to relieve heat conditions. Stretch a line to the 1st floor hallway. Take line into cellar or 1st floor as determined by reconnaissance operations.

11. Suppose the fire had extended into the store. You have been sent into the store to conduct post-control overhaul. What would be the best way to confront the tin ceiling and the overhaul?

    A. Only overhaul the walls and around pipe chases and utility openings. The tin ceiling is a barrier that fire will not penetrate.

    B. Overhaul the ceiling area from above by prying up the baseboard and floor boards over the hot spot on the first floor.

    C. Use a Halligan hook and pry the ceiling apart at the seams. Then pull it down in sections to expose the ceiling joists above.

    D. Use a Halligan tool from a folding or combination ladder. Use fork of tool to get a purchase at the seams of the tin ceiling, then pull it down in sections to expose the ceiling joists above.

    E. Using a pike pole or Halligan hook, punch a good-sized hole in the tin ceiling over the hot spot. Drive a stream into the hole to flood the ceiling space.

ANSWER SECTION

Scenario 9–1
Short Answer Questions

1. This appears to be a cockloft fire. Lines must immediately be stretched to existing and potential fire areas. The first line, a 1³/₄" or 2" line, must be stretched to the top floor leeward side apartment of the fire building. This line must work in conjunction with the ladder company who must pull the ceilings in the apartment, hitting and extinguishing fire in the cockloft as it is exposed. The line is placed in the leeward apartment as it represents the probable direction of fire travel. If the line were placed in the windward apartment first, the fire may outrun the attack and spread to the adjoining exposure and chase the roof team from the roof. Be proactive and get the line ahead of the fire.

2. The second line must also be stretched to the top floor and should also be a 1³/₄" or 2" line unless the initial attack team reports heavier than expected fire conditions that are beyond the capability of the initial attack line. This line can be used to back-up or work in coordination with the initial attack line. If need be, this line can work in the windward apartment, which must also be opened at the ceiling level in an effort to expose and extinguish the fire. It should not be necessary to stretch a 2¹/₂" line at this stage, as it will be difficult to maneuver. Cockloft fires require a line that is mobile. Cockloft fires that are properly fought from below will require a lot of movement and flexibility with hoseline operations.

3. Ladder company operations will be very labor-intensive at this operation. Obviously, a primary search must be extended in all areas of the building, including the cellar. Remember that the primary search is not only intended to locate fire victims, but also to locate fire in order to better direct attack teams. A fire in the cockloft may have originated in another area and extended to the cockloft by way of building voids. The primary search and recon will reveal this.

   For this cockloft fire, ladder companies must get to both the roof and the top floor. Communication between the interior team and the roof team is critical. The roof team must immediately open any natural openings on the roof to alleviate any built-up and unignited hot gases. Many times, the scuttle in these buildings will be boxed to seal off the cockloft. At a cockloft fire, if the scuttle is opened and no smoke comes out, then the box will need to be broken out. It is usually nothing more than wood of small dimension, usually 1" x 4". Be extremely careful what you stick into that opening to break out the box. A long-handled tool such as a Halligan hook is best. Fire erupting out of this area can burn the firefighter who is too close when the box is opened. The same boxed-out condition may be present beneath a skylight. In addition, a draft stop may also be present. This is a piece of glass at ceiling level. If smoke is present inside the structure and does not vent when the skylight is opened, probe with a long-handled tool to break the glass out of the draft stop. (Fig. 9–1)

Fig. 9–1 A draft stop may be present at the lower end of the skylight opening. If you break out the skylight and the smoke seems to be "stalled" below, suspect the presence of a draft stop.

Top floor crews must pull ceilings after confirmation from the roof team that the natural openings on the roof have been opened. The interior team must also direct the roof team to the most advantageous place to cut the roof given both the fire conditions in the cockloft and the prevailing wind. Careful consideration must be given to the exposed wood frame walls of Exposure B, the leeward exposure. Fire venting from the roof hole and other natural openings must not unnecessarily jeopardize this building. It may, if the fire is in an area where the cut will endanger the exposure, be necessary to order a handline to the roof to keep the exposed wall wet.

Horizontal ventilation of windows both on the top floor front and rear must be accomplished as soon as possible. Once the ceilings are pulled to expose fire and streams are applied, the visibility will be seriously obscured. The same is true in the exposure. It is always safer and easier to operate when you can see what you are doing

It will also be the responsibility of the ladder company to ensure additional means of egress are available to men operating on the roof and on the top floor. The wires in front of the building will eliminate the use of the aerial, so other means of access and egress must be used. As there are no fire escapes, ladders must be used to access the roof. A ladder may need to be placed on the rear one-story extension roof to get to the main roof. If rear access is difficult, it may be possible to take a small roof or combination ladder through the fire building and out the 2nd-floor window to the rear extension roof where it can be placed to the main roof. At no time should the scuttle on the fire building be used to access the roof. It may also be possible to get to the fire building roof from the roof of Exposure D. If the Exposure D roof can be accessed, a small ladder may be placed from the roof of the exposure to the roof of the fire building. (Fig. 9–2)

Fig. 9–2 Photo by Bob Scollan NJMFPA
Due to wires, it is easier here to access the second floor and roof by raising a ladder to the one-story extension roof at the rear of the building.

Egress from the top floor can be secured by placing a ground ladder to both the front windows and the 2nd floor rear windows on the extension roof. Use extreme caution when raising the ladder at the front of the building due to the wires. It will probably be necessary to raise the ladder parallel to the building using a beam raise or flat raise and pivot.

4. Additional lines must be placed to cover the exposure on the leeward side. The fire raging through the cockloft will take the path of least resistance, which could be through the combustible wood frame wall and into the exposure. A line should be stretched to the 3rd floor. This is because the 3rd floor is adjacent to the cockloft of the fire building. (Fig. 9–3) It also may be necessary to stretch an additional line to the top floor in case the fire runs the wall studs. Ladder support should be provided at these positions, as walls will be required to be opened and examined as conditions demand. The damage

Fig. 9–3 Exposures that are taller than the fire building present a major extension threat. This problem will be compounded if both buildings are wood frame. Be prepared to provide both interior and exterior protection.

created by the recon effort should be tantamount to the amount of fire present. Examine the cockloft from the roof via natural openings. Pull ceilings and open walls as necessary. A chief officer should be assigned as a division supervisor to safely and efficiently organize and manage exposure operations. The key to this operation is to get sufficient manpower and equipment to these areas before the threat materializes. Be proactive and don't wait for fire to show from the exposure windows before you decide to address the problem.

5. Defensive operations should concentrate on keeping the fire from spreading beyond the building of origin. Collapse zones should be established and maintained. If they are not already operating there, lines should be stretched to the 3rd and 4th floor of Exposure B. Aggressive control efforts must be extended from the interior of the exposure or there will most likely be no exposure to defend later on. Exposure D should also be entered and checked, especially the top floor and the cockloft. It may be possible to use a line from the roof of Exposure D to keep the exposed wall of Exposure B wet, but it will be better to use a master stream device for this operation. Radiant heat and/or convection in the form of direct flame contact from the roof of the fire building must not be allowed to accumulate on the combustible walls of either exposure.

As for the fire building itself, lines should be directed from outside collapse zones through the windows to hit the fire burning in the cockloft. Using a master stream to pour water on the roof and through any vent holes will be ineffective. Let the fire burn out of the vent hole and use the master stream to protect the combustible walls of the exposures. The master stream may also be useful if the cornice can be opened and a stream applied into it. This will be the most direct method of hitting the fire in the cockloft. An aerial platform or even a Telesquirt will be most effective here. (Fig. 9–4) The wires will present a problem here and will most likely prevent access to this area. Have the power company shut down these power lines as soon as possible so the operation can be safely executed.

Fig. 9–4 Photo by John Hund
For untenable areas where fire has not broken through the roof, use streams from the exterior through windows to hit fire burning under the roof. A Telesquirt will flow 1000 GPM into the fire area. Be aware of the disposition of the stream runoff.

If the fire does burn through a major portion of the roof, then the master streams can alternate between the main body of fire and the exposed walls. It will be extremely difficult to protect the walls of the exposures, as they are taller than the fire area and will be exposed to a tremendous amount of both radiant and convected heat. Water applied to the walls in conjunction to aggressive interior operations in the exposure will be the only defense here. If the positions inside the exposure become untenable, the building may also be doomed for once the interior crews are chased out, there will be nothing to stop the fire from spreading into this building via the now-open walls and ceilings. This is why aggression and adequate manpower in this position at the beginning of the operation will pay off later.

6. It will not be unusual, given the location of the fire, for the top floor to be tenable to the point where SCBA may not seem to be required. Don't get lulled into this trap. The conditions will change very rapidly once the ceilings are pulled and streams applied from below. The resulting steam and smoke created will obscure vision as soon as the stream hits the fire. Horizontal ventilation of windows will be required if it hasn't been done already. Resist the temptation to introduce PPV into this area, as the strong air currents produced will likely push fire into uninvolved areas.

It is important to be aware that even though the conditions on the top floor may seem friendly enough, there could be an extreme amount of heat and potentially fire just above the ceiling. All it may take is for a relatively small amount of oxygen to be inadvertently admitted into the ceiling from below. The resulting expansion of the gases may envelop the area below the ceiling, killing those who are not able to escape. Firefighters have been killed due to ceiling void backdrafts, mainly because they did not coordinate operations with the roof team before opening the ceiling. It is imperative that firefighters know what is going on in all areas of the building that they are entering. A thorough size-up of conditions, especially the location and extent of the fire, as well as the likely direction of travel are important indicators of potential life-threatening situations.

Scenario 9–2
Multiple-Choice Questions

1.

A. **+1** A hoseline of appropriate diameter is properly stretched to the fire area, but as this is top floor fire, there is no need to hold the line at the stairs. Any occupants escaping will do so from the level on which the line is positioned. With the wind and current venting direction of the fire, the conditions in the hallway should not be too bad. Attack the fire from the upwind and unburned side and all the problems should go away.

B. **-2** These would be proper tactics if the hallway were untenable. Once the fire is darkened down, the interior line can move in and extinguish any remaining fire. These are poor tactics at the current stage of this fire. All of the factors that allow for a safe offensive attack are present. The operation is in the rescue mode and the fire is venting on the leeward side of the building. Any defensive action such as an exterior stream operating into Side A will likely drive the fire throughout the building, endangering both firefighters and occupants. This is strictly an offensive operation.

C. **+2** Attacking the fire as soon as the line is stretched to the top floor will offer the best chances of success for the operation. Attacking the fire in this manner will also place the line between the fire and any victims via the safest, most effective path of least resistance. We still must be thinking *victims*, no matter what the occupancy or hour. A well-placed attack line, properly coordinated with primary search and other support operations will blast the fire right out of the building.

D. **-1** Two objectives are being pursued. A line on the interior to protect egress and a line to attack the fire from the fire escape. Not only is the positioning of the fire escape line in violation of the safest, most effective path of least resistance, but splitting the crew to stretch two lines will likely result in neither objective being successful. All attack personnel of the first-arriving engine company should be utilized to place the first line into operation to attack the fire as soon as possible. In fact, it may take the personnel from the first two arriving engine companies to get the line in place. Splitting the crew and attacking via unsafe access paths is impractical and unsafe. The first line can accomplish both stairwell protection and fire attack.

E. **-2** Even chief officers should have a sense of humor.

2.

A. **+1** The second line can be utilized to protect the interior stairs as well as assist in reinforcement of the fire attack as long as the first line is already in place. This fire is routine and should be handled by the first line. The second line may only have to stand by as the extinguishment is accomplished. Too many lines in the apartment may cause congestion and injury. Having the second line in place is proper and a good insurance policy.

B. *-2* The first line is already in the apartment, having stretched via the interior stairs. This line, operating from the exterior, will cause an opposing stream condition, always a mistake and cause for injury. It is imperative that incident commanders have control over the positioning and operation of all lines. Discipline and strict adherence to orders on the part of the company commander are a large part of this control.

C. *+2* Although the 2½" line is probably not required, it is acceptable as a second line. These are excellent tactics as they show a willingness to communicate and coordinate with the initial attack team. Using the line as a protection against fire extension is also appropriate and shows an understanding of fire travel and extension prevention priorities.

D. *-1* It is critical that a second line be stretched to the fire area as soon as possible. If the first line encounters a problem, the second line will be the only thing standing between the fire and the rest of the building. In addition, due to the proximity of the fire to the street and the unencumbered access, the stretch, positioning, and attack will most likely be handled by the first engine company. Vent, enter, and search operations should be the responsibility of the ladder company. The point is given for secondary water supply.

E. *-2* Here, again, no second line is stretched in addition to the lack of establishment of secondary water. The tactics described here are Ladder operations and should not be routinely conducted by engine company personnel, especially the second-arriving engine company. If, due to conditions and a lack of ladder company personnel, later-arriving engine companies are given these assignments, that is acceptable, but the job of the first two-arriving engine companies must be directed toward water supply, fire attack, and attack reinforcement operations.

3. The incident commander's understanding of the strategic concept behind the action chosen is the backbone of all operations. It will have a direct impact on the proper coordination of those actions. Incidentally, all of the answer choices here are complimentary with attack actions chosen for the first two questions. If you chose improper tactics, you could probably justify them (in your head) by choosing the answer choice appropriate for your actions. Thus may begin the downward spiral into test abyss.

A. *-2* The potential life hazard present prohibits the use of exterior streams, at least in the initial stages of the fire. This action would be a classic case of what has been called "candle-moth" syndrome. The line hits the fire where it shows. This philosophy is acceptable on interior firefighting, but must be avoided when on the exterior and the fire situation is not defensive in nature.

B. *-1* Pincer attacks are useful in large area buildings where it is safe to attack using two simultaneous flanking movements in an attempt to "pinch" the fire off, thereby preventing spread. Pincer attacks may also be utilized in contiguous structures where the original fire area has been surrendered. It is not a safe alternative in a small area such as this occupancy. The pincer action here would most likely lead to opposing streams.

C. *0* This concept follows the philosophy of keeping the line at the stairs until occupants have escaped. While this may be a suitable strategy for a fire on the lower floor of an occupied multiple dwelling, it is not necessarily advantageous in this situation. The best strategy here is to attack the fire before it gets a chance to spread into uninvolved areas such as the cockloft.

D. **+2** This concept will place the line on the windward, unburned side, and push the fire in the direction it is already venting. The presence and direction of the wind and the location of the fire make this a virtual textbook attack. Any other set of actions will complicate the issue and probably have an adverse effect on the fire. (Fig. 9–5)

E. **0** This is unacceptable in all but large, rapidly spreading fires where a defensive/offensive strategy is being pursued. In this situation, to wait for the fire to reach a specific area is surrendering and causing unnecessary damage. The key to this incident is to hit the main body of fire with an aggressive attack from the unburned side and all the other problems should diminish.

Fig. 9–5 Photo by Bob Scollan NJMFPA
As fire will always take the path of least resistance, venting fire should be allowed to do so as lines are stretched to attack from the unburned side. Use the stream to push the fire right out the window.

4.

A. **0** While the primary search and attendant support activities of the fire and adjacent areas are essential, the importance of vertical ventilation over the stairwell cannot be overemphasized. Especially at a top floor fire, the opening of the bulkhead door and other natural openings such as soil and vent pipes as well as the subsequent examination of the cockloft for the presence of fire is critical. Failure to perform this critical operation can lead to the loss of control of the fire. (Fig. 9–6)

B. **0** This answer choice goes to the opposite extreme from answer choice "A." Here, no interior ladder operations are taking place at all. One of the most important positions for first-arriving ladder personnel is on the fire floor, providing primary search and other support activities which clear the way for extinguishment activities. If the answer choices for "A" and "B" were combined, the information would pretty much represent an effective ladder company operation at this fire.

Fig. 9–6 Opening the bulkhead door will exhaust much of the accumulated products of combustion in the stairwell, essentially burping the building. In the early stages of the fire, this single act will provide more relief to operating crews and escaping civilians than any other tactic.

C. **+2** This is the best set of actions to take. A crew is operating on the fire floor, conducting primary search and providing access and support for the attack. At the same time, a crew is on the roof, having accessed it via the safest path of least resistance, the bulkhead of the exposure, which is located on the opposite side of a firewall. Providing horizontal ventilation of the fire area from the roof will make operations on the fire floor safer. It is important to remember that this roof will most likely have to be cut due to the proximity of the fire to the cockloft. At the very least, examination holes should be made to ascertain the amount of extension, if any, to the cockloft. It must be stated that when in doubt at a top floor fire in these old build-ings, cut the roof to localize the fire before it gets away.

D. *-1* Pulling the ceilings to expose the cockloft prior to opening the roof, especially the nat-ural openings, may be asking for trouble. If the cockloft is packed with superheated gases waiting for an oxygen source, opening from below may provide it. In almost all cases on top floor fires, it is better to wait for topside ventilation before pulling ceilings.

E. *-2* What about primary search? The interior of the fire building, especially the fire occu-pancy is the most critical area of operation. It certainly doesn't take the entire ladder company, especially when they are the only ladder company initially responding, to complete roof operations.

5. The primary concern here is the hour of the day coupled with the amount of manpower on the scene.

A. *0* The hour of the day and the accompanying visibility problems in conjunction with the limited amount of manpower responding on the first alarm make the striking of a second alarm necessary. Most departments respond with this type of manpower. It is actually a luxury to pull up on the scene with twenty men and another few dozen or hundred at your beckoned call. Trying to tough out a fire with what you have on scene when you may have more men available unnecessarily endangers firefighters, civilians, and the operation. For these reasons and the potential problems present, requesting a second alarm is not only playing proactively, but also playing smart.

B. *-1* This is playing it too safe. A second alarm should suffice. It is one thing to play it smart, but another to abuse manpower. The scenario states that a firewall separates the two build-ings. For sure, the exposure should be thoroughly checked, but a third alarm is overkill in this situation. Strike a second alarm, place your personnel according to the requirements of the action plan, and evaluate their effectiveness before striking a third alarm.

C. *+2* The life hazard at this hour, not only to occupants, but also to firefighters in dark con-ditions, require extra manpower be summoned. A second ladder company is the most important reason to request the additional alarm here, as the location of the fire in rela-tion to the cockloft may make both roof operations and interior top floor tasks consider-ably manpower-intensive.

D. *0* This is indecision and is not acceptable on the fire ground. The fire is showing at two front windows. What more information is required on the location and extent of the fire at the outset than what is already given? Make decisions based on current and predicted con-ditions. Waiting for information sometimes proves too little, too late.

E. *0* Waiting until reports are issued on cockloft involvement will force the incident com-mander into playing catch-up. Don't wait until the reports are given. The reflex time may be long enough to change the operation from a winner to a loser when the required apparatus and manpower are still on their way at the critical juncture of the fire. You usually only get one chance at a fire. If you blow it at the beginning due to indecision, the game may be lost.

6. The key to this question is the words "in the proper order." Anything out of order will either affect the smoothness of the operation or cause firefighters to be injured or killed.

   A. **-2** The indirect attack will not be suitable in this situation. Injecting a fog stream into a super-heated atmosphere without sufficient ventilation opposite the nozzle will result in a rapid expansion of steam into the path of least resistance, in this case, downward onto the firefighters. These actions will likely result in steam-related injuries to the firefighters if they are not first killed by the expanding superheated gases potentially igniting into their area of operation.

   B. **0** The first and second items should be reversed. Get the windows open first, providing more horizontal ventilation and some relief for the team being beat up on the top floor; then cut your examination holes in the roof to locate the danger area. Finally, the roof is cut as directly over the hot spot as is safely possible. If the windows are not ventilated, the interior team will likely be forced to withdraw. Additional horizontal ventilation should provide some quick relief.

   C. **-2** The first two items are acceptable; however, the directing of a stream into the vent hole blew this answer choice out of the water. Under no circumstances should a stream from above be directed into a vent opening, especially when men are operating on the interior. Anyone who says or does otherwise should be banished to Fire Bogeyland forever.

   D. **+2** This is the proper sequence. Open the windows, cut the roof over the hot spot, push the ceiling down to vent the apartment as well as the cockloft. Just venting the cockloft may provide some relief, but not near as much as removing the ceiling, which is acting as a barrier to effective ventilation. This will be especially important if tin ceilings are present. Ensure proper communication and coordination between roof and interior teams prior to pushing down the ceiling. (Fig. 9–7)

Fig. 9–7 For top floor and cockloft fires, the roof must be cut as directly over the fire as is safely possible. After the cut is made, use the blunt end of a hook to push down the ceiling. If the top floor ceiling is not removed, this hole will provide no relief for the interior crews operating on the top floor

   E. **0** This answer choice is in the right sequence, but the ceiling is not pushed down. The question asks how to alleviate the heat condition preventing the advance of the attack line. The only way to do this is by total ventilation, meaning windows opposite the attack and removal of the ceiling, where the hottest gases may be trapped. If the ceiling is not pushed down, the relief on the top floor will not be realized by the interior team as quickly as if this barrier were removed.

7. This question can be best answered by addressing the location and extent of the fire along with what building features will be most affected by the heat produced.

   A. **-1** Floor collapse will not be the greatest concern, at least not initially. In ordinary constructed buildings, water usually finds its way to the most level location by seeping down to the lowest floor by way of pipe chases, utility openings, and other pokethroughs in the floor. Knowledge of this should be an aid in salvage planning. Pooling of water on the fire floor to the extent that the stability of the floor is in jeopardy is not usually a major consideration when the fire is in the offensive mode.

B. **-1** The main roof joists will be of substantial dimension in this type construction, usually 2" x 10" at the minimum and may be as large as 3' x 12". It will take a lot of fire to burn these joists to the point of failure. An exception would be where a previous fire may have already compromised structural stability. Ladder companies should know this beforehand or upon examination of the roof joists.

C. **0** The fire escape is not being exposed to fire at this point. It is blowing out of the windows on Side A, while the fire escape is located on Side B. Unless the fire escape was weakened prior to the fire, and there is no reason to believe this is true until proven otherwise, it should not be a great concern regarding collapse at this stage. Remember, for the sake of safety, fire-fighters should always test the fire escape before climbing and operating on it.

D. **+2** As the fire is issuing from the top floor windows on Side A, the parapet wall above those windows is being subjected to the most heat exposure. The parapet is a freestanding wall and therefore receives no support from any other portion of the building. Fire may cause uneven expansion of the bricks in the wall, causing it to distort and lean to a point where gravity may finish the job. In addition, there may be steel girders present which reinforce the roof construction. As the fire heats these girders, they may expand and also push out the parapet wall. It must also be assumed that there is lateral reinforcement of the parapet wall and any isolated failure of the wall could pull the entire wall down with it. Any fire that is exposing a parapet wall will require that the wall be monitored for indications of collapse. (Fig. 9–8)

Fig. 9–8 Photo by Bob Scollan NJMFPA
Parapets, such as this old decorative stone wall above a taxpayer, are subject to failure from fire destruction of its supporting elements and/or by firefighting operations. When in danger of collapse, the area below should be cleared for the entire horizontal width of the wall.

E. **-1** The collapse of the stairs should not be an issue unless fire had extended to them and destroyed their load-carrying capacity. This will not be a threat in the early stages of this fire when the fire is contained to the occupancy of origin and venting on its own. If it spreads out of the occupancy and up the bulkhead stairs, then this may become a real concern and the stairs will need to be avoided.

8.

A. **+2** Fire will take the most effective path of least resistance to the floor above. Know where these areas are dependent on the building construction and specific layout of the occupancy. This most direct path for upward fire travel will be via the pipe chases both in the sinks, which will likely be located in the main area and in the bathroom.

B. **+1** This possibility cannot be overlooked, especially if the fire is blowing out of the front show windows. If this was the case, especially on arrival, position a line or operate a deck gun at the front of the building and use it to wash the wall above the barber shop and darken the main body of fire in the barber shop. The water should not enter the 2nd floor from the exterior at any time. The former action will dissipate some of the radiant heat while the latter quenches the parent body of fire. Waste no time in stretching lines to the floor above the fire to cut off and extinguish any extension.

C. *+1* Fire may also extend upward via light fixtures in the ceiling, but probably not as quickly as the pipe chases. Once in the ceiling, it may have to seek out an opening in the floor before showing on the 2nd floor. This will not be the case with the pipe chases, where the fire will have a direct path to the floor above.

D. *-1* Windows will act as firestops in the stud chases and will not be a likely path of fire travel to the upper floors.

E. *-1* The partition wall between the barbershop and the interior stair should keep the fire from spreading up to the 2nd floor via the stairwell, at least initially. The pokethroughs and other manmade openings in the store are much more likely paths of fire travel.

9. Question #4 addressed the assignment for the initial-arriving ladder company. Your orders for this ladder company should reinforce those actions taken by the first ladder company and cover areas yet to be addressed.

A. *-1* Recon of Exposure D is a sound assignment and should be one of the duties of the second-arriving ladder company. However, the first ladder company should have conducted the primary search of the fire building. This action will be redundant and a waste of manpower.

B. *+1* Exposure D operations are necessary at this fire, even though the scenario states that the buildings are separated by a firewall. The presence of a firewall is never a guarantee. Recon and evacuation must still be accomplished on behalf of civilian and operational safety.

C. *+2* These tactics are also sound and are a more efficient use of manpower than the other answer choices. One crew covers Exposure D. The fire escapes, which may be problematic if occupants are attempting to use them for escape, should be attended to by another crew.

D. *+1* This is another efficient use of manpower from the second ladder. Utility shutdown should be a concern, but the men can be more useful operating in recon and evacuation of Exposure D. Reinforcing the roof operation may be necessary due to the location and extent of the fire. Assigning manpower to this task will increase the chances of success in this operation.

E. *-1* Stretching a line is not the job of a ladder company. Crew 1 can be better utilized elsewhere. Crew 2 is used properly to recon for any fire extension in the exposure.

10. All firefighters should be trained in self-rescue and fireground survival. When lost or trapped, chances of survival are often dependent on the firefighter's ability to not only summon help, but also to help himself. The first of these abilities is the ability to remain calm and think. Once panic sets in, air supplies are depleted more rapidly and rational thinking ceases. All firefighters should be equipped with a portable radio. It is a verbal lifeline to the outside world. In addition, a working and armed PASS unit is mandatory.

In North Hudson, an emergency transmission can be issued by anyone on the fireground in distress. The firefighter turns his PASS alarm to "on" and holds the microphone of the portable radio to it, emitting the activated alarm over the radio. This stops all other transmissions. The firefighter in distress then sends out a Mayday call for help, giving out as much information as possible regarding his situation and present location. This emergency call must be practiced to be effective. The problem is that many firefighters do not turn on their PASS alarms. This is like not wearing one at all. So, the fire service comes up with alarms that are armed as soon as the SCBA is pulled out of the mounting bracket. Firefighters get around that by wrapping the activating pin and retaining cord around the waist straps. The best solution to this problem would be a PASS alarm that activates when the donning switch is placed in the facepiece of the SCBA mask, starting the flow of air to the wearer. Most firefighters are smart enough to wear their mask when working in a hazardous atmosphere and most departments have mandatory mask rules. If a firefighter refuses to wear his mask in a hazardous atmosphere, he is a liability to himself and his brothers to begin with. We are sometimes our own worst enemy on the fireground.

This brings us to the proper wearing of gear. Improperly worn, it could lead us to unnecessary injury and death. How many times are firefighters seen wearing SCBA without their waist straps buckled? I was working at the Fire Department Instructor's Conference in Indianapolis. We had put together a mask confidence course where the student had to negotiate a diabolical maze of hanging wires, cut-out floors, missing stairs, diminished clearances, and maniacal instructors pulling ceilings on them. The course was negotiated while the student had his protective hood turned backward over his mask, negating visibility. I am pleased to say that most students who properly wore their personal protective equipment did a good job negotiating this tough course. However, every student who had his waist strap unbuckled got hung up and trapped. In addition, every student who had his chin strap wrapped around the back of his helmet (I hate this; it's one of my pet peeves with firefighters), lost his helmet through a hole in the floor. (Fig. 9–9)

Fig. 9–9 Photo by Bob Scollan NJMFPA
The firefighter entering the basement does not have his waist strap fastened. A basement is one of the most entanglement-prone areas in a building. On the exterior, one firefighter is not geared at all. The other two outside members (possibly the RIC Team) also have waist straps unfastened and are not wearing hoods. The chin straps are also wrapped around the back of the helmet. This is unsafe and irresponsible.

Another favorite claim of students is that they don't wear their protective hoods because they like to use their ears to tell them how hot it is or if they are into the building too far. This is nonsense of the sheerest nature. A trip into the flashover container would certainly change their thinking. Heat and plenty of it can be felt through the hood, mask, gloves, turnout gear, and boots. Firefighters must realize that it is the signs of flashover and other dangerous phenomena that they should be looking and feeling for, not waiting until their ears melt off their heads. I wonder if these same firefighters test to see if coffee is hot by sticking their tongue in it or checking the heat of an iron by holding it to their face. Body parts should never be used as heat-sensing elements.

We have to practice proper protective equipment habits and know that when in a jam, sometimes the only protection we have is properly donned turnout gear.

A. **0** While keeping one's cool is a virtue, the firefighter who is trapped or lost must attempt to do something on his own behalf to rescue himself, especially if his air is running low. Using the radio and the PASS unit is a start. Other ways are trying to retrace your steps, looking for a wall and a window, shouting out, and looking for a hoseline.

B. **+1** While refraining from panic, it is a good idea to attempt to attract attention to your situation. Finding a wall is the first move as windows and doors are not usually found in the middle of a room. Look for light or apparatus lights showing through windows. Even at night in relatively dense smoke, it is sometimes possible to make out a window from a wall once you are close. Break it out and make yourself known. A hand or flashlight can be used to attract someone's attention (See chapter 12 of the text for more).

C. **+2** This is the best way of rescuing yourself. The male coupling leads out of the building, which is where you want to end up. Some departments paint, in fluorescent or dayglow paint, red next to the female couplings and green next to the male couplings. This eliminates remembering which coupling leads out of the building. The only problem here is if you find yourself in a pile of spaghetti. It may be confusing, but at least you have a better chance of both getting out and/or finding a fellow firefighter than if you were operating blindly. (Fig. 9–10)

Fig. 9–10 Follow the male end of the joined couplings to the exit. Following the female will lead you deeper into the building. A hose wrap indicating which way to go or painting the couplings will minimize the guesswork.

D. **0** Following the female coupling will take you further into the building, possibly all the way to the nozzle team, which could be a hundred or so feet away. Remember the situation states that you are running out of air, so you had better follow the hose in the right direction. If the hose is painted to give the direction out, your chances of mistakenly going the wrong way are minimized.

E. **-2** If the smoke condition is so bad that you can't find your way out, going up will only get you in further trouble. If anything, you must attempt to go down, toward the exit point that you came in. The scuttle area will probably be the hottest area of the building, as the rising superheated gases will also be using that exit. In this case, it is likely that you may never get through the scuttle. If you do, you should consider yourself the luckiest firefighter on the face of the earth.

Passing Score for Scenario 9–2 = 14 Points

SCENARIO 9–3
MULTIPLE-CHOICE QUESTIONS

The guide to proper tactics often is dependent on proper and timely reports. Knowledge about fire conditions on all sides of the building is imperative. Ladder personnel should be covering areas of the building that will provide this information. This information will help fill in the gaps regarding conditions not visible from the command post at the front of the building. Remember that sometimes the obvious will often be changed by reports from areas unseen. Likewise, on a test scenario, it is critical that you read all the questions before answering. Information not in the scenario is often found in questions.

1.

A. *+1* The duckbill lock breaker and a striking tool such as a maul or flat head axe is an efficient way to remove a padlock. However, there are, according to the scenario, at least half a dozen padlocks. Using this method would take an extensive amount of time. It is not the quickest way to get past the gates.

B. *+1* This is also an effective method; however, with the amount of locks present, this too would take an extended amount of time.

C. *-1* The inverted "V" or teepee cut is an effective way to cut a door in the gate. (Fig. 9–11) This would be used if quick access is required to knock down heavy fire, but it is only effective on gates covering entranceways. These gates cover the windows. According to the scenario, there is no gate over the front door. Cutting the V will not provide sufficient area for ventilation of the show windows or access into the building. It is better to force the gates and get them out of the way completely. Critically reading the information, just like listening to a size-up report, is one of the keys to success

Fig. 9–11 The inverted "V" cut is a quick way to get water on a hot, fast-spreading fire, but is not the ideal firefighter-created opening for access into a structure. When forced to cut a metal gate, remove as much of the gate as possible.

D. *+2* (Fig. 9–12) Cutting the locks using the aluminum oxide blade or its equivalent is the quickest way to get the gates up. Blades used for roof operations such as a carbide tip blade should not be used for this operation under any circumstances. While making the cut, it will be necessary to secure the locks so they don't move. This can be done with a pair of vise grips and a chain or by placing the fork of the

Fig. 9–12 There are seven padlocks on these gates. Operated properly, the power saw with an aluminum oxide blade could cut these locks in only a few minutes. Be cognizant of the signs of a potential backdraft in this closed up store, especially at night.

Halligan tool down over the bow. Use extreme caution in this operation, especially when the lock is above chest height. The best tool in the case of high-mounted locks is a torch; however, most departments do not carry them, so the saw will have to do. Also, be careful of the sparks and any projectiles cut loose by the saw. Full protective clothing must be worn, including eyeshields. Lines should be positioned, charged, and ready for action.

E. **-1** The reason that this is a **-1** is that it is not a backdraft situation. This is evidenced by the report from the rear of fire issuing from two windows. A fire venting from a window is an indication that backdraft conditions are not present. Remember that the fire load in this and other occupancies such as florists will produce heavy black smoke as many plastics and other petroleum by-products are present. Don't be fooled by the presence of black smoke, stained windows, and roll-down gates. Check for excessive heat and ensure a thorough recon of the perimeter is completed before backdraft tactics are executed. Backdrafts are rare, but they do occur. If there is any indication of venting fire, a backdraft is not likely.

2. The information given with this question indicates that this is not a backdraft condition. This information is here for a reason. Incident commanders must attempt to have as complete a picture of conditions as possible before a strategy is chosen and an action plan developed. Operating without complete information, while usually necessary, may lead to a strategic error. Reports from personnel providing coverage of the building are a major part of this picture and should round out preliminary information.

A. **-2** You will notice that either hydrant is acceptable here as is the apparatus positioning depending on from which direction the apparatus may have approached. The tactics accompanying the hydrant selection will make or break the choice. These are backdraft tactics. Backdraft conditions are not present here. The fire should be attacked in coordination with ground and roof support activities; however, making the attack through the show windows leaves much to be desired, especially when there is a door present.

B. **-1** The basic difference between choice "A" and "B" is the hydrant chosen. In addition, once the line is advanced, it is correctly stretched via the front door. Nonetheless, the same mistake of employing backdraft tactics is chosen. The action plan here is incorrect due to the inability to properly react to reports from perimeter positions.

C. **0** This answer choice correctly identifies this as an offensive fire, however, the placement of the initial attack line is incorrect. While the life hazard is on the 2nd floor, if the parent fire is not quenched, the attack team and the occupants may be trapped above the fire. This is at least a three-line fire. With the first line, hit the main body of fire and then provide back-up with the second line. The third line should provide coverage on the floor above.

D. **+2** The best plan here is to attack the fire at its seat while protecting the primary search area. You may notice that the first line is 1¾" instead of a 2½". This decision may depend on manpower and on conditions. It is usually better to stretch the larger line into a store fire. It is essential if the fire is blowing out of the store windows. If manpower is at a minimum, it may be better to quickly get a smaller diameter line into the building to attack the fire at its source than to struggle with a 2½" line being stretched by insufficient personnel. At the very worst, the initial, smaller line can operate in a holding action while the search is underway until sufficient personnel arrive on the scene to stretch a larger line. If the line must be backed out due to the heavy volume of fire present, then so be it. If the search was completed or a rescue was made due to fact that the line was in place quickly, it is justified.

With a three man engine company, (one is usually the pump operator), having two men to stretch a 2½" may not get the job done at all. The 1¾" may provide quick water to the area where it is most needed in the least amount of time. If manpower is sufficient, by all means, go with the bigger line.

E. *-2* As stated before, this is not a backdraft situation and it is not necessary to wait for roof ventilation to take place. In addition, the use of a deck gun in conjunction with an interior line placement on the floor above is both dangerous and unnecessary. The master stream may push the fire into uninvolved areas as well as create a steam condition that may bake unprotected occupants on the floor above the fire. There are almost no instances where using a master stream into the same area as interior operations are being conducted can be justified. This is especially true if the interior personnel are above the application point of the stream.

3.

A. *-2* These are complementary to the first engine company's incorrect backdraft tactics; however, a further error in operations is the failure to establish a second water supply.

B. *0* The tactic of stretching a line into the exposure at this stage of the fire is incorrect. The line should be used for back-up or at least to cover the next priority, the floor above the fire.

C. *-1* Again, incorrect backdraft tactics, but at least a secondary water supply has been established and the floor above is considered. That saves you from a *-2*.

D. *+2* The back-up of the initial attack line with a larger diameter line along with the establishment of a secondary water supply are proper reinforcement tactics at this fire.

E. *+1* It's best to reinforce the initial attack line in case something goes wrong. The next priority will be the floor above the fire. A line should be stretched there as soon as possible to cover both the primary search and fire extension checks above the main body of fire.

4.

A. *-2* There is a lot of space between the fire area and the library. Waiting for the library to be directly exposed means that the fire would have to spread to two adjacent buildings first. Delaying an additional alarm this long is the sign of an impotent incident commander.

B. *+1* A suspected life hazard is always a good reason for requesting an additional alarm. Manpower will be required to search and evacuate both the exposed and fire building. Don't get caught short if the fire unexpectedly extends. Send people in to get people out early.

C. *+2* The sound judgment of a competent fire officer should never be second-guessed. When a building that is contiguous for a whole block is threatened, it is best to be proactive and summon additional manpower. Possibilities sometimes turn into probabilities and then into realities. The proactive incident commander stays one step ahead and requests enough manpower to address the potential problems.

D. *0* This is wishful thinking. If the initial attack lines are not successful, there may be no one to put alternate action plans in place. Keep in mind that if you are covering all of the vital areas of the building, there will likely be no tactical reserve. This fact alone should be impetus to order additional alarms.

E. *-1* If you are still in the backdraft mode by now, you are about 4 points in the hole.

5. Ladder operations should always cover as much area as possible, providing primary search in addition to critical support functions that will clear the building and make the advance to the seat of the fire possible. I've never seen a ladder put out a fire, but I'd have to say that 90% of the fire-ground operations depends on the success of the ladder company operations.

A. *+2* These are all the right actions covering as much of the building as possible. The crew is correctly split and operates both on the interior and on the roof. On the interior, cover as much area as possible in your primary search efforts. On the exterior, if possible, always ladder the upwind exposure so a safe egress point can be maintained. Open natural openings and recon cocklofts, shafts, and adjacent roof openings for fire involvement. This is a tall order, but a disciplined and well-trained ladder company operating under solid SOPs should accomplish these tasks in a timely manner. (Fig. 9–13)

Fig. 9–13 Photo by Ron Jeffers NJMFPA
When the building must be laddered to access the roof, it is best to raise the ladder to the windward exposure. If possible, ladder two buildings away from the fire to the windward side to provide a margin of safety.

B. *-1* These are tactics for the backdraft scenario, but following that line of thinking has landed you further in the hole and probably hasn't exactly come up roses as a fire strategy here either.

C. *0* Again, as the backdraft condition is not present, points are not awarded, however, the crew is split and covers both the roof and the fire area, providing proper and coordinated support. If this was a backdraft scenario, this would likely be a *+2* answer.

D. *-1* No vertical ventilation is provided in this choice of actions. In these type buildings with open stairwells and numerous construction voids, an effort must be made to get to the roof and open the natural openings, providing relief and visibility to the crews below.

E. *-2* Forcing all the stores is a good tactic as it is likely that they will all have to be entered for recon, but using the stream to convert what the reader has interpreted as a backdraft condition into a free burn is extremely risky. First, it writes off anyone who might be living in the back of the store, which is common to these type stores. Second, it may push fire into uninvolved areas, turning what may be a routine fire into a fiasco. Often, one misplaced stream can do more damage than the good that several properly placed streams can accomplish.

6. The fact that this question is asked is a clue that a second alarm is necessary as these two companies are first-due on the second alarm. If, at this point in the book, you are not reading all the questions before answering, you should be ashamed.

A. *+1* To use the entire complement of manpower of these two companies in the primary search of the downwind exposure is not necessarily the best use of manpower. It is probably only necessary to commit one company to this task, freeing the other company for other duties. The point is given for being proactive.

B. *+2* This choice provides reinforcement of the operations at the fire building as well as in the leeward exposure. Splitting the ladder company will allow for the accomplishment of more tasks than if they were to remain intact as one group. The assignment of the rescue company will provide much needed reinforcement in the fire building.

C. **+1** The ladder company assignment is identical to that of answer choice "B." The using of the rescue as the FAST or RIC team accomplishes nothing tactically. By the time these second alarm companies are in place, a FAST team should have already been dispatched and should already be in place. The safest action to take on the part of the FAST team response is to dispatch the team as soon as the incident commander confirms a working fire. Second alarm companies should be committed to fire suppression activities.

D. **0** At this stage of operations, keeping a ladder company and rescue company as a tactical reserve is foolish. There is much to be done at this operation and these companies will be better utilized in the game than on the bench.

E. **-1** Sending the ladder company to the leeward exposure is an acceptable and sound assignment. However, using the rescue company to check conditions in the library is a waste of manpower. The library is not even remotely threatened at this time.

7. Recall that areas that the incident commander cannot see from the command post will usually cause him the most problems. Shafts are one of the largest culprits in regard to this because not only are they not visible from the front, they are also not visible from the rear and sides. This leaves the interior of the building, which is often not the best place to visualize this feature during a fire, the interior of the exposure, and the roof. This information will only be made available to the command post if the roof and exposure reports are complete and timely enough to address the shaft before it becomes a major problem.(Fig. 9–14)

Fig. 9–14 Note the presence of enclosed shafts between Side D of the taller multiple dwelling and Side B of the shorter mixed-use occupancy. Note also the small diamond-shaped shaft with no enclosing parapet between the buildings at the extreme right. Roof crews should report the presence of and continuously monitor conditions in these shafts.

A. **-1** First, the distributor stream in the shaft will interfere with any venting action taking place in the shaft. This may drive the fire back into the apartment or worse yet, into the window adjacent or directly above. Second, radiant heat from a fire issuing from one window will pass right through the water curtain and likely ignite combustibles in or near the exposure windows.

B. **+1** Removing fuel by taking down curtains and draperies as well as cutting down on the transmission of heat by closing windows on the shaft will reduce the likelihood of ignition in the exposure. The best insurance policy against extension via the shaft, however, will be the positioning of a protective hoseline.

C. **+2** The placement of a hoseline into the exposures will be the best method of safeguarding the exposed area from extension via the shaft. It must be assumed that the fire in the store will at least expose the shaft with heat. If the fire is of any major significance, flames. Recon reports from the roof and the exposures must be issued as soon a possible, and the shaft must be monitored from these vantage points. It is better to stretch the line to exposed areas early than be sorry later when the fire is already in the exposure by way of the shaft and no line is in place to stop it.

D. **-1** Operating a line from the exposure into the fire building will result in opposing streams. Opposing fire streams set the stage for injury and loss of control. The incident commander must guard against unauthorized stream placement. The best way to accomplish this is to have operational control over the incident and disciplined officers who carry out orders effectively and know the basics of firefighting. Opposing streams inside a building is always a mistake.

E. **0** The benefit of fire walls is negated by the presence of light shafts between buildings with windows in them. When heavy fire is present, the smaller the shaft, the greater the potential for fire spread to adjacent exposures via windows in the shaft. If there is a shaft present, be prepared for fire spread across it and position forces to head it off.

8. This is a stand-alone question regarding your ability to solve a problem. Be careful not to confuse it with your assignments from any other questions.

A. **-2** The incident commander is responsible for every person on the fireground. The potential for lateral fire spread, while small due to the presence of fire walls, is a threat nevertheless. If the fire suddenly escalates and it turns out this woman is several hundred pounds, the urgency of the rescue is compounded. Add to this the hazard of the oxygen as a fire accelerator, and the problem escalates exponentially. Don't take a chance unless the fire is definitely under control and the chance of the involvement of that exposure is zero. Removing her removes the potential problem.

B. **0** While removal of the woman is appropriate, leaving the oxygen cylinders in the apartment is not too bright. If the fire happens to spread to them, not only will there be an oxygen-fed fire to contend with, but also a BLEVE threat. They should also be removed.

C. **-1** This is unnecessarily jeopardizing the lives of emergency responders who are not wearing proper protective equipment, have proper training, and are unaware of the hazards of oxygen. The incident commander is responsible for these individuals as well. It is best to send fire department personnel to make the removal.

D. **0** This is an effort to both protect her in place if possible and monitor the fire spread profile at the same time. However, it places a company out of the action for an indefinite period. It should not be necessary to baby-sit an occupant. Remove her to the care of the proper authority and reassign the company.

E. **+2** Removing both the woman and the oxygen cylinders is the safest and best method of dealing with this problem. This action alleviates any and all potential problems regarding both.

9. This is an operational problem and will test your ability to adapt your action plan to the situation.

A. **-1** Evacuating the building and operating in a strictly defensive manner is giving up. In this case, that is unacceptable. The incident commander must be able to put Plan B and even Plan C into place when Plan A is not having the desired effect on the fire. If your only alternative plan is Plan "D" as in "D-efensive," you will be responsible for a lot of parking lots.

B. **0** While attacking from another direction is a good plan in this case, using the same crew and line is counterproductive and will take time. It will probably also be unsuccessful. The better plan is to get a line to the rear while the interior company attempts to keep the fire from extending toward the front of the store. Then, when the line is ready, pull the interior crew out and operate from the rear. This is a safer and better use of both manpower and strategy.

C. *+2* This is exactly what the last choice explanation is talking about. Given the fire situation, moving the attack to the rear will enable streams to be applied from the safest area and apply water from the path of least resistance in regard to the location of the fire. This situation requires strategy modification from offensive to defensive then back to offensive again as conditions permit. These actions will require a strong command presence and operational discipline and cooperation on the part of all the players.

D. *-2* Since the answer choice makes no mention of withdrawing the interior crews, it must be assumed that they were left inside as the exterior streams were operating. If you don't say that they were pulled out, you didn't do it. On the fireground, you better do it or you will either have injured or extremely agitated personnel to contend with, neither of which is particularly healthy for your career.

E. *+1* This may be acceptable if the master stream can blast through the partitions and hit the seat of the fire. It might be quicker, though, and cause less damage in the long run, to hit the fire from the rear.

10. The key to this question is to think big-picture. You are asked the best course of action regarding firefighter safety. This means all the firefighters, not just the ones operating in the immediate area. In the de-escalation period of the operation, firefighters frequently wander and are sometimes where they are not supposed to be. While this is not acceptable, it is reality. Notifying the entire fireground of the hazard at one time is the safest course of action.

A. *-2* Allowing a crew to operate in the area of the safe, which is stated to be hazardous, shows poor judgment and a lack of understanding of risk management. The fire is under control and it is an unacceptable to risk personnel for a piece of the building or contents

B. *+1* Keeping the overhauling crews out of the hazard area is a safe practice. It would be best to cordon off the area of the safe and below it. If visibility is poor due to either the hour of day or lack of windows in the area, the area should be well-lit. However, this focuses only on the personnel operating in the danger area. A wider spectrum of thinking is required here.

C. *-1* Assigning men to a hazardous area in this stage of the operation is an unnecessary risk. It is possible that the position the safe is in is holding other members in place. Drop the safe and you might dislodge other supporting members, dropping firefighters in the process. In addition, the impact load of the safe on the first floor and accompanying vibration may also cause a collapse of unstable members above it or a collapse of the 1st floor. The disposition of the safe is best completed when the fire is completely out, possibly by demolition or renovation crews. It should go without saying that whomever the fire department turns the building over to should be made aware of this hazard.

D. *+1* The assignment of the safety officer to the area to recon and monitor the hazard is proper delegation. His report back to the command post should guide your actions regarding the safe and the area.

E. *+2* The best method of making all fireground personnel aware of the danger is through the ordering of an emergency transmission. In this way, the hazard is broadcast over the radio to the entire fireground, making all personnel, both inside the structure and out cognizant of the hazard. Further action such as prohibiting access to the danger area and using the reach of the stream to overhaul the area will further guarantee the safety of personnel.

Passing Score for Scenario 9–3 = 14 Points

Scenario 9–4
Multiple-Choice Questions

1. Padlocks, especially on commercial occupancies, can be extremely difficult to force. Case-hardened padlocks with shackles of more than 1/4" cannot be forced using light-duty means such as twisting with a Halligan tool or driving the spike of the Halligan through the bow of the lock to force it open. If a duckbill lockbreaker is available, this may be effective; however, since the lock is at ground level, it may be more difficult to use this tool than if the lock was at a higher level as in a roll-down gate. There are other faster, more effective means of forcing heavy-duty padlocks. Firefighters, to be of a greater value on the fireground, should know how to defeat the many types of locks in a variety of ways.

   A. *0* Hinges on these steel cellar doors will usually have heavy-duty hinges which are machined and are not of the "pull the pin" type. Moreover, the hinges will likely be welded solidly into both the door and the frame. It will take a monumental effort to break these hinges.

   B. *+1* Using the saw with the aluminum oxide blade will quickly slice through the shackles of any padlock, even the tough hockey puck-type locks if the techniques are proper. (Fig. 9–15) It is important to have one member of the forcible entry team hold the lock in place using something other than his hands. A Halligan tool or a vise grip with a chain will suffice. Even the best saw operator will be relatively ineffective when visibility is poor. There is a better way to attempt to gain entry into the cellar given the scenario.

   C. *+2* It is important to pay attention to the details of the scenario. In other words, keep your eyes open, your head clear, and use all resources at your disposal. Do not be a victim of tunnel vision. The scenario states that the owner tells you that the store is empty. Well, if she is on the scene, there probably are few better people to ask for the key to the cellar than the owner. This would afford the attack team the safest, most effective path of least resistance into the cellar. If she doesn't have it, you are no better or worse off. At least you asked.

Fig. 9–15 Photo by Bob Scollan NJMFPA
If keys are not available, use a power saw with an aluminum oxide blade to attack the locks on a sidewalk cellar door. Beware that unignited, superheated gases may be just below the door waiting to vent and ignite. Have a charged line ready if this condition is suspected.

   D. *-1* Attempting to cut a case-hardened lock with bolt cutters will be futile. The jaws of the tool are not meant for this type metal. Always make every effort to match the tool with the job at hand.

   E. *-1* Chances are that if the effort to keep people out of the cellar with a heavy-duty, case-hardened padlock was made, the hasp and accompanying hardware will probably be of a stubborn metal as well. This brute force method is likely to be unsuccessful. There are times when the operation calls for the brute force method and times when you must outthink the lock. This is one of those cases. Take a second to survey your options before you put your head to the grindstone. You might find an easier, less crew-punishing method and gain valuable experience.

2. Line placement at this fire is a matter of priority in relation to the location and extent of the fire. Line placement at some time or another must address the cellar, the store, the top floor, the exposure to protect against fire spread via the air shaft, and the adjoining cellars to protect against lateral fire extension. In what order these lines are placed and if they are placed at all depends to a great extent on properly placing the initial lines to mitigate the problem in the quickest fashion.

   A. *+1* The path and direction of attack are correct. The exterior cellar door is the safest, most effective path of least resistance into the cellar to attack the fire. The trapdoor inside the store is too small, as most of these trap doors will be. Therefore, it is not the best place from which to mount an attack. The line size here, in relation to the heavy fire load in the cellar, leaves a lot to be desired. This fire is likely to be heavy and demands the punch, reach, and penetration of a larger diameter line.

   B. *-1* While the most severe life hazard is most likely on the floors above the fire, these areas are two floors and more above the fire. If water is not quickly applied to the parent fire in the cellar, problems will multiply. The line on the upper floor will not be able to remain there as the unattended-to fire escalates and spreads throughout the building. As a matter of fact, no other operation will be possible before long if the main body of fire is not attacked.

   C. *-2* It is possible that the fire at the cellar door entrance is vent point ignition, the proper mixing of oxygen with superheated products of combustion to produce flaming at building openings. Even if actual combustibles are being consumed in this entrance area, the angle of the deck gun and the ability for it to penetrate the interior of the cellar from the outside are next to zero. It is better to get a line in the cellar as quickly as possible. In addition, if there is a life hazard in the cellar, this action will certainly write them off. Many times, workers in stores who are illegally working or residing in this country will live in these areas. For this reason, a thorough primary and secondary search is mandatory in the cellar as well as all other areas of the building.

   D. *+2* A large diameter line aggressively advanced into the cellar will have the best chance of making any type of dent in this fire. The fire load in the cellar demands this type of stream. In addition, it will probably take the manpower of two companies to make the push into the cellar. It will be tough, hot, and obstacle-laden. Even though the 1¾" line may be more mobile, it will most likely not be effective against this heavy body of fire, when compared to the 2½" line, which has the advantage of reach.

   E. *-1* The trap door is only 30" x 30" square. That's only 2½ feet. This is akin to a scuttle opening in the floor. (Fig. 9–16) Not only will it be extremely difficult to get into the hole with SCBA and a hoseline, but also the venting heat will make it too dangerous to attack from. Recall the phenomenon of vent point ignition. The cellar ceiling will be the area of the greatest heat concentration. This is the most likely place for ignition of gases to occur. It is possible that these gases will not ignite due to lack of oxygen in the confines of the cellar ceiling. However, at the trapdoor, which is the path

Fig. 9–16 Descending this stairway into a burning cellar will be extremely difficult. In addition to the venting heat, the integrity of the combustible stairs may be compromised. If there is a safer direction of attack, use it. Keep the trapdoor closed and protect it with a charged line.

of least resistance out of the cellar in this area, the effect may be like opening up a roof over a superheated cockloft. Ignition of the gases may occur. If this is the firefighter's only way in, it may also be their only way out. If it is blocked by fire, the attack teams may be trapped. The best action to take at this fire is to attack via the exterior cellar entrance and keep the trapdoor closed to prevent extension into the store. A line should be stretched into the store at some point early on to guard against fire spread into the store via this artery.

3.

A. *0* If there was no exterior cellar entrance to this building, these tactics, along with answer choice "E" in Question #2, would be the best answer. However, the presence of the exterior cellar entrance allows the attack to be initiated from the same level as the fire, not from a descent through the interior of a building via a less than ideal access point. This answer does not score a *-1* because if the wrong tactics were chosen in Question #2, these would be the best tactics to protect the potentially ill-fated crew.

B. *+2* There are few places more critical than a cellar when it comes to the requirement for a back-up line. The egress is limited, as is the ventilation. If for some reason, the first line loses water, the back-up line may provide their only protection out of the cellar. The extreme fire load may require the need for more gallonage to advance through the barrier of heat, especially due to the fact that ventilation opportunities will be problematic at best. In most cases, you can never go wrong backing up the first line with a line of at least the same diameter, preferably larger. This means backing up a 1³/₄" line with a 2½" or backing up a 2½" fog nozzle with a solid bore tip. A 2½" line with a 1⅛" solid bore tip will flow 265 gallons of water minute where the same line using a 1¼" tip will flow 325 gpm. It will take a lot of fire and/or barriers to advancement to withstand the punch of this line, let alone if two are operating simultaneously. It is usually better, especially in a cellar, when using a 2½" line, or any line for that matter, to use a solid bore tip. The result will be less friction loss, less nozzle reaction, and more reach.

C. *+1* The size of the line is not appropriate for a back-up line unless the first line in the cellar was a 1³/₄". The point is awarded for the recognition that augmentation of the initial attack line is required.

D. *-1* Exposure D is certainly a concern due to the air shaft in between the two buildings, however, the positioning of the second line must address priorities in the fire building before turning to exposure protection. If the cellar was fully involved, this line placement may be acceptable, but the situation here is one of an offensive attack. The aggressive interior attack demands reinforcement.

E. *0* Like answer choice "D," it will be necessary to get a line to the apartment area above the store, but not just yet unless recon shows that fire has entered the access stairwell or the hallway and apartments above the store. The best plan is to hit the fire hard with the first two lines, then reinforce the operation in other critical areas with additional lines.

4. The placement of the third line will also be dependent on priorities relating to the location and extent of the fire. The aim of this line is to prevent extension via the path of least resistance.

   A. **+1** While the exposure is not the next priority, it is a priority nevertheless. The presence of the air shaft between the buildings may allow fire to spread into the exposure via unprotected openings such as windows. Lines must be placed to prevent this.

   B. **+1** Again, due to the search that must be conducted in this area, a line will be required to protect those operations. The best feature regarding the floors on which the apartments lie is that a separate access and egress is provided.

   C. **+2** This is correct positioning for this line. The store represents the floor above the fire, which may not only be almost as untenable as the cellar, but also represents the path of least resistance for fire travel inside the building. If the trap door has been left open, expect serious fire involvement in the store. If at all possible, the first company inside the store, usually the ladder company on a search and recon mission, should close the trap door if it can be done safely. If fire is issuing from the open trap door, it might be necessary to first knock the fire down, then close the door.

   D. **0** Extension into the adjacent store will not be as likely as exposure into the apartment windows bordering on the shaft. It is very likely that there will be no windows in the store bordering on the shaft. In addition, these old buildings are usually built with a fire wall separating each building terminating in a parapet at the roof which runs from front to back on both sides. However, the presence of the shaft negates the fire resistivity of the fire wall. Reconnaissance in the store along with line support will be necessary, but it is not the place for the third line at this fire.

   E. **+1** This is also a priority, but not the highest given the conditions. It is likely that fire walls separate the cellars of all the buildings on the block. However, remember that cellar layouts may be unusual and varied. The area just below any building may not encompass the whole cellar. Some smaller stores may only use half the cellar, while a larger, more storage-prone adjacent store may use the rear of the smaller store and its own cellar for storage. If renovations were done, the only separation between the areas may be a partition wall between the front of one cellar and the back of the same cellar and the adjacent cellar, where the wall, may have been breached to accommodate the renovation. Obviously, the best time to find this out is on pre-fire plan visits and inspections, but at a fire scene, it is critical that recon of these areas be completed as soon as possible. Surprises on the fireground are rarely beneficial to incident command.

5. Ladder company personnel must not only be resourceful, but must also be agile. Members assigned to the rear of the building will often have to resort to unorthodox methods of accomplishing tasks. Nothing should deter this mission, for the rear (and the unseen sides and shafts) will create a multitude of problems if ignored. The firefighters assigned the rear of the building had better get there or have a pretty good reason why they could not. Any areas that are not accessible should be immediately reported to the command post as an exception report so other orders can be given to cover the area.

   A. **-1** The diagram shows that all yards on the block are fenced and connected to the rear of Exposure C, the factory. Taking this route would mean having to climb several fences before the rear of the fire building is reached. A quicker way is necessary.

B. **-2** Accessing the rear via the cellar of the fire building is akin to suicide. This is like taking the interior stairs to the roof of a multiple dwelling during a serious fire in close proximity to or involving the stairwell.

C. **+1** Utilizing the fire building will eliminate the need to hop any fences, but if you are on the rear fire escape and the fire vents out the rear cellar windows, you will be like a piece of meat on a grill.

D. **+2** The exposure on the windward side is a safe way to access the rear. You will only have to scale one fence to get to the rear yard. Given the circumstances, this is the most desirable method of getting to the rear of the fire building.

E. **0** Not only will this take a long time accessing the roof and then descending the rear fire escape, but you may be exposed to venting fire from the rear windows.

6. When there is a heavy body of fire exists in the cellar of these old buildings, every possible effort must be made to provide ventilation to support the attack. This may require extensive secondary structural damage to open it up. If the attack teams cannot gain and maintain a position in the cellar, the building is likely to be lost as the fire extends through the myriad of voids and makes the positions on upper floors of the building untenable.

A. **0** There is no mention of providing horizontal ventilation opposite the attack line at the rear of the building. In fact, this is the most critical area of ventilation. In addition, vertical ventilation must also be performed or heat levels may build up and ignite secondary fires on upper floors.

B. **-2** The fire is located in the cellar. There is no need to cut the roof unless fire had extended to the cockloft or into a vertical artery such as a pipe chase. At this stage, opening the natural openings to prevent heat accumulation at the upper regions of the building will suffice. Cutting the roof will cause unnecessary secondary damage.

C. **+1** These tactics will certainly create as much ventilation as is possible by opening natural openings. This fire will likely require that something more be done to remove the heat from the cellar and allow for the attack to be successful.

D. **-1** First, there is no mention of horizontally ventilating at the front show windows. This will allow the heavy smoke and heat to remain on the first floor and continue to permeate the structure. Ignition away from the parent body of fire may be the result. Cutting a hole in the floor to supplement the natural venting operation is the right idea, but the location is improper. Locating a floor cut is different than locating a roof cut. While it is best to cut your vent hole over the main body of fire, it is better, when cutting a floor inside a structure, to cut it where the venting smoke can be released from the building. Even though the main body of fire is near the rear of the store, there will probably not be sufficient window venting opportunities to release the heat and smoke from the hole in the floor. This may endanger the rest of the building, especially if fire erupts from the hole with no exterior outlet.

The other problem is that in the "extremely heavy" smoke condition, there is a very great potential for the saw to malfunction. Gas-powered saw engines need the proper mixture of air to function. If the area where the cut is to be made is oxygen-deficient, as is likely the case at the rear of the store, the saw will sputter and not turn over. This actually happened at a fire very similar to this. A ladder company was ordered to cut the floor at the rear of the store. The saw would not start in the heavy smoke condition. As a result, they were nearly trapped and building was lost because the cellar could not be adequately ventilated. It is also extremely unsafe to attempt to vent in zero visibility, which will likely be the condition at the rear of this store. Remember: "Proper tactics protect people."

E. *+2* In this set of actions, the horizontal vent is accomplished on the opposite side of the attack line as is proper. However, notice that the front show windows are taken out and the hole in the floor is cut near these windows. This is the safest and most effective way to cut the floor given these conditions and building layout. The vent direction is established first by opening the rear. Then, opening the floor provides additional venting. The ideal would be to have windows on the rear of the 1st floor so the floor could be cut opposite the attack, but sometimes you must adapt when conditions are not ideal. It must also be noted that if it not already in position, a line must be used to protect the area surrounding the floor vent hole. It is forbidden to direct the line into the hole unless the attack team is withdrawn due to heavy fire. Then it may also be an ideal spot for a distributor or cellar pipe.

7. These are essentially "bread and butter" search tactics.

A. *0* In this choice, the cellar is omitted. When tenable, the fire area must be a prime target for the primary search, without exception.

B. *+1* The building must be covered and all areas searched. Safety of personnel demands the use of a lifeline when searching dangerous areas such as cellars.

C. *-1* The company should be split to provide the most coverage in the least amount of time. Keeping the crew intact in this fire violates the principle of economy of forces. Economy of forces basically means that tasks should be assigned to companies that accomplish as much as possible given the manpower available without compromising the safety of personnel.

D. *-2* There is no way that the building can be confirmed clear of occupants unless a thorough primary search of all areas is undertaken. Information from civilians, and for that matter, the police, must not be taken as gospel. Concentrate on the fire building first, then attend to exposures.

E. *+2* This set of tactics accomplishes all the proper coverage assignments as answer choice "B," but takes it a step further by attending to the most threatened exposure after the fire building search is completed. Sometimes the answer choice that gives a little more information, providing it is proper, differentiates the *+2* from the *+1* answer.

8. Officers supervising companies in cellar operations must be cognizant of the many dangers that may be present.

A. *0* The floor joists in these old, ordinary constructed buildings will usually be a dimension of 2" x 10" or 12". They should remain intact for a relatively long period of time. If these joists fail due to fire destruction, offensive operations will most likely have ceased and firefighting operations will be conducted from outside of the building.

B. *+2* The scenario states that the cellar is loaded with piles of tied-up cardboard boxes and unused material. This will not only constitute a heavy fire load, but also may create a collapse hazard. (Fig. 9–17) Once this stock absorbs

Fig. 9–17 Stacked debris anywhere in the building not only creates a heavy fire load, but can collapse and trap firefighters as well as absorb runoff water. Note how this stack blocks the sprinkler head.

water from hose streams, they may list and collapse, possibly on a firefighter. The weight of this stock can be several hundred pounds. They may also fall and create an obstacle between the attack teams and the point of egress. Collapse may not only be caused by water absorption, but may also be caused by the movement of the line and other fire-fighting operations. As the line is advanced and takes the path of least resistance around corners, it may topple large stacks of debris, cutting off the escape path. Several firefighters may have to be assigned to hose line management to prevent this. If the piles are unstable to begin with, it may be necessary to use the reach of the stream to knock down the fire. If the stream is unsuccessful in reaching the seat of the fire due to barriers created by the stock, alternative tactics such as the use of distributors from the floor above may be the answer. Otherwise, the cellar may be lost.

C. *0* The cellar stairs are wood and will not stand up to the ravages of fire for too long. However, in this situation, the cellar stairs should not have been used for either cellar access or egress. If there was no exterior access to the cellar, and the stairs were the main means of attack, then their stability will be a major factor in the offensive/defensive decision.

D. *-1* Having sufficient hose to make the attack is crucial. As the attack here should be made from the front exterior, stretching short should not be a problem. If the attack was made from the interior, such as from an interior stairway, estimating the amount of hose required will be of major importance. Getting caught short on the hose stretch is danger-ous, especially in below grade operations. If the fire is in the front, but the access is in the rear, the required hose will be twice the depth of the cellar plus at least a length to oper-ate on the safe side. For example, if the cellar is 50' deep, the hose required will be twice the depth, plus at least one more length. This adds up to 150' to safely reach the fire. Fire officers estimating the amount of hose needed should always err on the safe side and take a little more than a little less. You can always find a place to flake extra hose. If you stretch short, you may be in deep trouble.

E. *+1* The high heat conditions associated with cellar fires is one of the most significant factors in the ability of the attack teams to remain in the building. Fortunately, we may have some control over the ability to release some of this heat. From horizontal ventilation to cutting the floor, the incident commander should try to provide as much ventilation as possible before surrendering the area. Many times, the only difference between staying inside and putting the fire out and withdrawing is the absence of effective and creative ventilation.

9.

A. *-1* The conditions on arrival suggest that heavy fire may be present in the cellar. In addition, the substantial fire load in the cellar provides a potential for fire spread to the upper floors and to exposures. The building is attached and it will require manpower to at least recon these areas. Mixed-use occupancies are extremely manpower-intensive operations. Provide support for the operation by ensuring sufficient forces are on hand as well as a tactical reserve.

B. *+1* The fire is in the cellar. This means that the entire building will be exposed to the fire and the products of combustion. Recognizing the many tasks that will be required to keep the fire where it is and ensure it does not spread upward will require more manpower than there is on the first alarm response.

C. *+2* This is a better answer than choice "B" because the shaft will be a major fire spread problem which, if not recognized early, may cause severe problems in fire control later.

Taking steps to recon and protect areas on the shaft exposed to the fire will assist in confining the fire to the building of origin. It is bad enough to lose one building, but sometimes this cannot be avoided. Losing exposures when steps could have been taken early to prevent their involvement is a sign of incident command mismanagement and lack of foresight.

D. *-2* The only way to ensure that no life hazard exists is to conduct a thorough primary search. The fact that the owner tells you the building is empty is immaterial. Did she mean the store was empty, the building, or both? Search to ensure that all possible occupants are out of the building.

E. *0* Requesting just one more company is counterproductive. It is always best to request a standard greater alarm than it is to piecemeal companies into the scene. I have seen many buildings (and exposures) lost because the incident commander was calling one company at a time and relying on recall of personnel from miles away. If there is a potential need, get the additional alarm response rolling as soon as possible.

10. When attached buildings are involved, especially if shafts are present between buildings, ignorance of recon and required protection of these exposures may lead to involvement, embarrassment, and injury. Take steps to protect these areas early. (Fig. 9–18)

Fig. 9–18 Photo by Bob Scollan NJMFPA
The heavy fire above the roof is fire venting from a shaft between the buildings. Don't wait for this sign to place lines in the exposed building. Reports from the roof early in the operation should reveal the presence of this shaft.

A. *0* The areas that must be checked include the adjoining cellar, the first floor, the shaft, and the cockloft. Do not assume there are no penetrations between cellars. A line must be stretched to this area as soon as manpower permits.

B. *+1* These are the first tactics that must be taken as far as protection of this exposure. However, an examination of the cockloft is mandatory as vertical fire spread may find its way to this area unnoticed.

C. *-2* Unless something extremely out of the ordinary is happening, the protection of the windward exposure is usually never given precedence over the leeward exposure. An example of this exception may be when the leeward exposure is a vacant building and the windward exposure is a dealer of fine antiques. In this case, the rule of thumb of always protecting the leeward exposure first may be overridden. If nothing out of the ordinary is going on, attend to the leeward exposure first.

D. *+2* These are the best set of tactics to protect the exposed building. Coverage is provided for all possible threatened areas. Heat dissipation techniques are being conducted as well as fire extension monitoring. If these tactics are performed early, which will require a very early additional alarm, the building should not suffer severe damage due to unattended fire spread.

E. *+1* This tactic is a good way of providing coverage of as much area as possible with one line. This may be the most effective initial coverage of this building if manpower is not available to stretch a second line into the exposure. The line here will operate as per conditions and recon reports. It must be stated that, if manpower is available to stretch a second line, it should be done and lines placed both on the 1st floor and in the cellar. In addition, a line should also be stretched to the top floor and any other exposed areas.

11.

Fig. 9–19 Tin ceilings are relatively impervious to vertical fire travel except in areas where utilities have pipes passing through to the upper floors. This will be a path of least resistance and the first place that should be opened. It is also the easiest place to gain a purchase with a tool to open the ceiling for examination.

A. *-1* Tin ceilings are relatively effective in confining the fire to the area of origin, but they are not impenetrable. The same problems inherent in plaster and sheetrock ceilings can be found in tin ceilings. These are openings through which pipes and light fixtures pass through. Fire will also pass through these openings. (Fig. 9–19) These areas should be checked as well as the entire area above the ceiling. It only takes a hole the diameter of a pencil to spread heat and fire to an adjoining area, in this case, to the ceiling space and the floor above. A tin ceiling will also hold and radiate the heat of a fire back into the room, decreasing the time it takes for flashover conditions to accumulate. If the ceiling area shows signs of rollover, a stream, either a solid bore or straight to tight fog, should be directed at the ceiling to absorb some of the heat. Stream application should take place from a doorway and adequate ventilation should be provided opposite the stream in coordination with nozzle operation. This means as soon as the nozzle opens or immediately thereafter. Any delay in this ventilation will cause the generated steam to seek the path of least resistance out of the area, which will likely be from where the attack team is operating. This steam can cause burns and drive the team out of the area.

B. *-2* Pulling up the floor boards above the fire will not only create unnecessary secondary damage, but if the fire is burrowing in the ceiling space, it will be given a free pass to the floor above by this action. Let the floor boards act as a fire barrier while the area below is opened and the fire attacked from there.

C. *+2* The ceiling must be opened from below. The best tool to accomplish this is the Halligan hook. Tin ceilings are commonly hung in 4' x 8' sheets from wall to wall perpendicular to the ceiling joists. Thus, to expose an entire joist channel, it will be necessary to pull down more than one sheet. There will be seams between the sheets of tin. These seams can be hammered up by the head of the tool. A purchase can then be made on the sheet, allowing it to be pulled down in one sheet. If you try to get these ceilings down in any other way, you will just wind up poking holes in the ceiling and never get enough ceiling to pull down. Be sure a charged line is in the area to extinguish any fire that shows when the ceiling is pulled. Any apparent extension routes to the floor above can then be identified and the above fire team notified of their location.

D. *+1* This is another method of opening the tin ceiling. By standing on a ladder or something that will hold your weight, use a Halligan tool and attack the seam. The exposed tin edge can then be grabbed with a gloved hand and ripped down. This is not as safe as using a Halligan hook and operating from a distance, but it will work if you know where to begin the opening.

E. *0* This method will not only flood the ceiling, but the weight of the water could cause the ceiling to collapse. To overhaul the ceiling of a room, it is best to rip down the entire ceiling, especially if the room has been involved in any substantial way. This way, it can be ensured that there is no fire confined above the ceiling space and the entire ceiling is available for stream application.

Passing Score for Scenario 9–4 = 15 Points

CHAPTER TEN
SCENARIOS

SCENARIO 10–1
MOE'S PLASTIC WORKS FIRE

Construction

The building, at 222 Jeep Street, housing a workshop which packages plastics, is a large one story structure, with concrete block walls and a lightweight steel truss roof. There is an HVAC system on the roof, which feeds a center duct air handling system. On the interior at the northeast corner is a network of cubicles set up for record-keeping operations for the business.

Time and weather

The temperature is 26°F. The wind is out of the Northeast at 28mph. The time is 0810.

Area and street conditions

The fire building is in an area undergoing a renaissance. Jeep Street is a two-way street, while Hegarty Avenue is a one way street running west to east. Flood Street is a one way running east to west. To the southwest is a gas station. Running between the workshop and the motel (above and adjacent to Hegarty Avenue and Flood Street) is Victor Boulevard, a two-way elevated bridge.

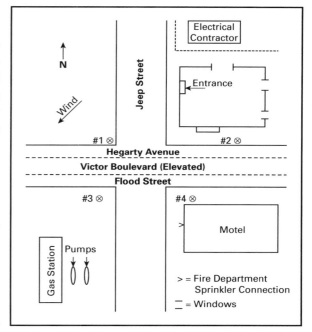

Fire conditions

As you arrive on the scene, you notice smoke issuing from underneath the loading dock door on the south side of the building. Some employees were seen stretching a garden hose into the structure prior to your arrival. There is a report from the onlookers that one of the people trying to extinguish the fire got lost. There are a number of people outside the building in a dazed and disoriented state.

Exposures

To the north of the workshop, there is an electrical contractor building surrounded by a high fence. The fence is topped with razor wire. On the south side of the building, across the street, is a motel which is also under renovation, but is still open for business. The three story motel, which faces on Flood Street is of wood frame construction.

Water supply

The hydrant spacing in the area is good, however, due to the ongoing renovation, one of the hydrants is out-of-service.

Response

Your response is three engine companies and one ladder company. Each is staffed by an officer and three firefighters.

Short-Answer Questions

1. Would there be a need to strike additional alarms for this incident?

_____

_____

_____

_____

_____

2. Where would the first line be stretched and what is the objective?

_____

_____

_____

_____

_____

3. Where would the second line be stretched and what are its objectives?

_____

_____

_____

_____

_____

4. How would a search of this structure be conducted?

_____

_____

_____

_____

_____

5. How would this building be most efficiently and safely ventilated?

_____

_____

_____

_____

_____

6. What are your concerns regarding exposures?

_____

_____

_____

_____

_____

SCENARIO 10–2
CARPET STORE FIRE

Construction

The fire building is #1055 Curtis Street, a one-story building measuring 270' along the Curtis Street side. The structure is 150' deep. At one time, #1055 Curtis St. housed a truck parts assembly plant. The building has recently been subdivided (equally in half by a fire partition, constructed of 2" x 4" metal studs with ⅝" sheetrock on both sides) and renovated, with some minor finishing construction still in progress. The northern portion of the building is fully sprinklered and occupied by a firm that manufactures toys. The southern half of the building is being retro-fitted with an automatic sprinkler system. Installation is not complete and the system is not in service. The southern portion of the building contains a large drug store and a carpet store.

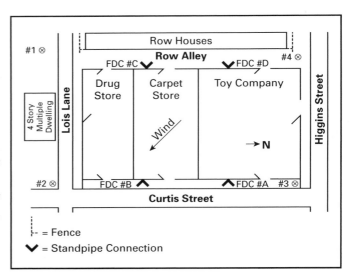

The building is of ordinary construction. The cellar has been sealed up during the renovation and its contents are unknown. The roof has been tarred over many times. The first floor joists are covered by 3" x 6" wood planking which were installed to support the heavy floor load of the previous occupancy. There is a three foot parapet encircling the building. The floor planking is oil-soaked from the previous occupancy and is now covered by floor covering and carpeting. Many portions of the fire partition are breached with unsealed poke throughs.

Time and weather

It is 0300, Sunday, May 25. The weather is clear, it is 65°F, and the wind is from the northwest at 15mph.

Area and street conditions

Noel Higgins Place, a two-way street, runs along the north side of 1055 Curtis Street. Lois Lane runs along the south side, while Row Alley separates the fire building from the C exposures, the row houses.

Fire conditions

A fire has been intentionally set in the carpet shop. A flammable liquid was used to accelerate the fire spread. Fire has control of the carpet store and is spreading rapidly in a horizontal fashion from north to south.

Exposures

A series of row houses are separated from the rear of the structure by a 12'-wide alley and an 8'-high cyclone fence. A four-story multiple dwelling is located south of 1055 Curtis Street across the 12'-wide Lois Lane.

Water supply

A water flow alarm is sounding. There are hydrants located on each corner, each on its own main.

Response

The response is four engine companies and two ladder companies. An officer and three fire-fighters staff each company.

SHORT-ANSWER QUESTIONS

1. Where would the first engine operate and what would be its orders?

   _____

   _____

   _____

   _____

   _____

2. Where would the second engine company operate and what orders would they be given?

   _____

   _____

   _____

   _____

   _____

3. What additional orders would be given to the third and fourth-arriving engine companies?

_____

_____

_____

_____

_____

4. What orders would be given to the ladder companies?

_____

_____

_____

_____

_____

5. What tasks would be given to engine and ladder companies arriving on the second alarm?

_____

_____

_____

_____

_____

Multiple-Choice Questions

6. As you arrive on the scene, you are to assume command by virtue of rank and department SOP. After receiving a briefing on actions taken thus far by the on-scene incident commander, what is your next action in regard to the incident command process? You have second alarm companies en route to the scene.

    A. Confirm the transfer of command with dispatch and all operating companies.

    B. Size up the fire situation.

    C. Notify all companies responding on the additional alarm to stage and await orders until you can assess the situation.

    D. Relocate the command post to a safer location.

    E. Request that additional chief officers respond to the scene.

7. How would you order a primary search of this fire to the ladder captain?

    A. Split the company into two crews. Crew 1 attempts to enter the carpet store to initiate a primary search. Crew 2 enters the drug store to conduct a primary search.

    B. Keep company intact. Enter the drug store to initiate a primary search. Use extreme caution due to the possible deteriorating conditions. Then, conduct a primary search of the Toy Company.

    C. Keep company intact. Conduct a primary search of the Toy Company. Conduct no search in the carpet and drug stores. These portions of the building are to be written off.

    D. Do not search this building. This is a defensive fire from the outset. Use crews to evacuate the row houses and multiple dwelling. Then set up a ladder pipe to protect the unburned Toy Company.

    E. Split the company into two crews. Crew 1 enters the drug store to initiate a limited primary search due to the deteriorating conditions. Stay in contact with the hoseline. Crew 2 conducts a search of the Toy Company

8. What is the major concern regarding building stability at this fire?

    A. The 3" x 6" wood planking will burn through and cause a collapse of the floor.

    B. Failure of the roof due to heavy fire conditions.

    C. Breaches in the fire partition will cause a fully involved building, spreading fire via radiant heat to the exposed row houses.

    D. Heavy fire will weaken the parapet and cause a collapse for the full length of the parapet.

    E. Building stability is not a factor at this fire.

9. A member of the press corners you during the latter stages of the fire and questions you as to the validity of the report of arson as the cause of this fire? What is the most appropriate response for you to make?

   A. "No comment."

   B. "These are the preliminary reports, but we will not know for sure until a complete investigation is undertaken."

   C. "We have reports that a flammable liquid was used to start this fire, but it is not proven at this time."

   D. "The fire is under investigation at this time."

   E. "Talking to the press is not my bag, baby."

10. A higher-ranking officer has arrived on the scene and designated you as safety officer. What is the most important step you will take regarding firefighter safety?

   A. Ensure all personnel are aware that you are the safety officer by radio announcement.

   B. Ensure all firefighters are in the proper protective equipment.

   C. Ensure all operating personnel are out of the collapse zone and/or operating in a flanking position.

   D. Ensure a Rehab Post has been established.

   E. Monitor the radio for reports of unsafe conditions and actions.

SCENARIO 10–3
CAESAR'S PLAZA STRIP MALL FIRE

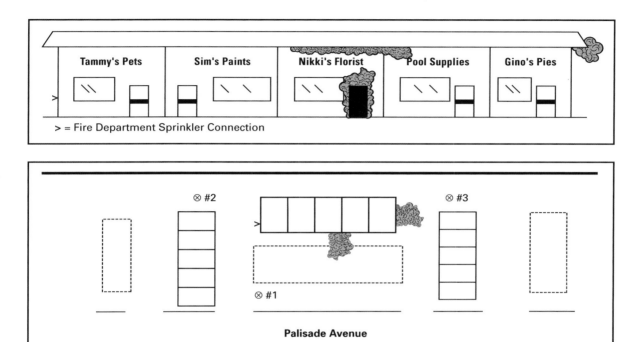

> = Fire Department Sprinkler Connection

⊗ #2   ⊗ #3   ⊗ #1

Palisade Avenue

= Parking Lots (Unfenced)

> = Fire Department Sprinkler Connection

Construction

A fire has been reported at the Caesar's Shopping Plaza. The one-story strip mall is concrete and steel construction with a quarter inch steel roof. Supporting the steel roof is a lightweight open-web steel parallel truss spaced eight feet on center. There are roof scuttles over each store. There is also a 2.5 ton HVAC unit above each individual shop. There are two-hour fire-rated walls between each shop. It is not known whether these fire-rated walls extend up to the underside of the roof or just to the top of the partition walls. There is no cellar. The ceiling of each store is made up of a removable-panel drop ceiling, above which wiring and HVAC components are present.

Time and weather

It is Saturday morning. The time is 1115. The temperature is 22°F. The wind chill makes it 9°F. The wind is blowing to the north at 12mph.

Area and street conditions

Traffic around the mall area is heavy. A recent snowfall has caused bulldozer-created snow drifts to dot the parking lot.

Fire conditions

Upon arrival, you see dark smoke emitting from the front of the Florist. There is no flame visible. The receptionist tells you that there is a "big fire" in the rear storeroom where most of the stock is located. She says that she thinks all customers and employees have escaped.

Exposures

The mall is U-shaped, with each strip housing five businesses. Each has a large show window over which a six foot parapet supported by a steel lintel is located. Covering the parapet at the front is a mansard-type facade made of 2" x 4"s and plywood covered with sheet metal. It extends for the entire row and was recently added to the structure. In the center of the row is Nikki's Florist. Exposure B is Sim's Paint Emporium. Exposure D is Frank's Pool Supplies. All businesses have front doors which are of tempered glass in metal frame set in a metal jamb. There is a bar across the center of each door to facilitate egress. At the rear, each store is served by a steel door set in a steel frame in a masonry block wall.

Water supply

There is a sprinkler connection at the south end of the row. All hydrants are on separate mains; however, through your preplanning visits, you have discovered that the hydrant adjacent to the fire department connection is on the same main as the sprinkler system.

Response

Your response is three engine companies and one ladder company. Engines are staffed by an officer and two firefighters. The ladder is staffed by an officer and three firefighters.

MULTIPLE-CHOICE QUESTIONS

1. Where is the best place for the Command Post to be established at this incident?

   A. On Palisade Avenue.

   B. In the North Parking Lot.

   C. In the South Parking Lot.

   D. In the Center Parking Lot.

   E. Inside one of the buildings in the South strip of the mall.

2. Will additional alarms be required at this fire?

   A. No, the fire should be contained to the rear of the store.

   B. Yes, the temperature extremes will lead to cold-related injuries if firefighters are not properly relieved.

C. Yes, due to the potential for exterior exposure spread.

D. Yes, due to the presence of involved hazardous materials in the two adjacent exposures, request both a second and third alarm.

E. Yes, due to the fact that there is smoke showing at the end of the row on the leeward side, indicating the probability of the open construction, more manpower will be required to get ahead of the fire.

3. What are your orders for Engine 1?

A. Establish a water supply at Hydrant #1. Position at the south end of the row. Stretch two lines to the fire department connection. Charge upon orders. Stretch a preconnected 2½" line to the front of the store. Wait for confirmation of vertical ventilation at the roof. Enter store to locate, confine, and extinguish the fire.

B. Establish a water supply at Hydrant #2. Position on the south end of the row. Stretch two lines to the fire department connection. Charge the system. Stretch a preconnected 2½" line around the back of the building to the rear of the store. After the door has been forced, enter store to locate, confine, and extinguish fire.

C. Establish a water supply at Hydrant #3. Position just past Florist on windward side. Handstretch two lines to the fire department connection. Charge system. Utilize deck gun in window to knock down fire at rear of store. Then, shut down deck gun and advance preconnected 1¾" line into store via front door to finish extinguishment of fire.

D. Establish a water supply at Hydrant #1. Position just past the Florist on the windward side. Handstretch two lines to the fire department connection. Charge the system. Stretch a preconnected 2½" line through front door to locate, confine, and extinguish the fire.

E. Establish a water supply at Hydrant #3. Position in front of Florist. Utilize deck gun in window to knock down fire at rear of store. Stretch preconnected 2½" line through the front door of the Pool Supplies Store to head off any extending fire on the leeward side.

4. What are your orders for Engine 2?

A. Establish a second water supply at Hydrant #1. Stretch two lines to and supply the fire department connection. Stretch a 2½" back-up line into Florist to reinforce Engine 1's fire attack.

B. Establish a second water supply at Hydrant #3. Position the apparatus in a defensive manner in anticipation for defensive operations. Stretch 1¾" back-up line into store to reinforce Engine 1 attack.

C. Establish a second water supply at Hydrant #1. Stretch a 2½" line to the Pool Supplies Store to protect against any fire extension.

D. Establish a second water supply at Hydrant #2. Position on the windward end of the row. Stretch a 2½" line into the Florist to reinforce Engine 1's fire attack.

E. Establish a second water supply at Hydrant #3. Position the apparatus in a defensive manner in anticipation for defensive operations. Stretch a 2½" back-up line into the store to reinforce Engine 1's attack.

5. Where would a third line be stretched and what would be the objective?

   A. Into the Paint Store to prevent against extension on the leeward side.

   B. To the rear of the building to advance through the back door to back up the attack line.

   C. Into the Pool Supplies Store to prevent against fire extension on the leeward side.

   D. Into the Florist to reinforce the initial attack.

   E. To the roof to protect operating personnel.

6. What are your orders for Ladder 1?

   A. Split the company into two crews. Crew 1 enters the Florist to conduct a primary search and conduct horizontal ventilation of the fire building. Crew 2 enters the Pool Supplies Store to vent as required and to expose extending fire by pulling the ceilings.

   B. Keep the company intact. Ladder the roof of the Florist. Open scuttles on roof. Also cut a 4' x 4' hole at the rear of the building to channel heat and gases out of structure. Push down drop ceiling to vent stores. Monitor the cockloft. In addition, use a heavy tool on a rope to swing into the front show window to horizontally ventilate the store.

   C. Split the company into two crews. Crew 1 enters the Florist to conduct a primary search and conduct horizontal ventilation operations of the fire building. Crew 2 operates initially at the rear. Breach the rear wall to allow ventilation opposite the attack line. Then ascend to the roof via a ground ladder. Open the scuttle on the fire building and adjacent buildings. Push down the drop ceilings to vent the stores. Monitor the cockloft for fire extension.

   D. Split the company into two crews. Crew 1 ladders the roof of the Pool Supplies Store. Open the scuttle of the leeward exposure to determine the extent of fire in the cockloft. Then open the scuttle on the Florist and windward exposure. Push down the ceilings to vent the stores. Crew 2 enters the Florist to conduct a primary search and conduct horizontal ventilation of the fire building.

   E. Split the company into two crews. Crew 1 ladders the roof of the Paint Store. Open the scuttle of the Florist to determine the extent of the fire in the cockloft. Push down the ceiling to vent the store. Then open the scuttle on the leeward side and the windward side exposure. Monitor the conditions. Crew 2 enters the Florist to conduct a primary search and conduct horizontal ventilation of the fire building.

7. What orders would be given to Ladder 2?

   A. Split the company into two crews. Crew 1 ladders the leeward exposure, opens a scuttle at the roof, pushes down the ceiling to vent the store, and monitors the cockloft. Crew 2 conducts a secondary search of the Florist and the exposures, while also checking for fire extension.

   B. Keep the company intact. Operate at the rear of the building. Breach the wall to establish a more favorable venting direction for the attack line. In addition, from the interior, open the doors at the rear of the adjacent exposures to secure easier access to all stores and additional ventilation. Enter the adjacent stores to check for extension.

C. Split the company into two crews. Crew 1 operates at the rear of the building. Breach the wall to establish a more favorable venting direction for the attack line. Crew 2 operates in the leeward exposure to conduct a primary search and to check for fire extension.

D. Split the company into two crews. Crew 1 operates in the leeward exposure to conduct a primary search and to check for any fire extension. Crew 2 operates in the windward exposure with the same mission as Crew 1.

E. Split the company into two crews. Crew 1 operates at the rear of the building, breaching the wall to establish a more favorable venting direction for attack line. Crew 1 then accesses the windward exposure to conduct a primary search and to check for fire extension. Crew 2 operates at the roof to assist in roof operations.

8. If the fire store was closed at the time of arrival and forcible entry was necessary, what would be the most efficient way of gaining entry to this area?

   A. Breach the sheetrock via an open, uninvolved store.

   B. Use a K-Tool in the "through-the-lock" method of entry.

   C. Use a pick-head axe and strike the tempered glass in the bottom corner.

   D. Use a Rabbit tool (hydraulic forcible entry tool) to pop the door.

   E. Use a Halligan tool and force the door at the hinges by removing the pins.

9. In the initial stages of this fire operation, what is your most immediate concern regarding the mansard-type façade?

   A. Collapse of the façade can pull down the parapet wall with it.

   B. The fire will heat the steel I-beam acting as a lintel over the show window. This can cause collapse of the parapet wall and secondary collapse of the mansard façade.

   C. Fire burning into the façade space will spread with the wind across the front of the stores. Openings in construction may allow spread into exposures in this manner.

   D. Fire blowing out of windows could expose overhead utility lines, causing them to drop, endangering operating personnel and other structures.

   E. Water accumulation inside the façade could cause collapse.

10. What is the most significant collapse hazard at this fire?

   A. Collapse of the parapet wall

   B. Collapse of the drop ceiling

   C. Collapse of the lightweight roof system

   D. Collapse of the mansard roof

   E. Collapse of the steel lintel over the show window

Scenario 10–4
Monmouth Strip Mall Fire

Construction

The strip mall is of ordinary construction. The roof is wood joist with a built-up tar covering. Recently added to the roof of the strip mall is a 2.5 ton HVAC unit above each individual shop. There are fire walls between each shop which rise above the roof line. There are no common cocklofts, however, there is a parapet located at the front and sides of the building. At the north end of the strip mall is Jen's Beauty Salon. It has been closed since 1500 Saturday. Jen's Beauty Salon has a tempered glass door set in a metal jamb at the front and a steel door at the rear. The front show windows are plate glass.

Time and weather

It is Sunday evening. The time is 1745. The temperature is 18°F. The wind chill makes it 5°F. The wind is blowing to the east at 12mph.

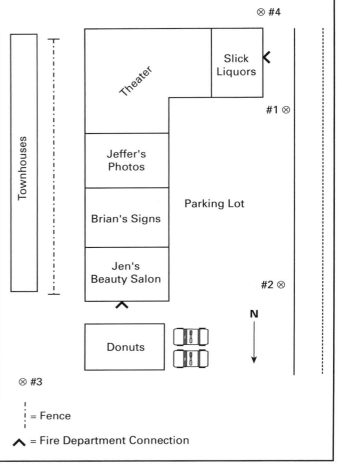

Area and street conditions

It is two weeks before Christmas and a strip mall containing shops and a triplex movie theater is crowded with shoppers and movie patrons. The same popular children's Christmas movie is playing at all three theaters. The theaters are completely sold out and the area is very congested. The mall is located on Highway 9. Traffic is very congested around the mall.

Fire conditions

The alarm was phoned in by the police department who are located on the north end of the complex. Puffing gray smoke is issuing from the top of the front doors. The show windows are stained black.

Exposures

To the east of the strip mall is a new lightweight wood truss townhouse development. A 7' wood fence separates the strip mall from the townhouse development.

Water supply

There is a fire department connection on the north wall of the salon.

Response

Two engine companies and one ladder company respond. An officer and two firefighters staff each engine company. The ladder company has an officer and three firefighters.

MULTIPLE-CHOICE QUESTIONS

1. What would be the most important size-up factor to consider at this incident?
   A. Construction
   B. Fire phase (backdraft condition)
   C. Street conditions
   D. Apparatus and manpower
   E. Life hazard

2. Will additional alarms be required at this fire?
   A. Yes, the theater must be evacuated. A major commitment of manpower will be needed to accomplish this objective.
   B. No, due to the fact that the store is on the end of the row, and there is firestopping present, on scene companies should be able to handle this fire.
   C. Yes, due to the fact that the roof may collapse under the weight of the air conditioning unit, this will be a defensive operation, and many exterior handlines will be needed to contain this fire.
   D. Yes, the weather, the slowed response due to the holiday traffic, and the problem of forcible entry on steel doors will allow the fire to spread beyond the capability of first alarm companies. With this in mind, a second alarm should be transmitted even prior to arrival.
   E. Yes, a second alarm should be struck due to the potential enormous life hazard and backdraft condition.

3. What are your orders for Engine 1?

    A. Establish a water supply at Hydrant #3. Supply sprinkler siamese with two lines and charge. Stretch a 2½" line to the front door of the Beauty Salon, standby until the ladder company forces entry. Locate, confine, and extinguish fire.

    B. Establish a water supply at Hydrant #4. Engine 1 positions at the front of the theater. Use manpower to evacuate the theater. Stretch two lines to the sprinkler siamese on the theater. Charge the sprinkler system. Protect the evacuating patrons with a water curtain, should convection heat from the Beauty Salon fire cause a sprinkler head to activate.

    C. Establish a water supply at Hydrant #2. Position just past the Beauty Salon. Stretch two lines to the sprinkler siamese on the Beauty Salon and charge the sprinkler system. Stretch a 2½" line to the front of the Beauty Salon. Stand by in a flanking position in a defensive mode until the ladder company confirms that vertical ventilation has taken place. Then, enter the Beauty Salon to locate, confine, and extinguish the fire.

    D. Establish a water supply at Hydrant #3. Position just past the Beauty Salon. Stretch two lines to the sprinkler siamese on the Beauty Salon. Charge the system. Stretch a 1¾" line to the front of the Beauty Salon. Stand by in a flanking position in a defensive mode until the ladder company confirms that vertical ventilation has taken place. Then, enter the Beauty Salon to locate, confine, and extinguish the fire.

    E. Establish a water supply at Hydrant #2. Position just past the Beauty Salon. Stretch two lines to the sprinkler siamese on the Beauty Salon. Charge the system upon orders. One firefighter stretch a 1¾" line to the roof of the Beauty Salon to protect the ladder company performing vertical ventilation to alleviate backdraft condition. Work under the command of the roof sector officer. Also have the captain and one firefighter stretch a 1¾" line to the front of the Beauty Salon. No offensive operations are to take place until vertical ventilation is complete. Upon confirmation of vertical ventilation, enter the Salon to locate, confine, and extinguish the fire.

4. What are your orders for Engine 2?

    A. Establish a secondary water supply at Hydrant # 1. Stretch a 1¾" line up a ground ladder to the roof to protect the ladder company engaged in vertical ventilation operations.

    B. Establish a secondary supply at Hydrant #3. Stretch a 2½" line to the rear of the Beauty Salon. Stand by in a defensive mode until vertical ventilation of the roof has been completed. Then, enter through the rear door to locate, confine, and extinguish the fire. Work the line in coordination with Engine 1's attack line entering from the front.

    C. Establish a secondary water supply at Hydrant #4. Stretch two lines of appropriate size to the sprinkler siamese on Slick Liquors. Charge the system. Stretch a 2½" line to the front of the mall area in between the Theater and Beauty Salon. Use the line in a defensive mode while vertical ventilation is being performed on the roof of the Beauty Salon. Use this line as a protective line if possible to protect evacuating theatergoers.

    D. Establish a secondary water supply at Hydrant #1. Stretch a 2½" line to the front of the Beauty Salon. Stand by in a flanking position in a defensive mode until the ladder company confirms that vertical ventilation has taken place. Then, enter the Beauty Salon to back-up Engine 1's attack line.

E. Establish a secondary water supply at Hydrant #1. Stretch a 2½" line to the front of the Beauty Salon. Stand by in a flanking position in a defensive mode until the ladder company confirms that vertical ventilation has taken place. Then, take the line into Brian's Signs to cut off any extending fire in the hanging ceiling space.

5. What are your orders for Ladder 1?

A. Position in front of the Beauty Salon. Raise the aerial to the roof of the Beauty Salon. Two firefighters perform a vertical ventilation of the Beauty Salon roof to alleviate the backdraft condition. After a vertical vent is completed at the roof level, the captain and one firefighter force entry at the front door. Horizontally vent at the Beauty Salon doors and windows, then perform a search of the Beauty Salon.

B Position the ladder truck in front of Brian's Signs. The captain and one firefighter force entry at the front door for Engine 1. Perform a primary search in the Beauty Salon and horizontally vent the doors and the front show windows. Two firefighters perform forcible entry and ventilation at Brian's Signs, checking for fire extension in the hanging ceiling space.

C. Position the ladder truck in front of Brian's Signs. Split the ladder company. The captain and one firefighter force entry at front door of Beauty Salon. Perform a primary search and ventilate the windows and doors. The other two firefighters assist Engine 1 in evacuating the theater.

D. Position the ladder truck in front of the Brian's Signs. Split the company into two crews. Crew 1 raises the aerial to the roof of Brian's Signs. Perform a vertical ventilation of the Beauty Salon roof to alleviate the backdraft condition. After the vertical ventilation is completed at the roof level, the captain and one firefighter force entry at the front door. Horizontally vent at the Beauty Salon doors and windows, then perform a search in the Beauty Salon.

E. Position the ladder truck in front of Brian's Signs. Split the company into two crews. Crew 1 raises the aerial to the roof of Brian's Signs and opens a natural opening above the theater to check for the presence of fire in the cockloft at south end of the strip mall. Crew 2, after a report from Crew 1, opens the roof above the Beauty Salon to alleviate excess heat and smoke from the fire building. Communicate via radio the conditions at the roof level and in the cockloft to the attack teams (Engines 1 and 2).

6. Your chauffeur has informed that both swivels on the sprinkler siamese are frozen and will not spin. What orders will you give him to remedy this situation?

A. Hook up to the sprinkler siamese on Slick Liquors and charge the system. This is the quickest way to supplement the sprinkler system.

B. Use the mallet and hydrant wrench to attempt to loosen the connections. Advise the command post of the results.

C. Place double males on the sprinkler siamese. Connect your supply lines and charge.

D. Place double males on the sprinkler siamese and double females on the male ends of your supply lines. Charge the system.

E. Get a feed from Engine 2 (on a secondary water supply), move the suction line to discharge and pump into Hydrant #2. This will increase the water into the main feeding the sprinkler system.

7. The ladder company, after vertically ventilating, has taken out the front show windows. Heavy fire is blowing out the front show windows. What is your most serious concern regarding building stability and/or fire spread?

   A. Fire will burn through drop ceiling, exposing the roof joists, which could cause roof collapse.

   B. The fire will heat the steel I-beam acting as a lintel over the show window. This can cause collapse of the parapet wall.

   C. Fire can now burn into hanging ceiling space via concealed spaces in the front wall of the store. This will cause lateral spread of fire to other stores.

   D. Fire blowing out of windows could expose overhead utility lines, causing them to drop, endangering operating personnel and other structures.

   E. Water accumulation on the roof from master streams could cause collapse of the roof. This will cause collapse of the side and rear walls.

8. Despite your best efforts, the fire has broken through the roof and is venting upward. What is the best course of action regarding line placement?

   A. Cut a small examination hole in the roof of Brian's Signs. Stretch a line to the roof and place a distributor to cool superheated gases in the cockloft.

   B. Place a deck gun or elevating platform at the sidewalk level in front of the Beauty Salon. Use the stream to attack the fire from below, allowing the products of combustion to vent out of the roof.

   C. Utilize the elevated master stream to extinguish the fire burning through the roof and to alleviate the flying brand danger.

   D. Stretch large diameter handlines into Brian's Signs to cut off any fire extending in the cockloft.

   E. Make a trench cut on the roof adjacent to the Beauty Salon, utilizing handlines on the safe side of the trench to drive the fire back and to prevent extension to the rest of the mall.

9. Assume that this fire occurred during business hours. The fire started at the rear of the store and was rapidly advancing toward the front. Upon arrival, many people are exiting the store by way of the front exit. How would you order the first line deployed to protect the people exiting?

   A. Break out the show window and advance the line through the show window. Use no water until you have passed the people exiting the store.

   B. Keep the line at the front door. Do not initially break out the show window. This may pull the fire toward the victims. Advance line into store via front door after people have exited.

   C. Stretch a line to the sprinkler siamese. The operation of the sprinkler should keep the fire in check and allow the people to escape.

   D. Break out the show window. Direct the stream over the heads of the people to keep fire back until people have exited. Then advance a line through the show window to attack the fire in the rear.

   E. Force the rear door. Attack the fire via the rear entrance so as not to impede the egress of people from store.

10. You have been ordered to perform post-control overhaul. You find a gas can in the rear of the store. You also find that all file cabinets are open, as is the safe. What is the best set of actions to take?

    A. Use water sparingly in the area of the safe and filing cabinets. Move the gas can into a safe area so as not to contaminate its contents with overhaul water. Notify the arson squad.

    B. Use a salvage cover to protect any evidence. Overhaul store contents. Keep a keen eye out for any other suspicious conditions. Notify the command post.

    C. Do not disturb any evidence. Place a positive pressure blower at the storefront to expel any lingering "dead" smoke. This will allow the arson squad to operate in a safer, clearer atmosphere.

    D. Notify the command post. Close the safe and filing cabinets so as not to disturb the contents with secondary damage caused by overhaul. Place a salvage cover over the gas can.

    E. Notify the command post of the situation. Suspend overhaul operations until an investigation can be initiated. Set up lighting to allow for a safe investigation. Keep a charged line ready.

ANSWER SECTION

Scenario 10–1
Short-Answer Questions

1. This situation poses many existing and potential problems. All must be factored into the decision to request additional alarms. The building is relatively large and the location and the extent of the fire is as yet unknown. There is a report of at least one person missing upon arrival. Add to this the possible unreliability of hydrants in the area. Another factor to consider is the presence of plastics which can be classified as "frozen flammable liquids." Plastics will burn twice as hot as ordinary combustibles, generating huge volumes of dense, black smoke. In addition, there is the matter of the wind, which is blowing across the structure at 28mph. This wind may present some big problems for fire control. In general, any wind in excess of 15mph must be considered a severe threat to exposures. This may be reduced to 10mph or even 5mph when closely spaced combustible buildings are present, both as the fire building and as exposures. (Fig. 10–1)

Fig. 10–1 Photo by Ron Jeffers NJMFPA
Wind-driven fires have created some major fires, especially when closely spaced structures are involved. This fire in Hoboken, NJ, occurred on a bitterly cold, extremely windy day. When it was over, several large factories and other buildings and nearly 100 automobiles were destroyed. The wind was so severe that flying brands ignited a pier almost a mile away.

The proactive incident commander not only sees the situation as it exists, but has the foresight to take into the account the potential developments given the circumstances of each incident. In this particular incident, the requesting of an additional alarm is a prudent action. The potential problems that may surface outweigh the risk of "waiting to see" if they materialize. In addition, resist the inclination of requesting one piece of apparatus at a time. This piecemeal resource pool is counterproductive.

A structured additional alarm response is far more effective than requesting one engine company or one ladder company by itself. As mentioned before, what you don't need, you can stage or send back. Companies work better as task forces, such as two engine companies and a ladder company. If possible, wait until you have a task force assembled at the command post before assigning them. This will eliminate redundancy of orders, reduce communication, help decentralize command and maintain a proper span of control. If possible, place a chief or senior company officer in charge of each task force to further increase efficiency. This structured task force approach may not always be possible at a rapidly escalating fire. However, keeping this structured format in mind can help manage the fireground more safely and effectively.

2. The first line should be stretched by the personnel of the first two engine companies. This line must be of a diameter to handle the size of fire either present or expected. As the fire's size and location are unknown at this time, it is prudent for the first-arriving engine officer to attempt to determine the fire's size and location. If this is possible, then the proper size line can be stretched to extinguish the fire. If in doubt or borderline in regard to the fire size/line size decision, stretching a 2½" line is the best and wisest action. (Fig. 10–2) Remember that while the line is being stretched, the fire is growing. In these open-area structures, there will usually be an ample

Fig. 10–2 Photo by Bob Scollan NJMFPA
In large area commercial buildings, unless the fire is of very minor nature, stretch a 2 ½" line with a solid bore nozzle. Fires in these structures can grow and spread rapidly.

supply of air to feed the fire. It is not a good idea to stretch an insufficient line and be outmatched by the fire. If you're not sure how big it will be when you get there, go with the bigger line. If the fire is small, the nozzle shutoff can be adjusted to flow sufficient gallonage without causing unnecessary water damage. In almost all cases, it is best to size up the volume and location of the fire in comparison with the potential fire area before any hose is taken off the apparatus.

The presence of the steel truss roof should also be a guiding factor in strategy determination as well as line size. While the larger line has the greater reach to knock down fire from a safe distance, it may take longer to stretch, especially if insufficient manpower is present. Stretching the smaller, more mobile 1¾" line may allow for protection of the primary search, knock down a small fire, and evaluate the tenability and feasibility of continuing an interior attack. If the truss is found to be compromised by the fire, the offensive strategy will have to be abandoned.

The next consideration is where to stretch this line. There are two choices here, the main entrance on Jeep Street and the overhead door on Hegarty Avenue. While both are paths of least resistance, the main entrance on Jeep Street is the safer point of entry. This is due mainly to the direction and velocity of the wind. A 28 mph wind will be difficult, if not impossible, to advance into. The heated products of combustion will likely drive the attack team from the building. It is better to use the overhead door as a vent point and stretch to the fire via the main entrance, keeping the line between the exit and the fire, and hopefully between the fire and any victims. (Fig. 10–3)

Fig. 10–3 Photo courtesy of Hartz Mountain Industries
If overhead doors are present, a decision must be made whether to use them for attack or for ventilation, or both. This decision will be influenced by wind direction, location of fire, and accessibility.

Water supply can be established at Hydrant #1 on Jeep Street. This can be easily accomplished by the pump operator of Engine 1 and the chauffeur of Engine 2. If the apparatus is properly positioned just past the main entrance on Jeep Street, leaving room for the ladder company, these two men can handstretch LDH to the hydrant. Since the hydrant is close, there will be no need to place an engine on the hydrant.

The positioning of the second or third engine should be, depending on SOPs and/or the orders of the command post, in proximity to Hydrant #3 under the bridge. This can be utilized in case the hydrant chosen for the attack (Hydrant #1) is out-of-service. The engine can establish a water supply at Hydrant #3 and hook up to the LDH supply taken off Hydrant #1. Even if Hydrant #1 is a good hydrant, Hydrant #3 can be used as a secondary water supply. Always have a contingency plan for water supply, especially when it is known that hydrants in the area may be unreliable.

3. The second line should be stretched by the personnel of the third engine company. This crew may be assisted by members of the second engine company who may not be required on the first line or by an additional alarm company. This line must be at least the same size and preferably larger than the initial attack line. If the first line is having trouble advancing and controlling the fire, the second line can provide a one-two punch or, if equipped with a solid bore nozzle, can provide reach and penetration to knock down fire well ahead of the advancing line.

It is imperative to coordinate the operations of these two lines, along with the operations of ladder personnel searching and providing support both on the interior and the exterior. For this reason, SOPs should establish clear lines of authority inside the building when tasks present a potential coordination problem. A good rule of thumb is that the first-arriving officer be designated attack director and both lines work under his supervision. A better plan is to assign a battalion chief to the interior to coordinate all interior firefighting actions. This frees the officers on the attack lines to supervise their respective companies.

4. There are two areas of concern in this building: the cubicle area and the processing area. A rope-guided search is essential in both areas as the heavy smoke conditions produced by the burning plastics may cause personnel to become disoriented and lost in this large and confusing occupancy if a rope is not used. A large floodlight should be placed at the entrance to the building to guide searching firefighters to the proper egress point. It is imperative that members operate in pairs, however it would be better to work in teams of three or four due to the complexity of the search. Additional alarm companies must be utilized to reinforce the search operation as required. An alternative to the lifeline is the use of the hoseline. Personnel providing support to engine company operations as well as the engine members can search in the immediate proximity of the line as it advances into the structure. Thermal imaging cameras for all search teams is an absolute necessity.

There is also the problem of the missing worker. Valuable information can be acquired from the workers themselves who were attempting to fight the fire as to where the worker was last seen. This is important in this large structure as time is running against survival probabilities and the area to be searched is considerable. It is also critical to conduct a thorough secondary search as all workers and possibly customers were not accounted for in the initial victim estimation.

5. Ventilation at large area buildings can either make or break the operation. More buildings have been lost by improper, inadequate, or non-existent ventilation operations than by any other factor. Ventilation is a major priority at all fires in the offensive mode. Even fires in the defensive mode can benefit from judicious venting of, as yet uninvolved, exposed areas. Smoke generated by burning plastics gives off extremely dense and toxic smoke. It is crucial to immediately and thoroughly ventilate this building due to this lethal fire load.

A critical task in the ventilation plan is the shutdown of the HVAC system. The system can channel fire and the accompanying products of combustion into undesirable areas. Shut it down as soon as possible. Consider its use after the fire has been placed under control. A representative from the building can assist in this tactic.

Fig. 10–4 Photo by Bob Scollan NJMFPA
Buildings must be vented both vertically and horizontally. Windows may be bricked or boarded up in many commercial buildings, even those that are still in operation. Lack of ventilation usually leads to lack of interior operations.

Horizontal ventilation of the overhead door is a key tactic in this fire. It is downwind and already showing smoke. If fact, with the given velocity of the wind, opening this door may allow attack crews to advance relatively close to the main body of fire without encountering excessive smoke and heat, much like a positive pressure fan. As the fire is being attacked, other windows on the north and the east side should be taken out. (Fig. 10–4)

Vertical ventilation should also be conducted if it can be done safely. Roof crews should look for the presence of natural roof openings. Buildings with steel deck roofs may have some type of skylight or scuttle. Ensure that opening these will not spread the fire into undesirable areas. Openings if possible, should be as close to directly over the seat of the fire as possible. If they are not, the horizontal ventilation may have to be sufficient until the fire is knocked down. Cutting these lightweight truss-supported steel roofs is not recommended unless it can be accomplished from the safety of an aerial platform or other supporting device, such as a roof ladder. The roof ladder will span several joists and distribute the weight of the firefighter making the cut. The reason for this support is it may not be possible to locate the truss beneath the cut. Trusses may be spaced as much as eight feet on center in these roof systems. There is the very real danger of the firefighter performing the cut sliding into the vent hole when the unsupported steel deck bends down into the cut before it is finished. In addition, a review of truss construction reveals that the truss is only as strong as its weakest portion. As the steel roof deck is only $1/4$" thick at best, the saw can slice right through the top chord of the truss, destroying its integrity. Immediate failure can then be expected. For these reasons, cutting a steel roof deck with a lightweight truss support system should be a last resort. Venting at natural openings is a much better and safer route.

6. The fire building is of non-combustible construction. As such, the contents of the building will constitute the main fire load. It should be expected that any fire of significant proportions will cause the collapse of the roof with the walls remaining intact for longer periods. The exposure hazard presented will be one of flying brands and convection and radiant heat at the roof level. Most of this generated heat should readily dissipate into the surrounding atmosphere, but flying brands will probably present the brunt of the problem.

The electrical contractor shop is upwind and not a serious exposure problem unless the walls of the fire building collapse at a later stage in the fire and pose a radiant heat problem. This is not the most likely scenario, causing this exposure to be placed on a lower priority.

The flying brand hazard as well as any other hazard presented by this fire will be worst on the leeward side. The wind is out of the northeast, causing the worst of the exposure hazards to be the gas station, which is in the most direct line of danger, and the wood-frame motel, which by its combustible nature, will present an ignition hazard.

The best action to take, of course, is to put the fire out and all the exposure problems are eliminated at once. The question asks what your concerns are. It is best to answer what the question is asking. The gas station, directly downwind, will be exposed to dense smoke produced by plastic burning. The station can be evacuated, temporarily closed for business, and the pumps shut down to eliminate any possible ignition source such as fumes. The point to make here is to eliminate the problem before it becomes a problem. The motel, being combustible, should be kept under surveillance, especially if the fire intensifies, escalating the situation, and possibly prompting a defensive mode of operation. The fire department connection, not in the narrative but shown in the diagram, should be supplied and handlines used to keep the exposed areas wet. This need not be done initially, but the incident commander must definitely consider these actions.

In addition, the elevated roadway, Victor Avenue must be closed to traffic, as dense smoke will certainly make this road impassable. It is the incident commander's responsibility to protect all life in and around the area. This includes bystanders and any traffic moving through the area, in this case on the elevated roadway, where smoke drift might be a factor

Scenario 10–2
Short-Answer Questions

1. Engine 1 will position on Curtis Street out of the collapse zone, but as close to Connection A as possible, leaving room for the ladder company. A water supply should be established at Hydrant #3. Engine 1 must stretch two lines of at least 2½" diameter to fire department Connection A and charge the system to supply the sprinklers in the north end of the building. Connections B and C would not be supplied as the system served by these appliances is not yet in service. Supplying them would be a waste of water.

   In addition, a 2½" line should be stretched into the drug store to protect the primary search and attempt to keep the fire from extending into the drug store. If there is too much fire and the store is sure to be lost, then all lines must be withdrawn as soon as the search is either completed or abandoned and used in a defensive posture from a flank to attempt to keep the fire from consuming the drug store, a formidable task.

   If the hydrant will supply the required water, a deck gun stream can be directed into the window of the carpet store to darken down the parent body of fire. This is unorthodox, as the engine company is already supplying handlines and the fire department connection, but if water is not applied to the main body of fire, the drug store is sure to be lost. (Fig. 10–5)

Fig. 10–5 Englewood, NJ, Fire Department file photo
When fire threatens to extend both through the roof and to adjacent exposures, whether they be interior or exterior, the best engine position is at the flank where it is possible to not only establish a water supply, but also use an apparatus-mounted master stream.

2. Engine 2 should also position out of the collapse zone on Curtis Street. As the wind is blowing the heat in the direction of Hydrant #2, no apparatus should be positioned there. However, a supply line of LDH can be stretched to this hydrant while the engine remains in a safe position. This supply can be used to supply both the deck gun of Engine 2 and a tower ladder if required. As a matter of fact, if a Telesquirt is available, Engine 2's position would be a good place for it. (Fig. 10–6) The boom can be placed right into the store front to blast fire. In order for this to happen, however, some ventilation must be performed or the resulting steam may enter the other stores where operations will be underway trying to keep the fire from spreading into them. Fortunately, by the time the boom is

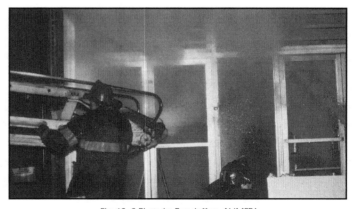

Fig. 10–6 Photo by Ron Jeffers NJMFPA
A Telesquirt is versatile and may be used to both protect exposures and to hit the main body of fire at the same time. The boom may be extended right inside windows and doors at ground or elevated levels to blast away ceilings and knock down heavy bodies of fire.

set up, the skylight on the roof should have been opened by the ladder company. Communication is essential here. No master stream should be used in the fire building indiscriminately and without orders from the command post.

In addition, Engine 2 should place in service an additional line stretched from and supplied by Engine 1 into the drug store. The operation in the drug store will be very manpower intensive. The only way to keep the fire from spreading into this exposure will be by aggressive ladder company work opening the ceiling and efficient and equally aggressive hose stream application. Due to the size of the drug store (150' deep), at least three lines will need to be stretched there.

3. Engine 3 must establish a water supply at Hydrant #1. The positioning of this company should be on the flank at the corner of Lois Lane and Row Alley. This positioning will put the company out of the collapse zone while allowing operations to protect the most severe exposures and still hit the fire in the carpet store.

   A line of 2½" diameter must be stretched into the drug store to protect the primary search and endeavor along with the lines from Engines 1 and 2 to keep the fire from spreading into the drug store from the carpet store. As the operations in the drug store are marginal at best due to the wind and the heavy fire in the carpet store, it is critical that a supervisor, preferably a chief officer, be assigned to supervise the interior operations in the drug store.

   As soon as possible, a deck gun stream should be established to wet down the four-story multiple dwelling. This stream can also be used, if necessary, to wash the exposed row houses as well as hit any fire issuing from the roof of the carpet store if the fire becomes defensive and all interior forces are withdrawn.

   Engine 4 should also establish a water supply, this one at Hydrant #4. The apparatus should position in the safest area possible given the potential fire spread profile. Two lines of at least 2½" diameter must be introduced into fire department Connection D at the rear of the Toy Company to supplement the supply established at the front of the building into Connection A. These lines must be charged immediately. Engine 4 must stretch from the Row Alley side into the toy store to assess conditions in that exposure and work to keep fire from spreading to that area.

   In addition, a deck gun can be utilized to protect the north side of the row houses and, if necessary, cover any fire issuing from the roof of the carpet store.

4. The two ladder companies responding are totally inadequate for this operation. However, properly positioned, assigned, and aggressively operating, they can make great headway on this fire. Then, proper reinforcement of key positions by additional alarm ladder companies will give the incident commander the best chance for holding this fire to the store of origin.

   As two engine companies are assigned to the front and two to the rear, so should the ladder companies be split up as well to support the engine company operations. Ladder 1, for a lack of a better term, should position out of the collapse zone at the front of the building on Curtis Street. If there is a tower ladder available, the best place for it is at the front of the building as the basket can be placed at the sidewalk level and used to direct a stream into the carpet store. The streams from both the tower ladder and the Telesquirt or deck gun should be able to blast the ceiling away and extinguish any fire in the cockloft. The crew of Ladder 1 must conduct a primary search of the drug store in as quickly a manner as possible as conditions may deteriorate rapidly given the existing body of fire in the carpet store and the wind velocity blowing the fire toward the drug store. If conditions permit, the ladder company personnel must work aggressively in the drug store to pull ceilings and expose any extending fire.

Ladder 2 will position at the rear. Since the alley is only 12' wide, it will not be possible to safely position on Row Alley, so the ladder can be positioned on Higgins Street. The roof should be laddered, either by aerial or ground ladder and the roof accessed via the Toy Company. This is a safe area. To ensure several ways off the roof, ground ladders should be raised to the southernmost end of the Toy Company, both at the front of the building and the rear. The task of this company is to effectively vent at the skylight on the fire building and, if conditions permit, cut a large hole in the roof as directly over the fire as is safely possible. (Fig. 10–7) The reason for this is to localize the fire and slow its horizontal spread into the adjacent exposures. A check should be made of the skylights on the expo-

Fig. 10–7 Skylights directly above or in direct proximity to the fire must be opened first, before adjacent, more remote skylights are opened. Knock out the boxed-out area beneath the skylight to vent the cockloft.

sure roofs to ascertain the fire extension profile in the cockloft. Communication of progress, conditions, and needs is critical. If the roof team is unable to complete the tasks assigned, the structure is likely to be doomed. Generally, if the ladder company cannot stay on the roof to complete ventilation assignments, the building and the successful outcome of the firefight will be seriously jeopardized.

5. At this incident, additional alarm companies should be utilized to reinforce operating positions already established. The amount of fire present and the amount of tasks to be completed requires additional alarms to effectively cover the building, provide relief, and keep a tactical reserve for contingency plans. However, it must be stated that the degree of this reinforcement and the decision to strike additional alarms will be dependent upon both the viewpoint of the incident commander at the command post and reports coming from around the fireground.

As the fire situation stands at the present time and thinking proactively, at least one ladder company will be required to support the roof operation, one to support the operation in the drug store, and one to operate in the Toy Company, opening ceilings and checking for fire extension. In addition, the utilities will need to be secured. If things really go bad, additional elevated master streams will be required. (Fig. 10–8) Further, the exposures on the B and C side will need to be evacuated as required by the escalation of the incident. All these tasks take manpower.

Fig. 10–8 Photo by Bob Scollan NJMFPA
At serious fires in large commercial structures, initial apparatus positioning should take into account the potential fire involvement profile. Keep them out of collapse zones and/or flanks to allow for incident escalation.

Additional engine company personnel may be required to man additional exposure lines, additional lines into the exposed stores, and to establish additional water supplies for master streams.

Also, don't forget that the more men you have on the scene, the more difficult it will be to control the operation. For this reason, additional chief officers will be required to properly maintain the span of control and best ensure the safety of those operating on the fire scene. (Fig. 10–9)

It is critical to remember that the operating personnel will need relief, requiring a constant rotation of companies. Fatigue leads to injuries—always be thinking "rehab."

Fig.10–9 Photo by Bob Scollan NJMFPA
When it comes time to surrender the building, it is best to assign a chief officer to supervise each side of the operation. This reduces radio communication and span of control while increasing overall effectiveness and safety.

It is safe to say that if this fire is not knocked down in a short time, this fire could reach as many as five or six alarm proportions, and maybe more, taxing the entire department and surrounding areas. Don't forget relocation to cover the city as well. The key is to be proactive and not put yourself or your personnel in a position of having to constantly play catch-up.

MULTIPLE CHOICE QUESTIONS

6. The establishment and enforcement of a structured incident management system is essential to the safety of operating personnel and the overall coordination of the incident.

   A. *+2* Announcing the command transfer via radio to all operating companies is the first step that should be taken once the new incident commander is in place. Especially in the early, bustling stage of the operation, confusion will be reduced and firefighters will feel more confident in the on-scene management system if structured command statements are made. This will include progress reports at regular intervals to let the troops know "where we are" at this time. Keep the players informed of who is at the helm, and the game will run more smoothly.

B. **0** Each player in the game must make an independent size-up when arriving and throughout the incident. This size-up is a mental action intended to organize your thoughts and put your head in the game. It should be begun upon dispatch, continued while en route with further information culled from radio reports, and reinforced upon arrival as the first view of the scene is absorbed. With this in mind, by the time that command is transferred, the new incident commander should have at least developed a preliminary size-up prior to taking command.

C. **+1** This is a good move as you don't want companies rolling into the scene without first figuring out what you want to do with them. It is very difficult to decommit companies once they have committed to a position. This incident requires a wide and varied positioning profile. Some companies are at the front, others at the rear, while still others should be staged and their manpower utilized to reinforce or complete some assigned task. The question, however, asks what the next step is in the incident command process. Notifying companies to remain uncommitted is not the next move after assuming command. However, it will be one of your first actions, so a point is given.

D. **-1** Unless there is an imminent danger, the command post should not routinely be relocated. There are many other problems to take care of without worrying about where the command post is located. A command post that requires relocation is indicative of poor initial positioning and planning.

E. **+1** A large incident such as this will require a strong command presence. In addition, the large manpower commitment operating in varied areas will require that a proper span of control be maintained. Requesting the response of additional chief officers will help reduce the span of control and allow officers operating at the task level to concentrate on the assignments at hand. Incidentally, the response of additional chief officers should be a part of the established additional alarm response. Establishing a chief officer response as part of the system will allow the incident commander to forecast sector/division chief positions and create a safer organization on the fireground.

7.

A. **-2** While entering the drug store to conduct a primary search is necessary, search of the carpet store, which is heavily involved in fire, is too great a risk to firefighters. It should be assumed that anyone inside the carpet store, which has been torched, is already dead. No primary search operations should take place in an area where fire is of these proportions. An extremely thorough secondary search should be conducted after the fire is placed under control and a safety check regarding the integrity of the building is completed. (Fig. 10–10)

Fig. 10–10 Photo courtesy of Mike Borrelli FDJC
A primary search will not be conducted in a fully involved area. Only a secondary search will occur here. The incident commander must conduct a risk analysis in deciding whether to conduct a primary search where live victims are unlikely to be found.

B. **+2** The drug store must be searched as the fire is heading in its direction. This search must be done quickly with a keen eye kept on conditions. Using all personnel assigned to the ladder company to accomplish this will get it done in half the time it would take if the crew were split. Lines should be stretched as quickly as possible to support the search. Only a rapid establishment of lines and related ladder support in this store will save the area from destruction.

C. **-1** A search must at least be attempted in the drug store. It is not yet involved and is the area of greatest life hazard at this time. It must be assumed that no life hazard exists in the carpet store due to the volume of fire there.

D. **-1** Even at the hour that this fire breaks out, a primary search must be initiated. There may be workers finishing the renovation job, night watchmen, and possibly the arsonists who may have become victims of their own game. This is not a totally defensive fire, but a defensive/offensive fire. This is because no operations are conducted in the carpet store, a defensive posture, and aggressive control tactics are conducted in the adjacent stores, an offensive operation. The evacuation of the exposures, if required, can be assigned to later-arriving companies.

E. **+1** This is basically the same tactic as employed in answer choice "B." The difference is that using only half the manpower of the company to search the drug store will cause it to be accomplished in twice the amount of time as keeping the crew intact will. The job may not get done if the conditions deteriorate. In a large area with the potential for deteriorating conditions, it is best to keep the crew intact and cover the area as quickly as possible. Using a lifeline and thermal imaging camera will speed up the process even more.

8. Answering this question requires some knowledge of how ordinary constructed buildings fall apart.

A. **-1** Wood burns through a rate of 1" every forty minutes. Even though the wood is oil-soaked from the previous occupant, it will still take a considerable amount of time for strength to be lost. Offensive operations will have been abandoned long before the floors collapse.

B. **+1** This roof, most likely supported by joists of at least 2" x 10" and constructed of wood planking, has some structural integrity. In the early stages of this fire, it should not pose a collapse threat and should not be the main concern, at least initially.

C. **0** While breaches in the fire partition will likely allow fire to spread to adjacent stores, it is also not the major concern regarding building stability at this fire. Before this becomes a factor, other less stable building features will likely collapse.

D. **+2** Fire issuing out of the front show windows may cause the parapet wall to expand, buckle, and collapse to the sidewalk. If steel reinforcement rods brace the parapet, the entire wall may be pulled down in one section. This factor is the main reason for positioning apparatus on the flanks in narrow areas and the establishment of collapse zones around the building.

E. **-2** The incident commander must always consider the stability of the building, especially when heavy fire conditions are present. Collapses such as those involving parapet walls can occur in early stages of the fire when men are moving about the area or in the overhaul phases when operating personnel may have let their guard down. For this reason, it is critical to establish a safety zone beneath any area supporting a parapet that is being exposed or has the very real potential of being exposed to heavy fire.

9. Inappropriate responses to the press by unauthorized personnel have caused the department both unneeded embarrassment and botched arson investigations. The rule of thumb to follow is if you are not authorized to speak to anyone outside the department on official department business, it is best to keep quiet if you are not in command. Referring the person seeking information to the incident commander is the best thing to do. Most departments have a public information officer who handles these details. (Fig. 10–11) However, if you are the incident commander, you will have to say something. Let's check out possible replies.

Fig. 10–11 Photo by Bob Scollan NJMFPA
In most departments, someone is authorized to speak to the press. Fire personnel without that authorization should not give out information to anyone, but should refer them to either the incident commander or to the designated public information officer (PIO).

A. **0** This is both a disrespectful reply and a possible cause for speculation on the part of the fire department to cover up what may be construed a scandal. Be as up front as possible without overstepping your boundaries. Disrespectful statements injure both your reputation and the reputation of the department. It is best to be firm, fair, and friendly when dealing with the press. Respect them and their rights and they should respect you and your duty to observe proper protocol regarding the release of official information.

B. **0** This is giving more information away than you are probably authorized to do. Remember that the press will take a statement such as this and declare it as arson before the preliminary investigation has even been started. Further, you can bet they will attach your name to the statement on the evening news, something your superiors will not be too pleased with.

C. **-1** Again, none of this is definite and is speculative on your part. Official information should not be released until fire investigators have had a chance to complete at least a cursory look-through. Even if it is known that a flammable liquid was used to start the fire, you are probably not authorized to offer this information.

D. **+2** This statement is short and sweet and says it all. It is the truth. You may say that the department spokesman will contact you when further information is available. This will not get you into trouble and will hopefully satisfy the press for the time being.

E. **-2** Leave the smart-aleck statements to Austin Powers.

10. At large-scale incidents where many responders are operating, one of the best ways the incident commander can address scene safety is the designation of a safety officer. Although the incident commander will still retain overall responsibility for the safety of all responders and bystanders, delegating the authority for safety to another officer will strengthen the organization by decentralization of duties, providing for better accountability on the fireground, and allow the incident commander to concentrate on mitigation of the problem. (Fig. 10–12)

Fig. 10–12 Photo by Bob Scollan NJMFPA
The assignment of safety officer has authority on par with the incident commander to "alter, suspend, or terminate" activities judged to be unsafe. This high standard of judgment demands that the safety officer position be chosen wisely.

The assignment of safety officer should not be taken lightly by the incident commander. A department-assigned ISO (incident safety officer) who responds to fires and other emergencies simplifies this decision. However, the ISO assigned on-scene must be someone who can be trusted to do the right thing all the time, who has a very strong knowledge of the job, and who possesses good leadership and decision-making qualities.

NFPA 1521, Standard for Fire Department Safety Officer, 1992 ed., regarding the Incident Safety Officer reads:

> *At an emergency incident, where activities are judged by the safety officer to be unsafe or involve an imminent hazard, the safety officer shall have the authority to alter, suspend, or terminate those activities. The safety officer shall immediately inform the incident commander of any actions taken to correct imminent hazards at an emergency scene.*

This statement basically puts the authority of the ISO on the same level of the incident commander. Remember that NFPA Standards are not requirements, but recommendations. However, failure to follow these recommendations has resulted in hefty liability. These standards are usually upheld in courts of law as the "standard of care and action" to follow. Likewise, the incident commander who fails to use the ISO properly as a partner-in-command is asking for trouble, both on the fireground and possibly in the courts.

A. *0* While this announcement is an established way to inform operating personnel that a safety officer has been assigned, it is not the most important step regarding firefighter safety.

B. *+1* This is one of the most important duties of the safety officer. However, in a disciplined department, the proper use of all personnel protective equipment should be a strictly enforced SOP. If the SOP is not followed, the safety officer may have to operate as a safety cop and focus on such minutiae as firefighters wearing gloves or having their chinstraps properly buckled. Lack of adherence to such routine details of personal safety on the part of the participants will force the safety officer to micromanage. This will put the entire fireground organization at a great disadvantage. Firefighters should properly follow personal protective procedures to allow the ISO to focus on more critical matters.

C. *+2* The safety officer must be aware of and monitor any unsafe features such as the fire's effect on the structure and how the particular structure will react and fail when assaulted by fire. This takes a good deal of fireground experience and knowledge. Ensuring that collapse zones are maintained and operational positions are safe is a major responsibility of the ISO.

D. *+1* A rehabilitation post is a crucial and required part of any major emergency scene. Fatigued firefighters are more likely to operate carelessly than rested firefighters. Constant monitoring of firefighters by company and chief officers will make the ISO's job easier when assigning companies to the rehab post; however, as the safety officer, it is his responsibility to ensure personnel get proper and regular relief. This may include advising the command post to request additional alarms to continue the firefight when companies are sent to rehab.

E. **+1** The ISO cannot be everywhere at once. Like the incident commander, he must rely on all operating personnel to be his safety eyes and ears and to both report and act on unsafe actions and conditions. It is axiomatic to say that any reports of unsafe conditions or actions should be immediately investigated further by the safety officer.

Note that there are no "negative" answer choices. This is because the tasks of the ISO are wide and varied. However, a "big picture" mentality must be established to ensure the safety of the most personnel. It would be a crime for a safety officer to be hollering at firefighters for not wearing their gloves while building parts are falling around them.

Passing Score for the Multiple Choice Portion of Scenario 10–2 = 7 Points

Scenario 10–3
Multiple-Choice Questions

1. The command post should always be established in a safe and conspicuous position. It should be easy for personnel to find, but not so close to the incident that it interferes with operations. To take this a step further, the command post should be in an area which is out of smoke drift (upwind) and located so that, if the incident escalates, the command post does not have to be moved. A command post that has to be moved represents poor planning and organization on the part of the incident commander.

   From the command post, the incident commander should be able to observe conditions and operations in the immediate vicinity. Ideally, the command post should be set up so the incident commander is able to see two sides of the building. If this is not possible, he must demand that recon reports be transmitted as soon as possible from the areas he cannot see. Recall that most problems on the fireground occur in areas out of direct sight of the incident commander. If the incident commander can see 50% of the building's exterior by virtue of effective command post positioning, then that is 50% of the picture he doesn't have to rely on others to get for him.

   A. *+1* Positioning of the command post in the street in front of the building (side A) is generally the rule of thumb to follow. In this case, it is out of the way of operations and in an area where the fire building can be effectively monitored. However, it would be safer in this case to position the command post inside the plaza. There is plenty of room and it would ensure that the incident commander is out of possible traffic on Palisade Avenue. Even if the street were shut down, apparatus movement into and out of the plaza would be easier if the command post were not right in the main route of access.

   B. *-2* The scenario states that the wind is blowing to the north. Looking at the direction of smoke movement in the diagrams provided shows which way is north and which way is south. South is to the left and north to the right. Thus, the north parking lot would not only provide zero visibility of the fire as it is behind the strip of stores on the Exposure D side, but it is also downwind—never a good place for the command post.

   C. *0* The south parking lot is also not a good choice for the command post as it is behind the stores on the Exposure B side. While it is upwind, it affords no view of the incident. Incidentally, it is an excellent area to designate as the staging area as it is close to the operational area, but enough out of the way as to afford easy apparatus movement without operational interference.

   D. *+2* This is the best place for the command post location. It is off the main thoroughfare, but also out of the way of operations and apparatus movement. While it doesn't afford the best view of two sides of the building, it will allow the incident commander to observe conditions and operations not only in the fire store, but also in the adjacent exposures. If the smoke is not drifting in that area, the best possible place to position the command post would be in the northeast corner of the center parking lot as a peek at Side D of the building may be obtainable. However, this will depend on conditions. Safety should be the guiding factor here.

   E. *-1* While this may be out of the weather, this is not a good place to initiate and maintain operational control of an incident. For one thing, it will be difficult to observe operations. These buildings may be a good place to establish a rehabilitation post as they should be warm and allow men to get relief from the elements, but it is not an ideal place to position the command post.

2. Your decision to request additional alarms should be based on a quick and pessimistic size-up. Supplementing the size-up information should be other factors including your knowledge of the building, your observation of conditions, your reports from the interior, and your present complement of manpower. Unless you know for sure that the opposite is true, it is likely that the cockloft above the drop ceiling is open over the entire row, giving the fire a ready avenue of travel. This being the case, forces must be committed to many areas to not only localize and extinguish the main body of fire, but also provide coverage to exposures to check for and address fire extension.

   A. *-1* This is not a guarantee at this time. There is a possibility that the fire will travel into the drop ceiling. If this threat materializes, a much larger complement of manpower will be required. If they are not already on-scene or at least on their way, the probability of losing the entire row is very great.

   B. *+2* One of the most important size-up factors to consider at this incident is weather. For anyone who has ever fought a fire in sub-freezing weather, it can be sheer torture. Once the initial adrenaline surge of the battle is past, the harsh reality of the weather conditions creeps in. This is especially true during extended defensive operations, but also when it is time for overhaul, salvage, and particularly during demobilization when equipment is picked up, hose drained and repacked, and apparatus placed back in service.

   A wind chill of 9°F will certainly cause much discomfort and possibly cold-related injuries such as frostbite and hypothermia to firefighters. The safety of personnel should always be the top priority on the incident commander's agenda. A warm area for rehab must be established. As mentioned before, an ideal place would be in one of the stores on the unexposed side of the fire. However, to allow men to take a break, more men will be required. Don't wait until the first alarm companies start going down, whether it be due to cold, heat, or just plain exhaustion.

Fig. 10–13 Photo by Lt. Joe Berchtold, Teaneck, NJ, Fire Department
At fires in strip malls, manpower will be the name of the game, not only to attack and get ahead of the fire, but to provide relief of fatigued companies. Weather extremes will compound this problem exponentially. Summon assistance early.

   Have a complement of manpower at least the same size as the on-scene complement to continue the operation. If you don't, once the first string is out of the game, the game will be forfeited and the building lost. (Fig. 10–13)

   C. *-1* At this juncture in the operation, there is very little threat of exterior exposure spread. Interior exposure spread is much more of a critical concern at this fire. It is one of the biggest reasons for requesting additional alarms at this type structure.

   D. *-1* The answer choice states that the hazardous materials present are involved in the two adjacent exposures. While it is true that both the Paint Store (Exposure B) and the Pool Supplies Store (Exposure D) are concerns regarding hazardous chemicals possibly stored there by nature of the occupancy, they are not yet involved in this fire, thus the hazardous chemicals are also not yet involved. This is a matter of critical reading.

E. *+2* The diagram of the buildings clearly shows smoke at the end of the row on the leeward side. This is indicative that at least the mansard is open across the front of the building. (Fig. 10–14) Smoke, accompanying heat, and possibly flame that can travel via this mansard may also find its way into the cockloft. Proper size-up of the building along with pessimistic forecasting based on where the fire is and where it is likely to go will prompt the incident commander of the need to get ahead of this fire as quickly as possible. This requires manpower and equipment.

Fig. 10–14 The mansard on the front of this strip mall may be open for the entire frontage. If fire is evident in the florist on the near end and there is smoke showing from the mansard on the far end, expect trouble.

3.

A. *-1* Hydrant #1 is the most appropriate hydrant. The supplying of the fire department connection is also appropriate; however, it is not a good practice to wait to supply the system until ordered. At a working fire, the system must be supplied immediately by one of the first-arriving engine companies. In North Hudson, SOP usually assigns this to the third-arriving engine company as the first two engines are used to get the attack lines into service as quickly as possible. We have the luxury in that all the responding companies arrive on the scene within the first few minutes. If this is not the case, it may be best to use the first-arriving engine to supplement the system, as it may not be clear when the next engine will arrive. In any case, it is imperative to supply the system as early as possible.

In addition, the apparatus placement here is acceptable, as the strip is not large and the stretch of the attack line to the fire area will be relatively short, however, it is not good practice to wait until vertical ventilation has been accomplished. This is not a backdraft situation. Delay of the application of water on the fire can have disastrous consequences.

B. *-2* Not only is hydrant #2 a poor choice as it also serves the sprinkler system in the stores, but using it to supply attack lines will rob the system of water. This can allow the fire to extend and spread throughout the row. In addition, attacking via the rear is not the path of least resistance, as the rear door is steel set in masonry. Even if the door was easily forced, this direction of attack is incorrect, as it will push the products of combustion into uninvolved areas, unnecessarily spreading the fire.

C. *-1* The hydrant selection is acceptable as long as the engine and pump operator is not left at the hydrant due to its location on the leeward side of the fire. A "drop and wrap" tactic would make use of Hydrant #3 acceptable. The strategy utilized here is defensive in nature. This is not a defensive fire. A primary search must be initiated, demanding the advance of offensive attack lines into the store to protect the search and attempt to locate, confine, and extinguish the main body of fire. Using the deck gun to initially knock down the fire will endanger any victims inside the store as well as any firefighters engaged in the primary search. In addition, the deck gun may push fire into uninvolved areas.

D. *+2* The best hydrant is chosen, the apparatus is positioned in a safe and effective manner, the fire department connection is supplied, and a line of appropriate size is stretched to accomplish the job of aggressive, offensive firefighting. The fire must be attacked at its seat before any of the exotic strategic and tactical operations stated in the other answer choices may be attempted. The 2½" line is the best choice as the attack line due to the potential fire condition given the estimated fuel load and size of the building. If the amount of water capable of being discharged from this line is not required, the nozzle operator can back off on the nozzle. However, if a smaller line is stretched and the fire is beyond its capabilities to control it, the battle may be over. In larger buildings, it is always better to get the bigger line in.

E. *-1* The hydrant chosen is downwind and the apparatus position in front of the building does not leave room for the ladder company. It is always better to leave the front of the building for the ladder truck. It is easier to stretch more hose than it is the stretch a ladder. The strategy here is that of defensive/offensive. This is not warranted given the circumstances at this fire. The first line should be stretched into the original fire store and the deck gun left unused, at least initially. Don't be too quick to give up the fire building. Chances are the fire will gain great headway and chase you out of the exposures, in which case, you will lose the exposure too.

4. Second arriving engine company duties, as at all occupancies, will be focused on ensuring a water supply is established and providing a back-up line for the initial attack. If the first engine company has already established its own water supply, the second engine company may establish a secondary supply or at least position in such a manner that a second water supply can be secured with a minimal amount of effort and time. When the potential fire area is large, think big water and have it ready to go before you need it. (Fig. 10–15) Remember that the second engine company may be required to assist in the stretch of the initial attack line. In this case, the third arriving engine company stretches the second line.

Fig. 10–15 Photo by Bob Scollan NJMFPA
Know your water supply and its limitations. If heavy fire conditions are expected, ensure both numerous and adequate water supplies are secured. Exposure protection, auxiliary appliance supply, and firefighting handlines and master streams all require ample supplies.

A. *+1* Hydrant #1 is where the primary water supply should have been established. In addition, if the fire department connection has not been supplied by Engine 1, then Engine 2 should perform this function. This will most likely be pre-determined by department SOP. In any case, the system must be supplied early. The back-up line is correctly stretched and is of proper diameter. Even if the initial attack team stretches a 1¾" line, the back-up must be larger. The 2½" line with a solid bore tip will provide reach and penetration as well as adequate water flow in case the first line has problems or is unable to advance in the presence of a high heat condition.

B. *0* The positioning of the second engine in anticipation of a defensive operation is good thinking due to the inherent structural weaknesses which make for a fast-spreading fire. This defensive positioning should be out of the collapse zone, but in an area where the transition from an offensive to defensive strategy can be smoothly accomplished. The problem with the tactics here is that a 1¾" line is stretched as a back-up line. Heavy fire must be anticipated, thus the back-up line must be of a caliber to handle this anticipated fire condition.

C. *-1* While the Pool Supplies Store is a definite concern, one that must be addressed by the incident commander in the action plan, it is not the correct location to stretch the second line. Safety of operating personnel must be the highest priority on the list of command. This is the major reason for providing a back-up line for the initial attack team. The second line is used to work in unison with, provide protection for, and reinforce the initial attack. If the main body of fire is not hit hard in the early stages of the fire, forget about saving the exposures.

D. *-2* Utilizing Hydrant #2 to supply handlines will rob water from the sprinkler system. This system is crucial in the initial stages of the fire as it will provide a holding action of the fire and may even retard fire spread into adjacent stores. Take the water away by using a hydrant on the same main and you may lose control of the operation. Even though the position and diameter of the line is correct, it is absolutely unacceptable to take water from the auxiliary system.

E. *+2* Defensive positioning is a matter of forecasting the potential for a fire beyond the control of operating forces. Although Hydrant #3 is on the leeward side of the fire, it is the only other available hydrant other than Hydrant #1, which should be the primary hydrant. If additional water supplies are required, relays may need to be established from outside the mall area, possibly on the other side of Palisade Avenue. Knowledge of alternative and additional water supply sources is critical to the success of the operation at large fires. In addition, the back-up line is of the proper diameter and correctly stretched to a position to back-up and reinforce Engine 1's attack.

5. The third line may be stretched by the third or fourth arriving engine company. I usually like the crews from the first two engine companies to stretch the initial attack line, but as this is a ground floor fire, the stretch will not be too long, and should be able to be accomplished by one company. If there are any doubts, then the first and second engine crews should stretch the first line, with the third-arriving engine stretching the second line. This will depend mainly on available manpower at the scene. It may be possible to use two companies to get the first line in place, then re-assign the second company to the second line, especially at a ground floor fire. The key here is to use as many men as are required to get the first line in position and operating. Once this is accomplished, work on reinforcing this attack position and getting ahead of the fire.

A. *-1* This is a matter of critical reading. As evidenced by the direction of smoke movement in the diagram, the Paint Store is actually on the windward side and should not receive as much priority as the attached and exposed stores on the leeward side.

B. *-2* The rear of the building is not only an area of difficult access and most resistance, but operating a line from this area would cause opposing lines with the initial attack lines, never a good tactic. If you chose the rear of the store for your initial attack, you lost two points there too.

C. *+2* This third line must be utilized in a "head-'em-off-at-the-pass" fashion. The Pool Supplies Store is the most immediate leeward exposure and represents the most likely path of fire travel. As soon as the initial attack is established and reinforced in the Florist, a line must be placed in the Pool Supplies Store as quickly as possible. The ceilings must be pulled, and streams operated into the cockloft to extinguish any fire spreading via the cockloft.

D. **0** A third line into the Florist would cause line congestion and not accomplish much. If the two lines operating there are outmatched by the fire, it is time to seriously consider a defensive operation or at least a defensive/offensive operation where the attack lines would be backed out and a master stream utilized to knock down the heavy fire in the Florist. Once this is accomplished, the attack lines can attempt to advance again into the Florist. However, even while the master stream is operating into the Florist, additional lines can be operating in the exposures, keeping the fire in the area of origin. This takes a great deal of coordination and discipline. For this reason, an additional chief officer or two should be assigned to key areas on the fireground to effectively assist in the coordination of the operation. (Fig. 10–16)

Fig. 10–16 Photo by Joe Berchtold Teaneck, NJ, Fire

Even if control is likely, lines must be stretched into attached stores, especially on the leeward side. Fire can spread unnoticed via pokethroughs and be past your position before you know it. Don't take this chance. Too many areas covered are better than not enough.

E. **+1** Although the roof is not the best place for the third line, one will be required there eventually. Stretching a line to the roof to protect the roof crews venting the fire area is an acceptable safety-oriented action. It deserves a point, but not two.

This fire is by no means a three-line fire, especially if it is of any significance. The three-line placement described for this fire is aimed at the areas of top priority which should receive primary attention. Any lines stretched in addition to the first three must be positioned according to present and anticipated conditions. Any large fire in a strip mall will require many handlines in addition to the possibility of master stream devices.

6. Ladder company operations in strip malls are very labor-intensive. The more successful the efforts of the crew are in confining the fire through effective ventilation operations, the better the chance of keeping damage to a minimum and saving the building.

A. **0** Primary search is an absolute must at any offensive fire operation. The order to conduct a primary search of the Florist will give any victims still inside the best chance of survival. However, the efforts of the other half of the crew operating in the Pool Supplies store may cause the fire to be pulled in their direction by pulling the ceilings prior to the roof being ventilated. Roof ventilation will channel the products of combustion to the exterior via the most effective path of least resistance, the scuttle, and must be completed before any ceilings in the exposure are opened.

B. **-2** Keeping the crew intact may be counterproductive at this fire, especially when only one ladder company is on the first alarm response. While operating in pairs, try insofar as possible to cover as much of the fire area as you can. Get men into the fire area and above the fire to cover as much of the critical areas as possible.

In this answer choice, no primary search is conducted at all. In fact, no interior ladder operation of any kind is conducted in the fire store. Remember that life safety is the primary fireground objective, followed by incident stabilization, and property conservation. The only way to ensure life safety is maintained is by conducting a systematic primary search of the fire area. It is also unsafe in a mall with attached buildings to ladder the fire store. If conditions deteriorate, the egress point may be cut off. It is better to ladder the windward exposure and use that area as both a tool staging area and an area of refuge. Especially where a scuttle is present, it is not necessary and dangerous to cut a vent hole in a steel roof supported by lightweight steel, open-web parallel chord trusses. Also, the front show window can easily be broken from street level. It is not necessary in this fire to vent it with a tool from the roof. (Fig. 10–17)

Fig. 10–17 Many non-combustible buildings will be equipped with scuttle hatches over each store. Opening this hatch will help to localize the fire to the store of origin. This is much safer and quicker than cutting a metal deck roof.

C. **0** Splitting the company is best in most operations. One half of the company is conducting the primary search. However, the other half is operating on the rear. This is not effective. Breaching the rear wall will be time consuming. While this is being attempted, the fire will be spreading and it is probable that by the time the task is complete, the cockloft will be well on its way to becoming fully involved. The best action to take is to get to the roof and attempt to confine the fire first by vertically ventilating at the scuttle. Leave the breaching of the wall to a later-arriving company, if necessary.

D. **-1** Laddering the leeward exposure is potentially dangerous and may leave no path of escape as the fire spreads with the wind. It is best to ladder the windward exposure. The action of first opening the scuttle on the leeward exposure will almost guarantee extension to the cockloft of this exposure. This action is likely to pull the fire into the exposure. It is best to open as close to directly above the main body of fire as is safely possible. This means opening the scuttle of the Florist first. The fact that the primary search is properly conducted saved you from a **-2**.

E. **+2** The windward exposure is laddered. Proper confinement operations such as venting the scuttle of the fire store and horizontally venting to alleviate conditions on the interior are accomplished. The primary search is also being conducted. These are all proper actions by the first ladder company. The incident commander must monitor progress reports from these positions and reinforce them as required. Negative reports from the ladder crew either on the interior or the roof might be a cause for the incident commander to consider pulling the plug on the offensive operation and shift the strategy. (Fig. 10–18)

Fig. 10–18 Photo by Bob Scollan NJMFPA
The incident commander must monitor, both visually and via radio, the conditions in, above, and around the building. If the roof team cannot operate at the roof level, it is likely that the interior operation will be discontinued as well.

7. The fact that there is a question about a second ladder company should clue the student in to the necessity of a second alarm. That is why it is important to read all the questions first before you answer them. There is often information in the question which can lead you in the proper direction.

   A. *0* Laddering the leeward exposure puts the crew in danger if the fire suddenly escalates. However, this is an acceptable tactic if a ladder has already been placed on the windward exposure. The more egress points on the roof, the better. The extension of a secondary search cannot be undertaken until the fire is under definite control. It is also a good idea to use different personnel than those who conducted the primary search. As the fire has not yet been placed under control, this crew can be better utilized.

   B. *0* The operations in the exposures is crucial, especially those actions taken to determine and limit fire spread into adjacent occupancies. Breaching the rear wall should only be undertaken if the attack line is stalled and there is no other venting opportunity. It creates extensive secondary damage and is a time-consuming process. However, there will be times when this action is warranted. Sometimes, an action such as breaching is a last resort to support the advance of the line deeper into the structure and closer to the seat of the fire. If breaching becomes necessary, then it should be accomplished without delay. Any problems should be reported immediately to the incident commander. In addition, completion of the task should also be reported to the command post. The problem with this answer choice is that it is more productive to split the company to accomplish the many objectives required of ladder companies.

   C. *+1* This is basically the same tactics as answer choice "B." However, it gets the point because the company is properly split to accomplish the tasks. So long as men can work in teams, it is acceptable in most cases to split the company.

   D. *+2* Lines must be placed by engine company personnel in the exposures, both on the leeward and windward side. These companies will require adequate and timely support. This will include pulling ceilings and opening walls, horizontally ventilating, providing lighting, shutting utilities, and other related activities intended to confine the fire and control its spread. Splitting the company will provide two teams that can accomplish twice as much in two different places. As conditions warrant, and they probably will, reinforcement of this split company will be a necessity in both exposures. Fire spreads rapidly in these buildings. The only way to stop it is to get forces aggressively working on either side of the main body of fire as quickly as possible

   E. *-1* The poor tactic here is that operations in the windward exposure at the expense of the leeward exposure will put the entire leeward side of the structure in jeopardy. Operate leeward first, then worry about windward. The fire will take the path of least resistance. Be there to stop it.

8. If the store is closed, the first thing the incident commander should take notice of is the potential for a backdraft. If the indicators for a backdraft are not present, the door must then be forced. The methods which may be used have varying degrees of success based on the firefighter's knowledge of doors, his tools and their limitations, and his experience. It is not the intention of this text to explore the intricacies and nuances of locks and lock assemblies, however, a basic understanding of what works best and in what situations is essential to fireground operations.

A. **-1** Breaching the wall of an uninvolved store not only causes unnecessary damage, but needlessly exposes the as yet uninvolved store to fire spread via this firefighter-created opening in construction. When fire barriers are present, try to maintain their integrity as long as possible.

B. **+2** Most doors in these type occupancies will be constructed of metal set in metal frame. Some type of cylinder lock mortised into the door will usually secure the door. This is a very difficult door to force using conventional methods. In this case, the through-the-lock entry method is the most efficient and quickest way of entering the building. The K-Tool is slid behind the lock and the cylinder is pried up and out of the door. From there, dependent on which type of cylinder is present, a key tool is used to duplicate the action of the key in the lock. An A-Tool is also effective on these type doors and is actually a better tool to use because it works on both doors with flush-mounted and recessed cylinders. In any case, this through-the-lock method of entry is the most effective means of forcing these type doors. Firefighters using these methods must be aware of such conditions as backdrafts and ensure once the door is forced, it can be controlled. (Fig. 10–19)

Fig. 10–19 Both the K-tool and the A-tool will work well on tempered glass doors set in steel frames. Through-the-lock will be necessary to defeat this lock. Breaking the glass on the door may create more problems than it solves.

C. **0** Shattering the glass will provide a quick means of entry into the building, but it creates several problems. First, these doors are often equipped with tempered glass. It is very difficult to shatter tempered glass. The best way is to strike the bottom corner with the pick of an axe. This will result in a zillion tiny pellets of glass strewn around the building's entrance, which become a slip and laceration hazard. Second, it does not maintain the integrity of the door. Any superheated products of combustion are now uncontrolled as far as a barrier between them and the operating crew at the front door. Finally, there will still be the problem of the bar across the door used by patrons to push open the door (see the diagram). This bar is often very strong and extremely difficult to remove from the door. Bashing it down with an axe or similar tool often only further restricts the opening in the door. Usually, when the door glass is smashed, the door frame remains intact. Thus, the door is still not able to be forced. This is because the door is often locked by a key on both sides. Operating companies then have to enter the store under the bar. Smashing the glass is a quick way, but it is hardly efficient.

D. **-2** A door on a commercial occupancy must, by law, swing in the direction of egress. This will mean that the door will swing out of the structure toward the firefighter on the exterior. The rabbit tool, also called the hydraulic forcible entry tool or HFT, is only effective on inward-swinging doors. The tool will be useless on this type of door. The only use for it may be if the firefighter, out of frustration and desperation, uses it to smash the door glass. It is imperative that firefighters know the applications and, more importantly, the limitations of their equipment.

E. **-1** The hinge pins are usually machined to be constructed of a single unit and will not be available for pulling, unlike the pins commonly found on wooden swinging doors. This will also be a wasted effort.

9. The key to the question is the time frame. The question addresses the initial stages of fire-ground operation and the most immediate concerns regarding the façade. Focusing on the time frame should direct the reader to the best answer.

   A. **+1** While the collapse of the façade and the subsequent failure of the parapet wall should be a concern at this fire, it will, under most circumstances, not be an immediate concern. The fire is located in the rear and should not initially cause a major problem of this magnitude at the front of the store.

   B. **+1** The reasons for this answer are almost identical to the reasons in answer choice "A." To destroy the façade, distort the steel lintel I-beam by the action of heat, and bring down the parapet wall will all take time and should not be the focus of attention in the initial stages of the fire operation. If the fire condition significantly escalates and these areas become exposed and involved, the incident commander's focus on these dangers becomes a higher priority as they come closer to becoming a reality.

   C. **+2** The diagram shows that smoke has already entered and is traveling via the façade area at the front of the store as evidenced by the smoke showing at that level in the fire store and at the most leeward exposure (Gino's Pies). Influenced by the wind, heat and fire will likely follow the smoke into the areas of least resistance—the rest of the façade and the cockloft of the adjacent stores (if it's not already there). This indicates a need for a large commitment of personnel, as the requirements of pre-control overhaul operations will be extensive and manpower-intensive. (Fig. 10–20)

Fig. 10–20 This lightweight wood mansard is open over the entire perimeter of the store. Expect fire control problems and a long, manpower-intensive operation.

   D. **-2** Again, in viewing the diagram provided, the presence of power lines is not evident. They are therefore not a concern.

   E. **0** Any water accumulation inside the façade will not be problematic in the initial stages of the operation. Façade collapse due to water accumulation will most likely be a result of accumulated effects of interior handline or master stream operation, both of which will not make an early impact at his fire.

10. This is a straight knowledge question regarding your knowledge of the involved type of construction.

    A. **+1** The parapet wall is certainly a significant collapse hazard, however, it will only involve the immediate perimeter of the building. In addition, a lag time will be present where the parapet wall and the supporting steel lintel I-beam are being heated by the fire. As this becomes evident, collapse zones should be established extending for the full height of the building in addition to the width of the parapet wall. No operations should be permitted in this area. (Fig. 10–21)

Fig. 10–21 Photo by Bill Tompkins
The greatest structural threat in a non-combustible building heavily involved in fire will be the collapse of the unprotected lightweight steel trusses which support the roof. This collapse can occur in as little as 5 to 10 minutes.

B. **+1** Collapse of the drop ceiling can trap firefighters underneath, causing burn injuries and possibly fatalities. This will depend on such factors as the type and weight of ceiling and the depth of the store. However, this is not the most significant collapse hazard at this type construction.

C. **+2** The lightweight roof system can be expected to fail in as little as five to ten minutes of fire exposure. The presence of the 2.5 ton HVAC system on the roof represents a significant concentrated load that will contribute to and diminish the time for collapse. Failure to recognize the collapse dangers inherent in this type construction can cause the incident commander to allow firefighters to operate inside this structure for too long. Reports from the roof as well as progress reports from the interior should guide the strategy here. (Fig. 10–22)

Fig. 10–22 Heavy roof equipment represents a concentrated load and will decrease the time it takes for the roof to fail. Companies operating on the roof must make the presence of this load known to the command post as soon as possible.

D. **+1** The mansard, while it may pull the parapet wall down with it, represents a probable local collapse threat. It is probably more prone to burn-through and demolition by flame which will cause it to fail in smaller pieces. It is not likely to fail as one unit, however, when establishing a collapse zone, the incident commander must be prepared for this to occur. It is always better to err on the side of safety when establishing collapse zones than to wish you had later.

E. **-1** The steel lintel will not collapse, but will attempt to elongate and possibly buckle and twist due to the effect of heat on unprotected steel. This may cause the collapse of the structural members it is supporting. The prudent incident commander recognizes these deficiencies in construction and takes measures to ensure the safety of the operating personnel.

Passing Score for Scenario 10–3 = 14 Points

SCENARIO 10–4
MULTIPLE-CHOICE QUESTIONS

1. To properly address the situation that exists in this scenario, the first officer on the scene must be able to recognize the potential for backdraft. Many of the classic indicators are present here. The store is sealed up and has been closed for over 24 hours. The built-up wood and tar roof will hold the products of combustion inside the store and keep them from burning through the roof, at least initially. The cockloft is not common to the rest of the building so heat that would usually dissipate by spreading through a common cockloft cannot do so. In addition, the presence of puffing gray smoke combined with window glass stained black is indicative of a third (smoldering) or decay stage.

    A. *+1* The building is of ordinary construction. Therefore, the roof and the supporting system is combustible. Add to this the presence of the 2.5 ton HVAC system on the roof and the collapse potential increases. However, the cockloft is not contiguous, which lowers it on the priority scale when addressing horizontal fire spread. The adjoining structures' ceilings will certainly need to be examined as there may be pokethroughs, but the compartmentalization of the cockloft is a major ally to the firefighting operation.

    B. *-2* The phase of this fire is not a size-up factor and cannot be considered as a valid answer.

    C. *+1* The streets around the mall are congested and may cause a delay in initial operations. This condition will also lead to a crowd control problem. Be prepared for panicky shoppers and take steps to control the area around the fire building.

    D. *+2* Apparatus and manpower are totally insufficient for this fire. Ten men and a chief officer cannot even begin to address the problems inherent in this situation. Priorities must be established. Lines must be stretched and immediate steps taken to address the problem at hand—the fire. Other concerns such as the theater should be assigned to other agencies and companies responding on additional alarms. (Fig. 10–23)

    Fig. 10–23 Photo by Bill Tompkins
    Strip mall fires, especially those of ordinary construction, will be extremely manpower-intensive operations. The major factor in whether the incident command system can control the fire may be directly proportional to the amount of manpower on hand in the initial stages of the operation.

    E. *+1* There is no life hazard in the fire building as it is closed and has been for some 24 hours. However, there is a potential for a major life hazard problem at the other end of the mall. Fortunately, the fire seems to be contained at this time to the end store. The presence of the fire wall between the stores that eliminates the common cockloft is the most significant factor in determining where to operate.

2.

    A. *0* Some manpower will indeed have to be committed to the safe evacuation of the theater; however, it is the fire building and the attendant exposure that should receive the most attention and the major commitment. The police under the guidance of and in coordination with the fire department can coordinate the evacuation of the theater. Liaison should be established with the police as soon as possible to carry out this assignment with the least amount of fanfare possible.

B. **-1** Even with the firestopping present, the on-scene complement of manpower is insufficient to conduct this operation safely and efficiently. The incident commander must ensure all positions are covered as well as providing for a strong tactical reserve. Additional alarms will provide this strength.

C. **0** The fact that the roof may be overloaded by the air HVAC unit is no reason whatsoever to pursue a defensive strategy at this time. While the backdraft condition will cause a lack of offensive action at the outset, once vertical ventilation is completed, the fire strategy will become offensive as lines are aggressively advanced into the fire area.

D. **-1** The reason that this is a **-1** answer choice is due to the fact that to call for a second alarm prior to arrival based on the possible expected conditions is both irresponsible and dangerous. This action must wait until the initial size-up has been completed and a better idea of scene conditions is ascertained. Especially in this congested area, calling for a second alarm before the initial companies are on the scene will only add to the already congested area and possibly hinder initial scene operations. Although the reasons stated for the additional alarm may be valid, especially the weather condition because cold-related injuries can be expected in this frigid setting, additional alarms should only be based on a proper size-up once on-scene.

E. **+2** The key here is the recognition of the backdraft condition and the attendant firefighter safety concerns. This explosive condition is a threat not only to the firefighters, but also to all the patrons in the area. Recognizing this and assembling troops in anticipation of a potential major incident will keep the incident commander one step ahead of the fire and provide for the safety of operating personnel in the most efficient way possible.

3.

A. **-2** Forcing entry and entering the building given the conditions will likely cause a backdraft explosion. It is imperative that both officers and firefighters immediately recognize and take steps to protect themselves from unsafe operations such as this. Both the strategy and tactics demonstrated here are incorrect and show a dangerous and uninformed sense of judgment.

B. **-1** The theater is not the main concern at this time given the amount of manpower available on the scene. Operating in the fire area will provide the best protection for patrons of the theater and the rest of the mall. Utilizing a localizing strategy (venting at the highest point and then attacking) and a cut-the-fire-off approach (ensuring no fire extends into the adjacent occupancy) will make the most efficient use of available manpower. As the fire area is stated to be somewhat compartmentalized, the urgency that would be found if the cockloft was common is minimized.

C. **+2** This is the safest action given the circumstances. The sprinkler system is supplemented, apparatus are placed in areas out of the potential blast area, appropriately-sized handlines are stretched to flanking positions, and then the ground operation stands by in a defensive mode until roof operations have been completed. Venting at the highest point is explosion control, and will channel the superheated, unignited gases into the path of least resistance—straight up and out of the building. There actually may be ignition at the roof level or just above it as gases that were too rich to burn in the cockloft are properly mixed with the atmosphere and flaming combustion results. Coordination and communication are critical at this operation. Once the gases and fire are venting from the roof, the front show windows can be taken out, the front door forced, and lines aggressively advanced into the store. The operation then becomes one of an offensive strategy. (Fig. 10–24)

Fig. 10–24 Photo by Bill Tomkins
Where heavy fire involvement is anticipated or already occurring, use of large diameter lines will be necessary. Small diameter lines will not have the cooling or penetrating power required to knock down the fire. Solid bore nozzles should be used for reach and penetration.

D. **+1** This is basically the same strategy employed in choice "C." However, once the topside venting has been completed, the decision to attack with a 1³/₄" line puts the attack team at a great disadvantage. It must be anticipated that once the roof is vented, the fire will almost immediately return to a fully developed fire. The fire load of a commercial occupancy can easily (and quickly) overmatch this small diameter line. It is best to go with the big water here. The lines can be initially laid out by the insufficiently manned first-arriving engine companies; but once additional alarm companies arrive, they should be assigned to assist on these lines. Remember that the roof venting will take some time. In this type fire situation, it is best to slow things down and get your attack plan organized and ready for action. The stalled combustion condition in the fire building should stay that way unless an improper opening is made at the wrong time and wrong place. Controlling the operation and the manpower will afford the incident commander the best chance for success.

E. **0** There are several things wrong with these tactics. First, at a working fire, if the company is going to take the time to connect to the fire department connection, they had better charge the supply lines. Otherwise, they might as well not hook up at all. Second, stretching two lines with the manpower of the first engine is counterproductive. Get the first line in position so that when the roof ventilation is complete, the line can move in without delay. A line to protect the roof team is important, but not the first priority here. Also, if the lines are being stretched by three men, who is on the pumps? Another poor decision is the line size. As mentioned above, the initial attack line should be of sufficient diameter to control the anticipated fire condition. It is also poorly positioned at the front door. Should the plate glass window blow out, the personnel would most certainly be seriously injured by the glass, the ensuing fireball, and accompanying blast of heat. Stand clear until it is safe to enter.

4.

A. **0** It is better in most cases, and certainly in this case, to provide back-up of the initial attack line. The fire situation, once the roof has been vented, can be anticipated to be severe. Provide safety and reinforcement of the initial crew, then worry about other areas.

B. **-1** While waiting for the roof ventilation to be completed is correct, stretching the second line from the rear will cause the line to be in opposition to the initial line. This is never a good tactic. When stretching a back-up line, it is almost always best to deploy and operate the line in the same manner and directional path as the initial attack line.

C. **0** This is inappropriate for several reasons. First and foremost, the second line in almost all cases should back-up the initial line. This is well established in this text. The stated use of the line here is a waste of manpower and water. Also, given the temperature, any water used in the vicinity of evacuating patrons would create an ice hazard. The hydrant choice is also not the most desirable, as it is the most remote. It is not the best place for the secondary water supply.

D. *+2* The company here works in coordination with the first team. It is best in this situation to order two lines standing by at the ready when it is time to attack the fire. It is the safest and most sensible action at this fire.

E. *+1* The initial line placement here is correct; however the destination of the stretch is not. Ideally, the adjacent exposure is where the third line should be placed. The hanging ceiling space or cockloft must be checked, but it is more important for the second line to provide back-up in the fire store.

5. It must be announced in the incident commander's initial status report that a primary search will not be extended due to the presence of a backdraft condition. This statement places the operation in the incident stabilization mode as the undertaking of activities normally indicative of the rescue or life safety mode are unsafe and will not be conducted at this fire situation, at least not in the fire area anyway.

A. *+1* The tactics performed here are correct given the fire condition. However, in a potential backdraft situation, no apparatus should be positioned directly in front of the show windows of the fire building for the same reasons that personnel should not be placed there: safety from possible explosion. An explosion will take the path of least resistance and blow out the front show windows. In addition, as the building is attached, it is also much safer to raise the aerial to the windward exposure to allow the roof team to access the roof in a relatively safe area. The windward exposure, when available, is also an ideal place to establish a tool staging area. It is unsafe to leave tools all over the roof of the fire building, which is likely to be an area of poor visibility. Tools not being used should be left in a designated staging area so they can be found easily when needed and so they do not become a trip hazard. This is true even if the building is unattached and the ladder must go to the roof of the fire building. In this case, a tool staging area in as safe a place as possible should be established and maintained. A good place in this situation may be next to the parapet wall in an area somewhere near the access/egress point. This aerial positioning to the windward roof also provides an egress point from a safe area should conditions deteriorate or the roof over the fire area appears to be in danger of collapse.

B. *-2* Can you say Boom!? Companies simply must recognize the indicators of a potential backdraft condition. To the student taking this test, if you pursued an offensive attack with the first two engine companies, then this tactic matched your strategy. Such are these type tests. The slide downhill is rapid, and rightfully so. Firefighters who cannot recognize this scenario as a backdraft probably do not deserve to get promoted anyway. Learn the lesson here and now before it is too late, or worse, the mistake is made on the fireground.

C. *-2* This is another "Boom!" scenario. However, sending two men to the theater, while not correct, may have inadvertently spared their lives. You still lose two points.

D. *+2* This is what can be termed a delayed two-pronged approach to support operations in this fire. First, the building must be opened to relieve the superheated gases and backdraft condition in the cockloft. This is roof support. Secondly, a forcible entry team must be standing by in safe position ready to be deployed once the roof team confirms that the topside vent operation has been successful. This is the ground support portion and will provide both entry and ventilation. Ventilation of the front show windows from below will supplement and further establish the venting direction of heated gases exiting via the roof while cooler gases follow the attack teams into the building. If it can be safely done, the ceiling of the store should be pushed down from the roof. Venting fire from the cockloft may make this impossible. In this case, the ceilings should be pulled from below as soon as possible to establish the proper ventilation flow. Notice here also, that the ladder is placed in front of, and the aerial is placed to the windward exposure. It is safer operation.

E. **0** These tactics are backwards. The reason that the roof is vented as directly over the fire as possible is to localize the fire and confine it. Channeling the products of combustion out of the building in as direct a manner as possible prevents horizontal extension under the roof. Even though the scenario states that the cockloft is not common, it doesn't justify opening an exposure roof before the fire building. The exposures can be opened and checked only after the roof of the fire building is opened and examined. This includes both natural openings such as bulkheads, scuttles, skylights, and soil pipes and forced openings such as cutting the roof. Work the fire building first, then worry about exposures.

6. This is a problem-solving question. One of the most distinguishing attributes of the competent firefighter compared to the ordinary civilian is the ability to solve problems and adapt to a situation during an emergency. This is the main reason that the fire department is summoned in an emergency. Something I was taught as a firefighter and then as a company officer was that if I was to go to a superior with a problem, I should also have a proposed solution. As a chief officer, I strive to instill this in my subordinates, for when problems can be solved or at least attempted to be solved at the lowest level from which the problem originated, it makes the entire organization more effective.

A. **+1** Hooking to another hydrant and feeding a different sprinkler connection is a definite solution to the problem, but it will take time if the engine has to be repositioned or another engine has to perform the task. There are other ways of quickly fixing this problem. However, as the problem is solved, a point is awarded.

B. **0** This is the brute force method and will probably not work at this juncture. It is hoped that before notifying the command post of the problem, this was already attempted and was unsuccessful. This shows a lack of adaptability and functional fixity.

C. **-1** While the idea is right, the puzzle is not complete. Know your equipment and how to use it to adapt to situation.

D. **+2** This is the right fix. The ends of the supply lines will be male. The fire department connections are female. Adding a double female to the supply line and then a double male to the fire department connection will allow the line to be connected and solve the problem in the least amount of time. (Fig. 10–25)

Fig. 10–25 If the female swivels on the FDC are inoperable, the quickest solution would be to attach a double male to the FDC and a double female to the supply line. Then the lines can be coupled properly, causing the least amount of delay.

E. **-2** Fire department connections are fed to supplement the water being supplied by the main feeding the sprinkler connection. That is their function. If for some reason the main OS&Y valve between the sprinkler system and the main is closed, any action intended to increase the flow of the hydrant main will be a wasted effort. To attempt to circumvent the fire department connection and increase the flow in the main is an end-around attempt at solving the problem. It is quicker to use adapters to solve the problem and correctly supply the system. These wordy answer choices usually lure in the uninformed. The effective problem-solver is an asset to fire department operations. Officers must develop their subordinates to be problem-solvers and solution-finders. Anyone can find problems. It is the valuable firefighter who routinely seeks and develops solutions.

7. To answer this question, knowledge of building construction and the attendant problems characteristic of each is required. The situation is that fire is blowing out the front show windows. Addressing this problem will lead to the most correct answer.

A. **+1** Wood joisted roof members should be of the dimensions of at least 2" x 10" and may be as large as 3" x 12". It will take some time for the fire to burn through the drop ceiling before it begins to attack the roof joists. These substantial joists will remain intact for a long period of time under the assault of fire. However, the presence of the HVAC unit on the roof may be a load the roof was not designed for and the roof may not have been properly reinforced to hold this unit. This factor may precipitate an earlier collapse than expected. Ladder crews on the roof should make the incident commander aware of the presence of this unit as soon as possible so it can consider it along with other strategic factors. (Fig. 10–26)

Fig. 10–26 Photo by author
This ventilation and processing equipment represents an extremely heavy roof load. Roof teams must make the command post aware of such concentrated loads. A strategy modification may be the warranted because of this hazard.

The problem stated here is that fire is blowing out the front windows and exposing the construction features at the front of the store, causing a more severe problem. Once again, the location and extent of the fire along with the path of least resistance will lead to the best answer choice.

B. **+2** In these old one-story taxpayers, there will often be large show windows at the front of each store. A steel I-beam that acts as a lintel is located above the show windows. A lintel is usually located over an opening such as a window or door. Above this I-beam, the parapet sits. The purpose of this load-bearing I-beam is to balance out and distribute the weight of the parapet wall constructed above it. Fire blowing out of the front show windows may cause expansion of this I-beam which can cause the parapet wall to collapse onto the sidewalk below. Many times, these parapet walls are also reinforced laterally. This means that a collapse of one section can bring down the wall for its entire length, like a wave. Horizontal collapse zones for parapet walls should be established for the entire length of the wall. (Fig. 10–27)

Fig. 10–27 Photo by Bob Scollan NJMFPA
Although they may withstand collapse for a longer period of time than their non-combustible counterparts, ordinary constructed strip malls are subject to complete fire destruction and collapse.

C. **0** The front walls are constructed of brick. There are no concealed spaces in the area of fire impingement. If the framing around the show window is wood, there may be some ignition in this area, but it will not be a major concern. In this scenario, the cockloft is not common, so lateral fire spread is not a major problem here. Remember that this is a test question. In the real world, the hanging ceiling spaces of the adjacent store must be examined, but on a test, use what is given: no common cocklofts. If you "what if?" this question, you may be led to the wrong answer.

D. **-1** If the diagram is examined, there are no power lines present. In addition, the question asks about building stability and fire spread. Fire exposing utility lines does not fall into this category.

E. **-1** Water accumulation from master streams is definitely a concern and one that must be monitored whenever master streams are utilized. However, this is not an initial problem and it is not as severe a problem as the other construction-related weaknesses stated above.

8. By this time, sufficient companies should be on-scene to not only get ahead of the fire, but also continue the attack. If they are not, a priority decision must be made. Defensive tactics must be planned for in advance, especially in initial positioning of apparatus. Forecast potential fire spread early and your chances of having to reposition apparatus later will be minimal. The tactics themselves must be well-grounded to keep the fire from spreading and must be based on what is burning versus what is left to burn. It may be necessary to write off some areas of the building to save the rest of the structure. Get ahead of the fire and hit it hard while at the same time continuing to attack the main body of fire with master streams.

A. **0** The tactic of using a distributor in such a large roof area will be ineffective. Distributors are generally effective over the main body of fire. In this case, the fact that there is no common cockloft makes this tactic somewhat of a waste. There are better places to stretch lines.

B. **+2** When fire is burning through the roof, let it continue in that upward direction. So long as a major flying brand hazard is not present, remember that fire venting vertically is not fire spreading laterally. Streams should be used to attack the fire that is still burning under the roof from below. This direction of attack is the path of least resistance for stream penetration that would actually hit fire and not add to the photo opportunity of having many master streams directed at and above the roof. (Fig. 10–28)

C. **-1** The combustibility of the roof will cause the generation of thick, black smoke and some flying brands. The problem with the tactics here is the application of a master steam into the roof vent opening. Vent holes, even those created by the fire, should not be the place where streams are applied until the fire has burned away a major portion of the roof. The action of the stream tends to force the fire back under the roof, eliminating the venting action and causing lateral fire spread under the roof, possibly to exposures.

Fig. 10–28 Photo by Bill Tompkins
Fire venting from the roof is a positive sign in that it may not be spreading laterally as rapidly. Use the exterior lines from below to attack fire burning beneath the still-intact roof. Heavy caliber streams may knock down a major portion of the problem.

I was in command of a fire in a lightweight wood-truss townhouse complex. Fire was heavy and was threatening to spread to the entire complex. The threat of lateral fire spread to adjacent units diminished greatly once the fire had vented itself through the roof. At that point we were able to get control of the fire.

D. *+1* Stretching a line into the adjacent exposure will be necessary to cut off any fire which has made it through the firewall between the stores. The scenario states that there is no common cockloft. As the building is old, the integrity of the firewall must remain suspect until proven otherwise and steps taken to cut off any lateral fire spread. While the scenario states there are firewalls, for the purposes of this test, you can assume they are uncompromised. In the real world, remember that the only true firewalls are in heaven. A line must be stretched and the ceilings must be pulled, in that order. The positioning of this line does not address the parent body of fire, which if left to burn without intervention, may not only extend to the adjacent store, but may make any operation in the area, including in the exposed store, too dangerous. While exposures are being protected, the attack on the main body of fire must also be continued.

E. *-2* The presence of the firewall will make the trench cut unnecessary. In addition, a trench cut takes a long time and will not necessarily be successful on such a large roof. Even if a common cockloft was present here, in the time it takes to complete the trench cut, the fire will be past the operation. A large amount of area will have to be surrendered to properly complete the cut. It is more practical on large roofs such as commercial buildings and taxpayers, to cut multiple large roof vent holes than to attempt a trench cut.

9. Action must be immediately taken to protect the evacuating occupants.

   A. *+1* Advancing the line through a different route than that of the fleeing patrons does not congest the egress point. However, use of water on the fire is required. Water used to push the fire back will be the only protection available at this time.

   B. *0* Failure to break out the show windows will hold the products of combustion in the store. You can be sure that a panic situation exists. Breaking out the window will alleviate smoke and heat conditions and provide some relief. Do not wait to apply water to the fire. The stream will act like a fan and push the fire and the accompanying products of combustion away from the people.

   C. *0* This is not an action which will solve the problem here and now. Obviously, the sprinkler system is not operating properly or has been overwhelmed. The fire is past the point where the sprinkler may do any good other than creating steam. Immediate intervention in the store is required.

   D. *+2* This is the best action. The line is kept outside the store and directed into the store via the window to keep the fire back until the patrons have exited. Then it is advanced into the store to attack the fire. This is the safest action to take for all involved.

   E. *-2* Attacking from the rear will push the fire right at the people. This tactic is not only poor, it is criminal negligence.

10. This question tests your knowledge of scene preservation.

A. **-1** Moving the gas can destroys the integrity of the scene and, if not documented properly, upsets the chain of custody required to move such items. It is the responsibility of all fire department personnel to preserve the scene if evidence of an incendiary fire is discovered.

B. **0** While protection of evidence is a main concern, when suspicious items are found, they should be left in place and overhaul suspended at least until a preliminary investigation can be conducted. Then overhaul can be completed in compliance with the orders of the arson investigator.

C. **+1** Making the scene safer for investigation is a definite concern, however, using a positive pressure fan without the authorization from the arson investigator may blow evidence around the room, thereby disrupting scene integrity.

D. **-1** Closing the safe and the filing cabinet irresponsibly destroys any possible shred of evidence such as fingerprints or evidence of tampering. Items should be left in place exactly as they were found. Overhaul should be suspended in this areas except that which is absolutely necessary to extinguish the fire. It should be carried out with extreme caution.

E. **+2** These actions take into account scene safety (lighting and charged line), scene preservation (suspending overhaul), and chain of command procedures (notifying command). These actions will give the investigation team the best chances of conducting a safe, thorough investigation.

Passing Score for Scenario 10–4 = 14 Points

CHAPTER ELEVEN
SCENARIOS

> *Note: The scenarios addressed in this chapter focus on incidents involving some of the more common hazardous materials that the fire department will encounter. These are mostly incidents dealing with some type of petroleum product, as this commodity is present in virtually every community across the country and the world over.*

SCENARIO 11–1
TANKER ROLLOVER

Initial arrival condition

A tanker carrying a flammable liquid has rolled over on Sherman Street

Time and weather

It is 1225 and the temperature is 87°F. The wind is blowing at 10mph out of the east. A strong thunderstorm has just passed, but it is very humid, as the storm did not provide much relief from the hot weather.

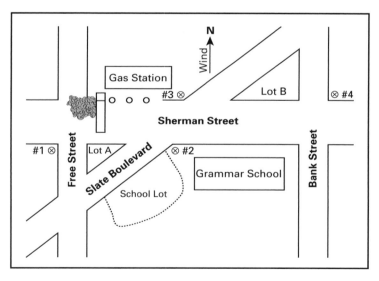

Area and street conditions

The rollover occurred while the truck was turning into the Sherman Street gas station to make a delivery. The truck is lying on its side about halfway into the gas station lot. Sherman Street is a two-way, four-lane street. It is crossed by several side streets. Bank Street is a one-lane, two-way street to the east of the gas station. Free Avenue is a one-way street to the west of the incident. Running on an angle crossing Sherman Street is Slate Boulevard. Slate Boulevard is a two-way, two-lane street. The streets are slick from the rain.

Fire conditions

Upon your arrival, you can see that there is product leaking from the large driver side fuel tank and has ignited. The fire appears to be confined to the area of the tank, but is still exposing the shell of the tanker. There is a "Flammable" placard on the truck, but you cannot ascertain anything else about the cargo. The driver is still in the vehicle. His status is unknown.

Exposures

The gas station is located in the center of the congested business section of the city. There are many civilians in the area.

Water supply

There are numerous hydrants in the area, each on a separate main.

Response

Your response is two engine companies and one ladder company. You do not have special agent capability on the responding apparatus. An officer and two firefighters staff each engine company. An officer and three firefighters staff the ladder company.

MULTIPLE-CHOICE QUESTIONS

1. In regard to firefighter safety, what is the best way of determining the contents of the tanker truck?

    A. Shipping papers

    B. Placards

    C. Information from the driver

    D. Shape of container

    E. Your senses

2. What resources would you request for this incident?

    A. EMS, Foam Unit, Haz Mat, and a Canteen Unit

    B. Utility company, police, Haz Mat, and CHEMTREC

    C. EMS, police, Haz Mat, and a Foam Unit

    D. Red Cross, EMS, utility company, and Haz Mat

    E. EMS, Foam Unit, police, and CHEMTREC

3. Would you request additional alarms for this incident?

    A. No, as the area is already congested. Calling additional alarms would further congest the incident.

    B. First determine the product. Request additional alarms consistent with the hazard presented.

    C. Yes, the potential for evacuation of the area requires the need for additional manpower.

    D. Only request the Foam Unit and the Haz Mat. No other companies are needed.

    E. Yes, rescue, evacuation, and suppression activities will require reinforcement of the initial alarm response.

4. Where would you set up the command post?

    A. In Lot A.

    B. In the school lot.

    C. In the school itself.

    D. In Lot B.

    E. Mobile command requires no command post be set up.

5. After ascertaining that the leaking product as well as the contents of the tanker is gasoline, what would be your orders for the crew of Engine 1?

    A. Establish a water supply at Hydrant #3. Position the apparatus in Lot A. Stretch a preconnected 1³/₄" attack line to knock down the small fire in the fuel tank. Keep the shell of the tanker wet.

    B. Establish a water supply at Hydrant #2. Position the apparatus at the intersection of Sherman Street and Slate Boulevard. Stretch two 1³/₄" preconnected attack lines. Advance on the small fire with a crew from Engine 2 using water in a wide fog pattern as protective shield. The captain of Engine 2 will attempt, under the protection of the hose line, to plug the leak. Retreat when this is accomplished and continue to keep the tank shell wet.

    C. Establish a water supply at Hydrant #4. Position the apparatus uphill and upwind. Use the deck gun from a distance to keep the shell of the tanker wet and to protect the rescue of the driver in the cab.

    D. Establish a water supply at Hydrant #2. Position the apparatus at the intersection of Sherman Street and Slate Boulevard. Stretch two 1³/₄" preconnected attack lines. Advance toward the cab with a crew from Engine 2 using water in a wide fog pattern as a protective shield. Protect the ladder crew engaged in the rescue of the driver, using the water fog to keep the fire from the victim. Retreat when the rescue is accomplished and continue to keep the tank shell wet.

    E. Establish a water supply at Hydrant #1. Place the apparatus in a defensive position. Isolate the scene and deny entry. Use the deck gun from a distance to keep the cab interior wet. Await the arrival of the Haz Mat team.

6. What would be your orders for Engine 2?

    A. Secure a secondary water supply. Use the deck gun to keep the shell of the tanker wet as a crew from Engine 1 attempts to plug the leak.

    B. Secure a secondary water supply. Position the apparatus in a defensive mode uphill and upwind. Use the deck gun to keep the gas station wet.

    C. Secure a secondary water supply. Approach from the rear of the tanker. A crew from Engine 2 advances a preconnected 1¾" handline in tandem with a crew from Engine 1 to push the fire away from the victim, allowing the ladder company to make the rescue. After this is accomplished, back out and protect exposures while keeping the shell of the tanker wet.

    D. Secure a secondary water supply. Approach the tanker in a flanking maneuver. Advance in tandem with a crew from Engine 1, using a wide fog pattern to push the fire away. Engine 2's captain will attempt to plug the leak.

    E. Secure a secondary water supply. Approach from both flanks of the tanker. Have a crew from Engine 2 stretch a preconnected 1¾" handline in tandem with a crew from Engine 1 to push the fire away from the victim, allowing the ladder company to make the rescue. After this is accomplished, back out and protect exposures while keeping the shell of the tanker wet.

7. What would be your orders for Ladder 1?

    A. Position the apparatus in Lot B. Evacuate the school through the rear doors onto Bank Street to avoid exposing any students to the incident.

    B. Position the apparatus upwind and uphill. Split the crew. Crew 1 makes a thorough search of the gas station. Crew 2 attempts to rescue the victim under the protection of the fog streams.

    C. Position the apparatus in Lot A. Use the ladder pipe to keep the shell of the tanker cool. Make the rescue of the victim as the engine crews attempt to plug the leak.

    D. Position the apparatus uphill and upwind. Under the protection of advancing fog streams in tandem, make the rescue of the driver. Then, conduct a search and evacuation of the gas station.

    E. Position the apparatus upwind and uphill. Stand by in a defensive posture until the Haz Mat team arrives and secures the scene. Then make the rescue of the driver.

8. What would be your orders for the Foam Unit?

    A. Utilize the roll-on method of foam application to cover the spill and extinguish the fire.

    B. Utilize the bank-down method of foam application to cover both the ignited spill and the tanker shell.

    C. Utilize the subsurface injection technique by plunging the foam into the ignited product to take advantage of the cooling power of the foam.

    D. Utilize the rain-down method of foam application to gently float the foam down on the tanker shell and ignited product.

    E. Utilize the rain-down method of foam application to cover the advancing personnel engaged in offensive plugging and rescue operations.

9. What action would you take in regard to the school and the gas station?

    A. Take no action in regard to these occupancies. Extinguish the fire, plug the leak, and all problems will disappear.

    B. Close the windows on the school and protect-in-place. Evacuate the gas station.

    C. Evacuate the school. Search and evacuate the gas station. Shut down the gas pumps.

    D. Take no action regarding the school. It is downwind. Shut down the pumps at the gas station.

    E. Search and evacuate the gas station. Shut down the gas pumps. Take no action regarding the school until the situation is properly sized up and a clearer picture of the magnitude of the incident is evident.

10. What would be the principle behind extinguishing this fire?

    A. Cool and quench

    B. Fuel removal

    C. Oxygen exclusion

    D. Inhibition of chain reaction

    E. Controlled burn

SCENARIO 11–2
INCIDENT ON THE BRIDGE

Initial arrival condition

A school bus carrying thirty high school students to a field trip has collided with an 18-wheel tractor trailer truck. The truck has overturned. Both vehicles are in the eastbound lanes of the six-lane Metro Highway bridge. There is a placard that reads "DANGEROUS" on the sides and rear of the truck. There are no other markings on the truck.

Time and weather

The time is 0500. It is Monday morning, March 17. The weather is clear and the temperature is 53°F. The wind is from the west at 15mph.

Area and street conditions

Although many are injured, most of the high school students, their teacher, and the bus driver have gotten out of the bus. Some of the injured are sitting or lying on the ground about 50' to the east of the accident. Some others are milling around, disoriented and confused.

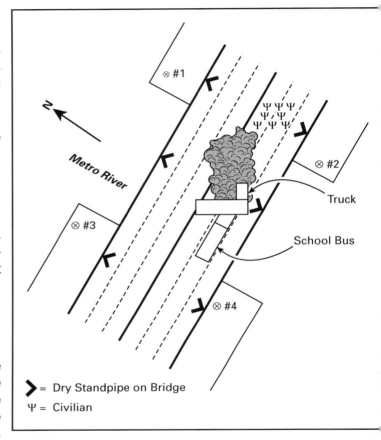

= Dry Standpipe on Bridge

Ψ = Civilian

Police are on the scene and have stopped traffic in both directions at the east and west ends of the bridge. Traffic has backed up, but the police have opened one west bound lane for emergency responders.

Fire conditions

The driver of the truck has evacuated from the truck and is not physically hurt, but is dazed and incoherent. The truck motor compartment and cab have caught fire and are exposing both the bus and the cargo area of the truck.

Exposures

Exposures are limited to the river below the bridge and the surrounding river bank.

Water supply

There are no hydrants on the bridge, but a dry standpipe runs along the guard rails on both sides of the bridge with outlets every 150'. There are hydrants on each side of the bridge located on service roads under the bridge from where the standpipes can be supplied.

Response

The response is two engine companies, one ladder company, and a rescue company. All are staffed by an officer and three firefighters.

MULTIPLE-CHOICE QUESTIONS

1. What is the most important size-up factor when assessing this scenario?
   A. Area
   B. Water Supply
   C. Auxiliary Appliances
   D. Apparatus and manpower
   E. Exposures

2. As you size-up the scene and give your initial radio report to dispatch, what are the most significant resources required at this incident?
   A. DPW, EMS, and the highway department
   B. EMS, the Coast Guard, and the utility company
   C. EPA, Haz Mat Unit, and EMS
   D. EMS, Haz Mat Unit, and a Foam Unit
   E. Haz Mat Unit, Foam Unit, and CHEMTREC

3. What are your orders for Engine 1?
   A. Proceed onto the bridge. Position the apparatus just past the bus in the westbound lane. Stretch two 2½" lines from the standpipe above Hydrant #3 to the intake of Engine 1. Stretch a preconnected 1¾" line across the median to the bus. Use this line to keep the bus wet and to protect any rescue attempts. Then move to the cab area of the truck and extinguish the fire.
   B. Proceed onto the bridge. Position the apparatus just past the bus in the westbound lane. Stretch two 2½" lines from the standpipe above Hydrant #4 to the intake of Engine 1. Stretch a preconnected 1¾" line to the cab area of the truck to confine and extinguish the fire.

C. Proceed to Hydrant #4. Connect to Hydrant #4 with a 5" LDH and feed the standpipe at Hydrant #4. Use a ground ladder to access the bridge. Bring a high-rise pack to the standpipe connection above Hydrant #4. Stretch a 1³/₄" line to confine and extinguish the fire in the truck.

D. Proceed onto the bridge. Position the apparatus adjacent to the standpipe above Hydrant #1. Stretch two 2¹/₂" lines from the standpipe above Hydrant #1 to the intake of Engine 1. Utilize the deck gun to knock down the fire in the truck. Stretch a 1³/₄" pre-connected line to the bus to protect any rescue attempts.

E. Proceed onto the bridge. Position adjacent to the standpipe above Hydrant #3. Stretch two 2¹/₂" lines from the standpipe above Hydrant #3 to the intake of Engine 1. Utilize the deck gun to knock down the fire in the truck and to keep the fire away from the bus. Stretch a preconnected 1³/₄" line to the bus to protect any rescue attempts. Also attack any fire in the cab area of the truck.

4. What are your orders for the captain of Engine 2?

A. Proceed onto the bridge. Position adjacent to the standpipe between Hydrant #1 and Hydrant #3. Utilize the deck gun to knock down the fire in the truck. Connect a high-rise pack to the standpipe between Hydrant #1 and Hydrant #3. Stretch a 1³/₄" line to the bus to back-up the initial attack line.

B. Proceed to Hydrant #3. Connect to Hydrant #3 using a 5" LDH. Feed the standpipe to supply Engine 1. Raise a ground ladder to the bridge. Stretch a second line off of Engine 1 to back-up the initial attack line and to protect any rescue attempts.

C. Proceed onto the bridge. Drop a feeder line off of the bridge to the crew waiting below at Hydrant #1. Receive the supply from Hydrant #1. Utilize the deck gun to protect any rescue attempts and to knock down the fire. Stretch a 1³/₄" preconnected line to the bus area to protect any rescue attempts and alternate between wetting down the bus and attacking the truck fire.

D. Proceed to Hydrant #4. Connect to Hydrant #4 using a 5" LDH. Feed the standpipe to supply Engine 1. Raise a ground ladder to the bridge. Bring a high-rise pack onto the bridge. Stretch a second line off of the standpipe above Hydrant #4 to back-up the initial attack line.

E. Proceed to Hydrant #4. Connect to Hydrant #4 using a 5" LDH. Feed the standpipe to supply Engine 1. Also stretch a 5" LDH under the bridge and feed the standpipe above Hydrant #3. Raise a ground ladder to the bridge. Bring a high-rise pack onto the bridge. Stretch a second line off of the standpipe that is not supplying the initial attack line. Back-up the initial attack line and protect any rescue attempts.

5. What are your orders for the captain of Ladder 1 regarding any rescue attempts?

A. Position the apparatus on the bridge adjacent to the group of civilians. Perform triage and prepare the victims for EMS treatment.

B. Position the apparatus just past the center standpipe in the westbound lane. Split the company into two crews. Crew 1 (the captain and one firefighter) searches for and removes any other victims from the bus. Crew 2 (two firefighters) checks the truck for any other victims, then assists Crew 1 in rescue operations at the bus.

C. Position the apparatus on the service road near Hydrant #3. Raise the aerial to the bridge to act as an access route to the bridge for the engine and ladder crews. The ladder crew will assist EMS in patient triage after searching the bus and vicinity for any other victims.

D. Position the apparatus just past the center standpipe in the westbound lane. Split the company into two crews. Crew 1 (the captain and one firefighter) searches for and removes any other victims from the bus. Crew 2 (two firefighters) uses axes and Halligans to remove all of the windows in the bus for rescue access.

E. Position the apparatus just past the center standpipe in the westbound lane. Split the company into two crews. Crew 1 (the captain and one firefighter) searches for and removes any other victims from the bus. Crew 2 (two firefighters) sets up the ladder pipe to knock down the truck fire and to protect any rescue attempts by wetting down the bus.

6. Would additional alarms be required at this incident?

A. No, as scene access is limited, summoning additional companies would only cause unnecessary scene congestion.

B. Yes, due to the numerous victims in the area and still inside the bus.

C. No, request additional medical personnel only, no other fire personnel are required.

D. No, the on-scene assignment will be able to handle this incident.

E. The battalion chief will decide the need for additional alarms when he arrives.

7. The rescue company has arrived on the scene and is awaiting your orders. What are your orders?

A. Stretch a third line off of Engine 1 to protect the rescue company.

B. Use extrication tools as necessary to gain access to the victims. Assist in the removal of the victims.

C. Assign them to the EMS sector to perform triage, treatment, and transportation preparation of the victims.

D. Find out exactly what the dangerous cargo is and take whatever steps necessary to prevent it from negatively impacting the incident.

E. Set up a dike operation to prevent any motor vehicle fluids from running off the bridge and contaminating the Metro River.

8. The foam unit has arrived on the scene and is awaiting your orders. What are your orders?

    A. Stand by at this time until an assessment can be made as to the need for foam.

    B. Take a feed from Engine 1. Lay a foam blanket around the accident scene to protect any rescue attempts and to suppress any escaping vapors.

    C. Have Engine 1 shut down and disconnect one of the 2½" feeder lines from the standpipe connection. Take this line and use it to supply the Foam Unit. Lay a foam blanket around the accident scene to protect any rescue attempts and to suppress any escaping vapors.

    D. Take a feed from Engine 1. Lay a foam blanket on the bus only. Allow the engine companies to extinguish the fire in the motor and cab area of the truck.

    E. Take a feed from Engine 1. Cover the rear of the truck with foam and protect the rescue company in their attempt to identify the cargo. Cover the crew with foam if necessary.

9. You are attempting to identify the cargo that is in the truck. What is the best way to go about this?

    A. Attempt to find out cargo information from the driver. Ask the driver for the manifest.

    B. Ascertain from the driver if the cargo manifest was in the cab, now involved in flame.

    C. Ask the driver what the name of the truck company is, and then call them to identify the cargo.

    D. Have the rescue company open the back doors under the protection of foam lines and investigate.

    E. Have the police run the license plate on the truck. Find out what company it belongs to, contact them, and inquire as to the cargo inside and its inherent hazard.

10. The standpipe you have been using for the initial attack line has blown apart due to years of deterioration. What is your next move to get water to the incident?

    A. Have a second alarm engine company feed another standpipe and take your feed from them.

    B. Have the second alarm ladder company raise its aerial device to the bridge. Utilize it as a portable standpipe to supply the attack lines.

    C. Disconnect from the standpipe at the top and the bottom of the bridge. Lower the supply line over the bridge to the supply engine below. Continue the water supply to the operation in this manner.

    D. Use whatever tank water is left in Engine 1 to protect any rescue attempts, then allow a controlled burn of both vehicles, keeping all personnel and victims uphill and upwind of the scene.

    E. Stretch a 5" LDH under the bridge to the standpipe on other side of the bridge. Feed this standpipe to supply the attack lines.

SCENARIO 11– 3
HIGH SCHOOL HAZ MAT INCIDENT

Construction

Lennon Memorial High School is a three-story, non-combustible building measuring 250' x 300'. There is an attached gymnasium on the east end, and an annex containing science labs on the west end.

Time and weather

It is 1345 on a Monday. The temperature is 72°F. It is cloudy and there is a wind blowing at 15mph from the east.

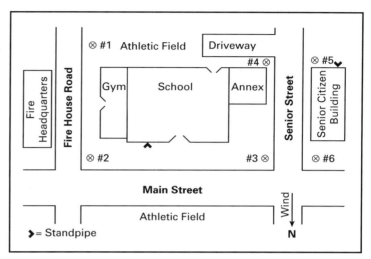

Area and street conditions

Streets are wide in this area, however, vehicle and pedestrian traffic during school hours congest this area.

Incident conditions

The school principal has come into headquarters and informed the house watch that a student has broken open a cylinder of concentrated liquid chlorine and has been overcome. It is not known how many students are in the classroom, there may possibly be up to 15 or more.

Upon arrival, you see a white vapor-like cloud migrating out of two windows on the west end of the annex. The HVAC system is in operation at the time of your arrival. Student evacuation of the school is in progress. Student population is 4000.

Exposures

The school is located on Main Street. There is a six-story, fire-resistive senior citizen home to the west on Mercer Street, and athletic fields to the south and the north. Headquarters for the fire department is located to the east, and has an adjacent parking lot.

Water supply

Hydrants are located on all corners, all on separate 12" mains. The science lab is sprinklered and is equipped with an HVAC system. All controls for the sprinkler system and HVAC system, as well as all school utilities, are located in the rear of the gymnasium in the main utility room.

Response

Your response is four engine companies, two ladder companies, a rescue company, a battalion chief and a deputy chief. The on-scene complement of manpower is 25 firefighters and officers.

MULTIPLE-CHOICE QUESTIONS

1. What is the least important size-up factor to consider at this incident?

    A. Location and extent

    B. Weather

    C. Water Supply

    D. Life hazard

    E. Auxiliary appliances

2. What would be your first action after establishing command and performing a scene size-up?

    A. Establish control zones.

    B. Set up an operational perimeter.

    C. Give an initial radio report.

    D. Attempt to identify the product and determine the associated hazard.

    E. Identify a safety officer.

3. Where would be the best place to establish a command post at this incident?

    A. In the school gym.

    B. In fire headquarters.

    C. In the athletic field on the south side of the school.

    D. At the corner of Main Street and Senior Street.

    E. In the athletic field on the north side of the school

4. What is the most important additional resource you would request to the scene?

    A. DPW

    B. EMS

    C. Haz Mat team

    D. Chief of department

    E. Police

5. What orders would you give the crew of Engine 1?

    A. Establish a water supply at Hydrant #3. Position the apparatus adjacent to the fire department connection (FDC) on Main Street. Stretch two lines to the FDC and charge the system. Stretch a preconnected 1³/₄" line to the flank of the annex. Use a fog stream to disperse the vapor cloud.

    B. Establish a water supply at Hydrant #1. Position the apparatus on Firehouse Road. Stretch a dry 1³/₄" line through the gymnasium door. Assist in the evacuation of the students.

    C. Establish a water supply at Hydrant #2. Position the apparatus near the FDC on Main Street. Stretch two lines to the FDC and charge the system. Stretch a preconnected 1³/₄" line into the school via the main entrance. Use a fog stream to cool and push the vapors toward the annex.

    D. Establish a water supply at Hydrant #2. Position the apparatus near the FDC. Stretch two lines to the FDC. Do not charge the system until ordered to do so. Stretch a dry 1³/₄" line to a safe area uphill and upwind and wait for orders.

    E. Establish a water supply at Hydrant #4. Position the apparatus in the parking lot. Stretch two lines across Senior Street to the FDC on the senior citizens' building. Do not charge the system. Stretch a dry 1³/₄" line to a safe position in the building and await further orders.

6. What orders would you give the captain of Engine 2?

    A. Establish a secondary water supply at Hydrant #3. Stretch a back-up line adjacent to Engine 1's line and use a fog stream to disperse the vapor cloud.

    B. Establish a secondary water supply at Hydrant #2. Stretch a dry back-up line adjacent to Engine 1's line and await further orders.

    C. Establish a secondary water supply at Hydrant #1. Stretch a dry back-up line adjacent to Engine 1's line and await further orders.

    D. Establish a secondary water supply at Hydrant #5. Take a high-rise pack into the senior citizen building. Attach this to the standpipe and use a fog pattern to develop a water curtain to keep the vapor cloud from entering the building.

    E. Establish a secondary water supply at Hydrant #1. Stretch a dry back-up line into the school via the gymnasium door. Assist Engine 1 with the student evacuation operation.

7. What orders would you give the captain of Ladder 1 regarding ventilation?

    A. Open all of the windows on west side of the annex.

    B. Shut down the HVAC system. Close all windows and doors on the north and south side of the school. Open the windows on the east side.

    C. Keep the HVAC system in operation, but place it in recirculation mode to keep the vapors in the annex and prevent their spread into the school.

    D. Don the proper protective equipment for the hazard. Place several PPV blowers outside of the west annex windows. Seal the school doors leading into the annex. Use the PPV fans to keep the vapors confined to the building.

    E. Shut down the HVAC system.

8. The Haz Mat officer has arrived on the scene and is awaiting your orders. What action do you take?

   A. Transfer command to the Haz Mat officer and work as part of the command staff.

   B. Direct him to where your personnel are operating so he can evaluate their effectiveness.

   C. Take him into the school to show him the layout of the area and how to access the annex from the interior.

   D. Establish a unified incident command structure, utilizing the expertise of the Haz Mat officer to assist in arriving at an effective action plan.

   E. Establish a hazard sector. Place the Haz Mat officer in charge of this sector. Brief him on conditions and what actions you have taken thus far. Provide support for the operation through liaison with the hazard sector. Provide manpower and resources as required.

9. What actions would you take regarding the senior citizen building?

   A. Send the second ladder company and rescue company into the building to undertake an evacuation of the building. Use the firehouse as a temporary shelter for the occupants.

   B. Shut down the HVAC system. Close all dampers to the exterior. Open windows on the unexposed side of the building.

   C. Allow the HVAC system to operate, but place it in a recirculation mode. Close all of the windows on the exposed side of the building.

   D. Shut down the HVAC system. Close all dampers to the exterior. Close all of the windows. Move the occupants from the apartments on the exposed side to the apartments on the unexposed side.

   E. Place several PPV blowers at the front entrance. Blow pressurized air into the building to pressurize the interior. Shut down the HVAC system. Open all dampers to the outside.

10. You have been assigned as safety officer. What would be your most important duty?

   A. Isolation of the hot zone.

   B. Mitigating the hazard.

   C. Safety of all participants, especially the emergency responders.

   D. Determining the severity of the threat of exposure to the students and the occupants of the senior citizens building.

   E. Ensuring all personnel working in the hot and warm zones are in proper protective equipment, and that the decontamination area is properly established and operated.

ANSWER SECTION

SCENARIO 11–1
MULTIPLE-CHOICE QUESTIONS

1. The best response to this question is based on two factors. The first and most important is the weight you place on firefighter safety, which must always be your highest priority. The second is your knowledge of potential hazards and how to best go about identifying them. This is one of the most important jobs of the initial incident commander at a hazardous materials incident.

   A. *-1* Shipping papers are usually kept in the cab of the truck. Identification of the product is always best accomplished from as far away as possible. There is an inverse relationship between safety and distance when attempting to identify hazardous materials. The closer you have to get, the more danger is involved. Retrieving the shipping papers as a first option is unsafe.

   B. *+1* Placards are usually visible from a distance. A good pair of binoculars will keep the person attempting to identify the product at a safe distance. A "Flammable" placard may not tell you exactly what is in the tanker, but it gives you more than enough information in regard to the main hazard of the product. From even this limited information, an initial action plan can be developed.

   C. *-2* The driver is still in the cab and his status is unknown. He may not be the best source of reliable information depending on his injuries, especially if they are fatal. You must physically access the cab to get to him. That is too close and an unnecessary risk to personnel. An attempt will have to be made at some point to assess the condition of the driver, but not during the initial product identification stage. Emergency responders have been killed or permanently injured because they rushed into a hazardous materials incident without using precautions.

   D. *+2* Shapes of containers, whether they are stationary at a tank farm or rolling down the highway, are a good indicator of what the product is inside the container. You can, at the very least, narrow down the possibilities and make some decisions based on this information. This is the safest method of product identification and should always be the first method attempted. The container shape of a vehicle is often a reliable indicator of the class of material being carried. (Fig. 11–1)

Fig. 11–1 Although this truck does not have placards, from the shape of the container, the incident commander may be able to narrow down the possible contents being carried. Until you are sure, always assume the contents are extremely hazardous.

E. *-2* Senses are both unreliable and dangerous. Attempting to use senses such as smell and touch is an indication of a lack of one of the most important senses, common sense. There are many methods that may be used to attempt product identification. The best ones are made from a distance. Play it safe and use them first.

2. You cannot expect to successfully mitigate an emergency such as this without some incident-specific resources. Knowing whom to request is critical to both the safety of the players and the successful outcome of the incident. There are basically four objectives that must be met to meet the strategy. These are:

    1. Firefighter safety—most important. This is the main reason for EMS response.
    2. Scene security/Civilian safety and control: This is the most important role of the police department.
    3. Vapor suppression: Foam must be used for this task.
    4. Mitigation: The leak must be plugged. Only a trained Haz Mat team should attempt this action.

Now let's look at the answers and scoring:

A. *+1* This answer covers three out of four. Failure to provide for scene control may cause needless casualties.

B. *-1* Foam is a necessity here. Unless overhead wires are involved, utilities are not one of the primary initial resources required. CHEMTREC can only give advice. They do not respond, nor do they provide resources to the scene.

C. *+2* All of the significant resources are requested. Given the choices, this list of resources will best provide for a successful conclusion to the incident.

D. *-1* Some limited, temporary evacuation will be necessary. There is no need for the Red Cross as widespread evacuation and relocation is unnecessary at this time. It is better to secure the proper resources so the emergency can be controlled without excessive civilian disturbance. With this set of resources, which does not provide for either scene security or vapor suppression, evacuation and permanent relocation may be the result.

E. *0* It is critical that a Haz Mat team be requested as soon as possible. The product is leaking and is ignited. Small problems, not handled properly, can quickly become large problems. Get the experts to the scene and things will get better.

3.

A. *0* This is not a valid reason for neglecting to request additional alarms. Proper command and control from the outset will ensure that scene congestion is avoided. A safe apparatus and personnel staging area should be established, most likely in the area of Bank Street. This area is upwind and will allow personnel to safely report to the command post without becoming exposed to the hazards of the incident.

B. *-1* If you even have the slightest bit of potential for their need, never wait to request additional alarms. By the time you do, it may be too late. Get them on the scene. If they are not required, they can be released.

C. **+1** The potential for civilian evacuation is a valid reason for summoning additional alarms. This will require manpower. The first alarm responders will be busied with many tasks. Having to conduct an evacuation, even a limited one, will spread the troops thin. Get extra personnel on the scene to accomplish these tasks.

D. **0** While the Foam Unit and the Haz Mat team are definite necessities, additional manpower will be required to assist and support the activities required to bring the incident under control. Some of these vital activities are listed in answer choice "E."

E. **+2** All scene activities will require reinforcement. In fact, the only activity that will be exclusively assigned to the Haz Mat team will be the plugging of the leak. That is the main point of this whole chapter. Even though most first responders are only trained to the operations level, there are still a great many tasks they can accomplish to improve conditions and scene safety. Activities such as rescue, vapor suppression, evacuation, and water supply must be addressed by the manpower on the scene. This will require reinforcement. To expect the three companies on the initial response to do all this is unsafe and displays a lack of respect for safety on the part of the incident commander.

4. The command post must be established as a control point for all scene activities. It must be announced over the air as soon as possible, be clearly marked, and positioned in a safe area. Choosing the best answer requires an analysis of the diagram provided. A poorly located command post will create more problems than it sets out to solve. (Fig. 11–2)

Fig. 11–2 Photo by Bob Scollan NJMFPA
The command post must be located in a safe area and allow room for expansion. A hazardous materials incident will require additional agencies respond. The command post will be the point for representatives of these agencies to report and possibly operate.

A. **-2** Lot A is situated on the wrong end of the tanker. No matter what they are carrying, a tanker is most dangerous at the ends. If they explode, they can become deadly missiles. If you have read the questions first, you would already know that the product is gasoline. Tank trucks that carry gasoline are of the non-pressure type. Vapor pressure in the tank is under 3 psi. In addition, these tankers are usually made of aluminum, which melts at about 1200°F. They are designed to melt down to the level of the liquid, which will alleviate any possibility of an explosion, but create a heavy fire condition. However, no matter what the pressure is or the explosion potential, the ends of the tank are no place to operate in proximity to. It is certainly not the place you would want to set up a command post. The tank is not yet ignited, but if it becomes involved and you were not prepared for it, this will not be your day. (Fig. 11–3)

Fig. 11–3 Aluminum containers are designed to melt down to the level of the liquid, thus minimizing the BLEVE threat. If the tank does begin to melt, be prepared for heavy fire and potential exposure problems. Nevertheless, operating at the end of this tanker is asking for trouble.

B. *-1* The school lot is the wrong location for the command post for the same reasons as Lot A. It is located at the ends of the tanker. It is, however, further away, so you only lose one point here instead of two.

C. *+1* The school would provide a command post that is in a protected area. This would be most ideal if the weather was inclement and the school was closed. If the school is open, placing the command post in it can add to the fear and confusion on the part of the students, who are pre-teen. The grammar school may also not offer the best view of the incident.

D. *+2* From a strategic standpoint, Lot B is the best place to position the command post. It is on the windward side, is a safe distance from the incident, and is an ideal place for companies to report for assignment from staging. Furthermore, placing the command post in this lot will provide for a constant evaluation of the incident due to the unobstructed view it provides.

E. *-1* The concept of mobile command always seems to confuse people, especially acting officers. There is no such thing as an establishment of mobile command unless the incident is in a Mobil gas station. Mobile command is understood to mean that the first-arriving company officer will be in the operational mode and not setting up a fixed command post. Even if this is so, the command must be named after the street, facility, or other area where the incident is actually occurring, such as "Sherman Street Command." The term "mobile command" is incorrect. At this and other "stop and think" incidents, a fixed command post must be established. A mobile command here is improper and denotes that action is being taken which the officer may not be trained to do.

5. We are in the fifth question and the product is only now being identified. Read the questions first to help round out the information in the scenario. These type incidents are often handicapped by incomplete information. Decisions may have to be made prior to product identification. Always make them on the side of safety.

A. *-1* Lot A, as already mentioned, is a dangerous position due to the orientation of the tank truck. The hydrant selection is acceptable, but the tactic of putting water on burning gasoline is improper and should be avoided. Water will not cool gasoline below its ignition temperature and only cause the fire to spread as the burning product floats away on the water. The only proper tactic in this choice is the application of water on the tank shell to keep it cool.

B. *-2* Again, the hydrant selected is acceptable, as it is downwind. The tactic employed of advancement using a wide fog is proper, but the objective is wrong. First responders should undertake no offensive hazardous material intervention. Use the fog stream to make the rescue, then get out and continue to operate in a defensive mode until the Haz Mat team can mitigate the leak. It may be possible to let the product burn itself out as long as the tanker shell is kept wet. The fuel tank has a finite amount of product, but if it is not controlled properly, it can spell disaster. (Fig. 11–4)

Fig. 11–4 Fires involving saddle tanks on tractors (located below the passenger door) may not only threaten the product in the container, but also make the rescue of the vehicle occupants more difficult and dangerous.

C. **+1** Hydrant selection is as far upwind as possible. Positioning is also proper. Using the deck gun to protect the rescue, however, may not be as safe as advancing with a fog stream. At worst, you may drown the victim in the vehicle cab.

D. **+2** In this answer, fog lines in tandem are used to protect the rescue. The pressure of the fog streams will drive the gasoline-fed fire away from the rescuers. (Fig. 11–5) Once the rescues are made, companies should retreat using the fog streams as cover. The streams are then utilized in a defensive manner until the leak can be plugged. It may also be possible, in small spills, to use a dry chemical or CO_2 extinguisher to knock down the flames. Remember that even though the flames are extinguished, the vapors will still be issuing from the product and present a major ignition potential. It is critical that all ignition sources be controlled. Foam should be applied as soon as

Fig. 11–5 BCFA file photo
Tandem fog streams can be used as a shield to approach and remove a victim, shut off a burning flammable gas cylinder, or sweep fire from a burning pit. Once the objective is accomplished, the lines are backed out. This operation requires strict control.

possible. If foam is not yet available, it may be prudent to let the small fire burn in a controlled manner, which will eliminate product and the excessive buildup of vapors. This decision will depend on conditions such as the size of the fire, the size of the leak, the potential exposures, and the availability of foam.

E. **-1** Hydrant #1 is a poor choice as it is upwind and will possibly place the apparatus in the path of both the smoke and the tank shell should it fragment. Using the deck gun to keep the interior of the cab wet may drown the driver. Deck guns are usually a minimum 600 gpm. If the stream is to be used, use it as cover to protect the rescue. Waiting for the Haz Mat team to rescue the driver may cost him his life. A rescue attempt must be made.

6.

A. **-2** While this is a safety-oriented action, the plugging of the leak by the first-arriving engine crew must not be undertaken. It is likely that their level of training is such that only defensive operations are permitted. Sending an untrained and ill-equipped crew in to take an offensive action is unjustifiable. (Fig. 11–6)

B. **0** The idea of keeping the gas station wet is proper if the tanker container becomes involved. This is not yet the case. It is better to keep the shell of the tanker cool. This will help prevent meltdown. The leak and fire is still small and may be kept manageable. Companies should make an effort, if it can be done safely, to dike areas ahead of the spill in an effort to contain it to one area. Flowing and possibly burning gasoline can wreak havoc on an area if it enters sewers, cellars, or flows under vehicles. Even though offensive operations are not permitted, actions can be taken so as not to make matters worse.

Fig. 11–6 Photo by Ron Jeffers NJMFPA
Protecting drains by diking ahead of the spill are defensive measures. They can be performed by members trained to the Operational level. These are just some of the support activities "unqualified" firefighters may accomplish while the HMRT do their thing.

C. **+1** These are the right tactics, but the wrong approach angle. Never advance from the ends of a tank. The flanks are the best place from which to approach any kind of tank. This rule of thumb, incidentally, also applies to vehicle fires as well.

D. **-2** Any attempt to plug the leak is the wrong strategy to take for first responders. Protect life, protect property, and let the experts worry about the leak. Doing nothing is better than doing the wrong thing here.

E. **+2** The second engine company should, like a structure fire, stretch a second line to reinforce the operation of the first line. In this tanker incident, the lines should be advanced simultaneously. They should only be used in the proximity of the tanker to protect the rescue. Once complete, the lines should back out of the area and keep the shell of the tanker cool to prevent meltdown and protect any threatened exposures.

7.

A. **0** Ladder company operations should focus on life safety in the immediate area of the emergency. This means an attempt to rescue the driver and any passengers in the truck. It also includes any civilians in the vicinity of the tanker. The evacuation of the school should be considered, but after the main priority in the hot zone is addressed.

B. **+1** It is not prudent or effective to split the crew in this situation. All hands will be required to make any rescues in the area of the immediate emergency. Even the rescue of one person in an overturned truck can easily eat up the entire ladder crew. Leave the search of the gas station to later-arriving companies.

C. **-1** Setting up the ladder pipe to keep the tanker shell cool will take time. That time can be best spent making the rescue from the vehicle. In addition, the offensive strategy of plugging the leak is mentioned in this choice. Even if you, as the ladder officer, are not involved in this operation, condoning its undertaking is as bad as doing it yourself. Know the capabilities and limitations of yourself, your assigned crew, and those on the response with you. It only takes one wrong action by one person or crew to land several companies in the hospital or the funeral parlor.

D. **+2** This answer choice gets the priorities straight. Using the fog streams for protection, first get the driver of the tanker out of the vehicle and to safety. Then, concentrate on the search and evacuation of exposures.

E. **-2** Waiting until the Haz Mat team shows up may cost the victim his life. If the truck container was fully involved, this would be a proper action as the risk versus gain would be too great. However, the fire and leak are limited at this time. Now is the time to make the rescue attempt. With proper protection from advancing fog streams in tandem, it is acceptable and proper to attempt the rescue.

8. It is critical that the foam blanket be properly applied to most effectively provide coverage of the tanker as well as the area. Gasoline has a flashpoint of about 536°F and an ignition temperature of as low as -40°F. This means that the vapors are always ready to ignite. The best way to prevent this is to control your ignition sources and apply a blanket of foam to keep the vapors from contacting the oxygen supply. Remember that to have ignition, all the segments of the fire tetrahedron must come together. The action of the foam blanket works to keep the oxygen portion of the tetrahedron from contacting the fuel portion of the tetrahedron, thereby preventing ignition.

A. **+1** The roll-on method of foam application will work well on the spilled foam. It is the preferred way of providing coverage of foam on a flat surface such as the ground. It is not the best for three-dimensional fires and spills such as a tanker on its side. This method will not provide proper protection of the tanker shell. (Fig. 11–7)

Fig. 11–7 BCFA Photo
The lob method is being utilized by this aircraft-type apparatus. No matter what method of foam application is used, the reach of the stream should be used to best protect personnel.

B. **+2** The bank-down method is the best application technique to use at this incident. The foam may be applied to the side of the tanker and vehicle cab, keeping it cool and covering product as it flows downward. This method will also allow the foam to roll over the product as it flows off the side of the truck to the ground. (Fig. 11–8)

C. **-2** The subsurface injection method of foam application is used in specific circumstances and with certain types of foam. This is not one of them. Plunging the foam stream into the ignited product will be disastrous. It will cause the ignited gasoline to splatter all over the area, exposing both firefighters and uninvolved exposures. This would be a classic "don't have a clue" action.

Fig. 11–8 Photo by Ron Jeffers NJMFPA
The bank-down method is most suited for overturned vehicles. The blanket can coat the tanker shell as well as run down its side to the ground where it can blanket and extinguish spills

D. **+1** The rain-down, also known as the "lob" method, will also provide coverage of both a ground and object spill. However, it will not be as efficient as the roll-on method on ground spills or the bank-down method on 3-D spills. It may be the method of choice if heat or some other obstacle prevents a close approach to the product and vehicle. Knowing what your options are will allow you to apply the foam using the most appropriate method given the situation.

E. **-1** First of all, we have established that offensive operations are not to be conducted by the first responders. If you went this way, you are probably about ten points in the hole and contemplating a career change at this time. In addition, deliberately covering any personnel with foam is of no value. Remember that foam is over 90% water. If the heat conditions are severe, that water may turn to steam, causing burns to the firefighters. Use the foam on the product.

9.

A. **-1** Plugging the leak is an offensive operation, one that should not be taken by the first responders. If things go wrong and the incident escalates to threaten and possibly involve these two critical exposures, it may be too late to protect the occupants. This is not a quick-fix incident. Mitigating this incident will take time. Eliminate as many problems as possible by addressing the exposures as soon as possible.

B. **+1** Protection–in-place is a definite consideration at this incident. If the evacuation of the school cannot be controlled, this may be the best alternative. There is nothing worse than several hundred children gawking at the incident. In addition, the evacuation of the gas station addresses the life hazard, but other measures can be taken to minimize the hazard presented by this occupancy.

C. **+2** These actions take incident potential into account. If potential problems are eliminated early, you will not have to worry about them if the incident escalates. Even though the school is on the windward side, the wind could shift. Imagine having to explain to parents why you didn't evacuate the school if the wind did shift and students were overcome by smoke exposure. Play it safe here and evacuate the school on the unexposed side. Before you do, it might be a good idea to have school buses dispatched to the Bank Street side. Load the students on and take them to a nearby school or other safe area where their parents can be notified of the closure. This eliminates the life hazard at the school. Shutting down the pumps in addition to evacuating and searching the gas station will minimize both the life hazard and ignition source problem. The thinking incident commander plans ahead and eliminates potential problems before they become reality.

D. **0** This answer tests the critical reading factor. The school is not downwind, it is upwind. If it were downwind, it would present an extreme exposure hazard, making it a major priority.

E. **0** The gas station tactics are correct, but waiting to take action regarding the school could prove to be a critical mistake. Never gamble with lives, especially those of children. Inaction could cost you dearly here.

10. Knowing the principle behind the extinguishment of any fire will allow the incident commander to choose the proper tactics as well as the extinguishing agent in proper form and quantity.

A. **+1** Foam is 94% to 97% water, depending on the foam-to-water proportioning. Foam and its main ingredient, water, will be instrumental in lowering the temperature of the gasoline below the ignition temperature and preventing re-ignition should the foam blanket be compromised. The foam application should continue until the leak is plugged and the product can be properly controlled through diking, diverting, or rendering it inert.

B. **-1** Fuel removal to extinguish this fire is not practical. These tankers can hold thousands of gallons of product. Fuel removal would entail offloading the product to another vehicle, which would take a long time. It is more practical to apply foam and attempt to plug the small leak using properly trained Haz Mat personnel.

C. **+2** Oxygen exclusion or smothering will be the primary method by which the fire is extinguished and stays that way. The foam acts as a blanket to suppress the vapors being released by the gasoline. This vapor suppression prevents the gaseous fuel source from contacting the oxygen source, the ambient air, which could allow ignition to occur. As long as the blanket remains intact, ignition of the vapors is unlikely. The key is to have enough foam to continue blanketing the product until the incident can be stabilized. Incident commanders must know where and how to contact foam reserves to maintain the current mitigation strategy. The degree to which the strategy will be successful is largely dependent on the amount of foam that can be secured. Don't forget to take reflex time into account.

D. **-1** The inhibition of the chemical chain reaction, which is the linking element in the fire tetra-hedron, does not apply to this situation. Chain reaction inhibition is more applicable to the use of Halon and is not pertinent to the principle of extinguishment in the use of foam.

E. **-1** A controlled burn is generally not acceptable in built-up areas of a city. Gasoline is a petroleum product. The byproduct of the burning of petroleum products is large volumes of acrid, black smoke. A controlled burn would not only expose the entire area to deadly toxins and soot accumulation, but may also risk the potential of tank failure, either result-ing in a boiling liquid expanding vapor explosion (BLEVE) or a tank meltdown, where a running fire may be the result.

Passing Score for Scenario 11–1 = 14 Points

Scenario 11-2
Multiple-Choice Questions

1. Many size-up factors must be considered and will play key roles in this incident. Apparatus positioning and water supply operations for this bridge should be preplanned in advance. This will allow the incident commander to concentrate on mitigation rather than apparatus directing. The more you can take care of before the incident, the more smoothly your operation will be and the more you can focus on critical, incident-specific problems.

   A. **+1** The area of the incident will present many problems. A bridge is the runner-up to the tunnel in incident-unfriendly environments. Only one lane is open for emergency vehicles, demanding a careful coordination of apparatus positioning. If a specific positioning SOP is not in place, companies must stage at the base of the bridge and wait for positioning orders from the incident commander. This will cause a delay in operations and force the incident command to focus on factors other than the critical requirements of the incident. It is always better to be prepared.

   B. **+1** The standpipe is dry. The only way to get water onto the bridge is by supplying this standpipe or by the use of tank water. Both should be utilized. Tank water should be used in the initial operation and be supplemented by standpipe supply. It is imperative that these standpipes be inspected and tested on a regular basis. It is not conducive to smooth operations if the water supply system fails when it is needed most.

   C. **+2** The existence and, more importantly, the serviceability of the auxiliary appliance will make or break this scenario. It would be a nightmare to have to stretch supply lines onto this bridge. The presence of the standpipe simplifies and speeds up the operation. The ability to effectively utilize this appliance will impact on the success of all other operations.

   D. **+1** Apparatus and manpower is always a consideration. The rescue problem created by this accident will tax the on-scene responders. There are more tasks to be accomplished than personnel on the scene. In addition, one company will be required to supply the standpipe system on the road adjacent to the bridge. The manpower of this company will not be immediately available on the bridge. This drops the manpower available to operate on the bridge by four. You will need more to get this job done safely and effectively.

   E. **0** The only major significant exposure to address will be the river below. It may be exposed to any leaking fuel or fire stream runoff. For this reason, it will be necessary to notify the agency in charge of this waterway. Temporary diking well ahead of spilled product may be required to protect the waterway.

2. Incidents that are beyond the scope of both the routine response and the initial responder's expertise should prompt an immediate request for incident-specific resources. This request should address both conditions and prerogatives based on the three fireground priorities. These priorities, of course, are life safety, incident stabilization, and property conservation. Keeping these in mind will guide you to the best resources.

   The first priority is life safety. Remember that the most important factor at any incident is firefighter and civilian safety. The response of EMS personnel, especially in this vehicle accident, should be the first resource that comes to mind. Many civilians are injured and some are possibly trapped.

The second priority is incident stabilization. Hazardous materials are involved, although at this time they are as yet unidentified. A "Dangerous" placard can be just about anything. DOT guidelines state that this placard is used when transporting two or more hazardous substances in one load exceeding 1,000 pounds. Explosives Class C, which include some types of fireworks, small arms munitions, and safety fuses are also required to bear the "Dangerous" placard. It is critical that a Haz Mat team as well as a Foam Unit be requested immediately. (Fig. 11–9)

Fig. 11–9 The "Dangerous" leaves a lot of room for trouble. Whenever unsure of the hazard, treat it as if there is nothing more harmful on Earth.

The third priority is property conservation. Not only is a sloped roadway involved which may cause a running spill problem, but also a waterway is involved. Proper notifications must be made to agencies responsible for the waterway. This will include the Environmental Protection Agency for inland waterways and the Coast Guard for waterways that serve ocean-going vessels. Both agencies have established action plans for such incidents and can provide valuable assistance in the form of technical expertise to the incident commander. On even smaller waterways, the local fish and game authority may have jurisdiction. This information should be known to the incident commander and logged into the dispatch center's resource index beforehand.

Now let's look at the answers based on the most significant resources required:

A. *0* The DPW and highway department are not of major significance.

B. *0* No utilities are involved; this is a waste of a request.

C. *+1* Covers all three fireground priorities, but not special extinguishing agent response (foam). The incident mitigation priority will take precedence over property conservation, especially when there is a severe life hazard. The Coast Guard or EPA should be on the resource list, but are not as significant as the resources that will address life and the incident, your top two concerns.

Fig. 11–10 Standpipes built into elevated roadways are an advantage for the incident commander. Companies assigned to supply this appliance should know its location and closest hydrant.

D. *+2* Covers life hazard and incident mitigation, the two top priorities. It just so happens that there are two significant resources required to stabilize the incident. They, along with the EMS response, are the most significant.

E. *-1* No EMS response. Life hazard must be addressed.

3. There are many problems here. First, there is the life hazard in and around the bus. The fire in the cab and motor compartment of the truck further threatens this hazard. This fire must be kept from spreading to the cargo. Base your strategy on the three fireground priorities.

A. *0* Hydrant #3 is the best hydrant and standpipe to utilize. It is upwind and in the lane opposite the emergency. The line is used to protect the rescues and rightfully so. However, while this is happening, the fire in the truck may be spreading to the cargo area. If this happens, the rescue effort may be seriously jeopardized. Further actions must be taken to both protect life and stabilize the fire condition. (Fig. 11–10)

B. **+1** Hydrant #4 and the standpipe above it is also upwind, but may be a little too close to the incident. Hydrant #3 is better. Here the line is stretched to attack the fire, which if allowed to continue to spread unchecked, could endanger the entire rescue operation. This is superior to keeping the bus wet. There is a way to address both while keeping the company intact.

C. **-1** This high-rise type of operation will take an unacceptable amount of time. It also results in less equipment on the bridge by having the first engine company at the ground level. It is best to position on the bridge and take a supply from the standpipe. Use tank water while waiting for the continuous supply.

D. **-2** This position and hydrant/standpipe selection places the apparatus and its crew in the direct path of the smoke drift. If the fire reaches the cargo area, the fumes from the unidentified cargo can be lethal. Never position downwind or downhill at a hazardous materials incident. This rule should never be broken except under the most extreme and limiting circumstances. The apparatus positioning here completely cancels out the proper strategy in regard to line and stream placement.

E. **+2** This strategy protects both the rescue and addresses the fire extension problem. The deck gun stream will be kept well away from the rescue effort and will be an effective method of knocking down the fire in the truck. The handline is used to both protect the rescue and keep the fire away from the cargo if possible. The best way to accomplish this is to stretch around the back of the bus to the south (passenger) side of the bus. From this angle, the line can be used to keep fire away from the bus. Reinforce this operation as soon as possible with additional lines.

4.

A. **-2** No matter what tactics are chosen here, they cannot overshadow the fact that the second engine failed to provide the water supply. When the lines run dry, the operation may go directly into a nose-dive, endangering both the fire personnel and the victims. A dry standpipe with no water in it is still a dry standpipe.

B. **+2** These tactics will ensure that a continuous water supply will be established as well as the reinforcement of the initial protective/attack line.

C. **-2** Positioning in smoke is a display of tunnel vision and endangers the crew and the operation. Dropping a feeder line over is inferior to the simplicity of operating off a standpipe. The only time this would be acceptable is if the standpipe is out-of-service or is known to be suspect as to its reliability.

D. **0** If the second engine company's apparatus is supplying the standpipe, and the engine company on the bridge is directing that supply to its intake, it is best to utilize a line from the engine. The pressure as well as the control over the water will be more effective. This action is totally different than the action taken in choice "E."

E. **+2** Here, two standpipes are supplied by the same engine company. This will act as an insurance policy in case one side of the interconnected standpipe fails. The other supply can be quickly utilized. The same goes with the choice of line stretch. Although the line is not stretched off the apparatus on the bridge, it is using a different standpipe. Again, if something happens to the first standpipe, the water supply will not be interrupted at the other standpipe. This is different than taking a line off the same standpipe in which the attack engine is being supplied.

5. Ladder company operations should focus on the life safety problem. This problem will be most severe in the hot zone.

   A. *-2* Positioning the ladder truck in the area of smoke drift is just as bad as positioning the engine there. Keep the rig out of the path of smoke. In addition, triage and treatment are the responsibility of EMS personnel, not the first-due ladder company.

   B. *+2* Splitting the crew is an efficient use of manpower. It is unknown whether there is another occupant in the cab of the truck. If there is, no time should be wasted in removing this victim. Once this is addressed, the crews can reunite and work on the removal of victims from the bus.

   C. *-1* Although the tactics once on the bridge are acceptable, the positioning is improper and causes the whole question to be blown. The best place for the ladder company is on the bridge. The ladder truck is basically a rolling toolbox. Leaving it out of the area of operation puts the rescue and support operation at a great disadvantage.

   D. *0* The positioning is acceptable as are the rescue tactics. However, the intentional act of breaking out all the windows in the bus will not only create a flying glass hazard, but a panic problem as well. These windows are designed to be removed without breaking them. Firefighters should be trained and familiar with these procedures.

   E. *-2* The use of the ladder pipe is not warranted in this incident, not at this time. This is an inefficient use of manpower. The balance of the ladder crew can better be utilized in the rescue operation.

6.

   A. *-1* Scene congestion will only be a problem for the incident commander who fails to control the fireground by setting up a proper and safely positioned command post and staging area. The potential for scene congestion should never be a reason to fear additional alarms.

   B. *+2* There are at least thirty victims to attend to, either dazed and disoriented in the area, or still on the bus—possibly trapped. In addition, the geographic location of the incident makes operations more difficult than if the accident occurred on a street. Summon additional manpower to assist in overcoming those limiting factors caused by the geographical location.

   C. *0* Just requesting additional medical personnel will not help in the rescue of the victims. It is firefighters that remove victims and deliver them to the triage and treatment area. Make sure enough personnel are on the scene to handle the rescue problem.

   D. *-1* There are only 4 officers and 12 firefighters responding to this incident. There are potentially more than 30 victims. Failure to summon additional assistance will tax the resources on the first alarm response, creating the potential for firefighter injuries.

   E. *-2* Inaction is the trademark of a poor leader and an incompetent incident commander. Decisions that directly affect the incident must be made in a timely fashion. Many times, there is only a small window of opportunity to make a decision that can make or break the incident and have dire consequences on the safety of the players. If you are unsure, always err on the side of safety.

7.

A. **0** Stretching lines is not usually the responsibility of the rescue company, except under an extreme circumstance such as a fire threatening a victim, and all other engine personnel are already engaged. This incident does not require the rescue company to operate a hose line. They can be better utilized assisting in the rescue.

B. **+2** The rescue company should be involved in aiding where the largest threat to the life hazards exists. In a building fire, this should be the reinforcement of the primary search. In this scenario, the victims still in the bus are the top priority. As many hands as necessary should be assigned this duty. Rescue company personnel should be well skilled in the nuances of bus and other vehicle extrications, including the safety precautions inherent to each. (Fig. 11–11)

Fig. 11–11 Photo by Ron Jeffers NJMFPA
Incidents involving school buses can be extremely complex. Coordination of triage and treatment, school notifications, press matters, and interagency operations are some of the areas that the incident commander must attend to.

C. **+1** It is advantageous for the incident commander to have rescue company personnel trained at least to the first responder level of medical response. The EMT level is preferred as the fire department is often on the scene prior to medical assistance. It is beneficial not only to have personnel who can address medical concerns of victims, but also can be available to provide rapid primary care to injured firefighters, which may have to be initiated inside a fire building. (Fig. 11–12)

D. **-2** Hazardous materials incident mitigation is not the responsibility of the rescue company. Most are not trained to operate in this fashion. More importantly, rescue company personnel can be more useful assigned to meeting and stabilizing the life hazard objective of the incident.

Fig. 11–12 Photo by Ron Jeffers NJMFPA
A medical group should be established at all hazmat incidents. Included in this group will be the triage, treatment, and transportation units. Mass casualty units will be required at incidents that involve a large number of victims.

E. **0** The task of diking ahead of spilled material is something that the operations level responder can accomplish. However, the rescue company can be better utilized in victim rescue and stabilization operations.

8. This scenario does not state anywhere that fuel is leaking or has been involved in the ignition.

A. **+2** As the foam unit is not on the initial response, water should be used to extinguish the motor compartment and cab fire. To wait until foam is available may allow the fire to spread to the cargo. While it is important to have a Foam Unit on the scene, it is equally important to properly size-up the problem. The truck is labeled "Dangerous." The cargo is unidentified. If the cargo is water-reactive, using foam, which is mostly water, may make the situation worse. Waiting to lay down a blanket until the cargo is identified is the best action. Get the unit on the scene. Use it if the situation warrants.

B. **0**   There is no mention of any vapors escaping. While the laying down of foam is a "better-safe-than-sorry" action, it may escalate the operation if the cargo is water-reactive. Do your best to find out the contents of the truck before applying an extinguishing agent to it.

C. **-1**   The lines supplying Engine 1 must not be disconnected. If the Foam Unit must be supplied, either let Engine 1 provide the supply or secure another supply from one of the unused standpipes. To shut down the primary water supply could endanger both firefighters and victims.

D. **0**   There is no need to apply foam to the bus. There is no fire in the area of the bus, nor are the fuel tanks threatened or involved. Keep the foam operation at bay until more information is known.

E. **-2**   Covering the crew with foam is ridiculous. Applying foam to an unidentified dangerous cargo is equally ridiculous. Allowing an untrained crew to attempt to identify a dangerous cargo by coming into close proximity to it is criminal. The effective incident commander must never subject personnel to these unnecessary dangers.

9. This question asks for the best way to identify this cargo. Best must also mean safest to both score well and safeguard crews

A. **+1**   This is acceptable as the driver has self-evacuated from the burning cab. However, it is stated that he is disoriented. The information may not be reliable or correct. In this and any other hazardous materials situation, you must be certain of the cargo. Almost certain is the same as somewhat uncertain.

B. **0**   Cargo information that burns up will never come back. This is not an effective way of identifying the cargo.

C. **+1**   This, like asking the driver what is in the truck, may be fruitless if the driver is unable to answer accurately. It is still worth a try. The incident commander must have an alternative plan for identification of the cargo.

D. **-2**   Untrained personnel should never attempt to make contact with product. In addition, the foam lines may compound the problem if the product is water-reactive, further endangering the ill-fated curious rescue company.

E. **+2**   Contacting the company via the license plates on the truck will result in the safest, most accurate cargo identification. The incident commander must be able to rely on outside agency assistance to help mitigate these and other non-routine incidents. To operate with an exclusive "home-rule" policy is asking for trouble.

10. Enter the problem. This is where Murphy's Law enters into the picture. The incident commander, whether a chief or company officer must be prepared to address these type problems and develop effective solutions to solve them. Anyone can run the show when things are going smoothly.

A. **+1**   This is an obvious choice and will be an effective way to establish another water supply. The fact that the standpipe failed may be indicative of the same problem with the additional standpipes. Can you trust them? It is more effective to develop a solution that solves the problem instead of putting a Band-Aid over it.

B. *+2* While this may take time setting up, it will effectively solve the problem and ensure that a continuous water supply will be re-established to the area of operation. Don't be a victim of functional fixity. Use the flexibility of your apparatus to your advantage.

C. *+1* This is also a feasible solution to the problem, however, the weight of the supply line may put excess stress on the attack pumper intake. It must be secured sufficiently to safely continue the operation. Utilizing a tower ladder as a standpipe is a more effective method of delivering water to the bridge from below.

D. *-2* A controlled burn of a dangerous, unidentified cargo is irresponsible and should be a last resort action. Imagine if the cargo was explosive, radioactive, or possessed some other dreaded feature and you decided to let it burn? You could wind up exposing the whole city to the hazard. Attempt to develop solutions to solve the problem before throwing in the towel.

E. *+1* This, again, is a feasible solution to the problem, but one has to question the integrity of the standpipe system if a part of it has already failed.

You may have noticed by now that poor apparatus positioning was test-death in this scenario. Whereas proper positioning will make the operation safer and run more smoothly, poor positioning will delay operations, negatively impact support operations, and make the scene more dangerous. In the absence of a specific SOP, the ability of the incident commander to effectively position apparatus will have a direct relationship on incident success and safety.

Passing Score for Scenario 11–2 = 14 Points

Scenario 11–3
Multiple-Choice Questions

1. This response was initiated by a verbal alarm—a walk-in. There will be times, where, because of the proximity of the fire station to an emergency, we will be called upon to respond. These responses include stabbings and shootings, technical rescue and hazardous materials, and, of course, the oddball response. We have all been to these. Most people call us because they do not know whom else to call. I once responded on a verbal alarm to a bat capture. The bat was trapped in between the screen and the storm window of a tenement. It was hissing and spitting. We donned full protective gear, including SCBA to protect our faces, and armed with no more than an empty coffee can, the capture was made. The bat was subsequently released on the cliffs overlooking the Hudson River. Sometimes, these calls will be beyond the scope of our expertise. We are obligated to respond and do whatever we can to gain control of a situation, without unnecessarily jeopardizing personnel. This will be our primary duty until the experts arrive to stabilize the incident.

   This spill at the local high school is a call that we would certainly be called to, even if it was not a verbal walk-in. We must take steps to safeguard our personnel, protect life, and isolate the incident.

   This question addresses the least important size-up factor. Failure to critically read the question could cause a –2 response.

   A. *-2* Location and extent is the most important size-up factor, thereby making it the least desirable answer choice here. The fireground priorities cannot be safely addressed until the location and extent of the incident, in this case, the chlorine leak, is known.

   B. *0* Weather will always play an important role in the disposition of a hazardous vapor. Wind conditions along with relative humidity can help dissipate a cloud into a safe area or keep it low to the ground and cause it to endanger exposures, both life and property. Here, the wind is blowing toward the senior citizens complex, which will compound the problem caused by the vapor cloud.

   C. *+2* The water supply is good, with each hydrant supplied by a separate 12" main. While the product may limit your use of water, water supply will be the least of your problems. It is the least important size-up factor.

   D. *-1* The life hazard is critical. There are many students in the school that will potentially be exposed to the cloud should the wind shift. There are also the students who may be overcome in the annex.

   E. *-1* When they are present, auxiliary appliances are also a chief concern. The area of the spill is sprinklered. The system should be supplied in case the ignition of product activates a sprinkler head.

2. The establishment of command is the first action taken in regard to scene control. It establishes a central figure that will be accountable for all actions from here on in. The scene size-up actually occurs prior to or at the same time as the establishment of command.

   A. *+1* Control zones must be established at every incident to regulate access to operational areas. Control zone establishment and identification should be applied to all incidents. This includes, but is not limited to structure fires, vehicle fires and extrications, and hazardous materials. Control zones are established to provide for control of the emergency scene and to safeguard both civilians and emergency responders.

   B. *0* "Operational perimeter" is an improper term used in place of control zones.

   C. *+1* The reason that this is not the *+2* answer is that the initial radio report is given at the same time command is established. The initial radio report is the medium by which the incident commander both establishes command and relays his size-up to other companies. Any other way of letting the companies know who is in command, such as whispering in each officer's ear, is ineffective.

   D. *+2* Once the initial radio report is given, establishing command, the product must be identified. All other actions at the incident will depend on whether or not the product is identified and, if so, what hazards are present. These hazards allow the incident commander to make a hazard/risk analysis and determine what is the best way to stabilize the incident.

   E. *+1* Identification of a safety officer is mandatory and will take some of the weight off the incident commander's shoulders. The safety officer can reconnoiter the area and report conditions, hazards, and other pertinent information back to the incident commander. It would be best if the department had a dedicated safety officer who responded on all such incidents, operating under an established set of roles and responsibilities. Dedicated safety officers are superior to on-scene appointments. They eliminate the potential of breaking up a company as well as the disadvantage of "springing" such a role on someone who may not be prepared or completely willing to accept such a responsibility.

3. To find the best answer here, analyze the diagram.

   A. *-1* The gymnasium is upwind and a good distance away from the spill area. However, with the exception of a high-rise fire, it is never a good idea to run the operation from the interior of the same building where the incident is occurring. A quick shift of the wind and the entire "brains behind the operation" will be endangered.

   B. *0* The fire department headquarters is also upwind and out of the spill area. In addition, if the weather were bad, this area would provide shelter for command. However, it will not provide the incident commander any type of view of the incident. Inasmuch as possible, the command post should afford a view of the operation from a safe distance, so a first-hand evaluation can be continuously carried out. This evaluation should be supplemented by accurate and timely reports from key areas. Do not rule out the possibility of using the firehouse as a support area or press area. It is a key area that might come in handy.

   C. *+1* The athletic field south of the school will give the incident commander a vantage point of the operation from a safe distance. The problem is that access into and out of that area may be problematic. The only safe access is via Firehouse Road. This is unfortunately the only egress. The point is given because it is acceptable, but not the ideal.

D. *-2* The corner of Senior and Main is too close to the incident. It is as close to downwind as you can get without being in the senior citizens' building. Hazardous Materials incidents, especially in a school, will be big news. Many agencies will approach the command post. It should be in a safer area.

E. *+2* Utilizing the north athletic field will expand the options of the incident commander. This position will allow companies reporting for assignment to access the command post in a safe, uncongested area. It will allow for an unobstructed view of the annex from a safe area. There is also room for apparatus if needed.

4. Recall the fireground priorities of life safety, incident stabilization, and property conservation. Following this guideline will guide you to the best answer.

A. *0* The Department of Public Works is not required at this incident. The role of this agency is one of property conservation, usually providing sand to be used for diking. This spill will not require this resource.

B. *+2* Life safety is the number one reason for requesting additional resources. A student is stated as being overcome. Moreover, there is the potential for at least fifteen students in the annex as yet unaccounted for. Providing EMS response is the most effective way to address the life hazard problem.

C. *+1* The Haz Mat team are definitely a requirement at this incident. However, the responsibility of this resource is incident stabilization. While stabilizing the incident will sometimes take care of the life hazard problem, it should never be a substitute for it.

D. *-1* Department protocols may require the notification of the chief of the department. However, unless he is a Haz Mat technician or specialist, the chief is not the most important resource required at this scene.

E. *+1* Scene control is a major factor at this incident. Traffic must be rerouted. Control zones must be maintained. The police will also be of great importance when distraught parents begin to arrive. They should be used to keep non-essential people out of the area of operation.

5. The product here is concentrated chlorine. Chlorine vapor will be about $2\frac{1}{2}$ times heavier than air, making it conducive for migration along the ground. It will seek out low spots. It is nonflammable, but highly corrosive. Chlorine vapor will react with water, including moist skin or moisture in the lungs to form hydrochloric acid. This can cause severe damage to both skin and the respiratory system. Full protective clothing and SCBA are absolutely mandatory. All skin must be covered to avoid chemical burns.

A. *-2* Hydrant selection is too close to the incident area. A wind shift could be calamitous. In addition, for the reasons listed above, water used on the vapors can be extremely hazardous. Severe burns will be the likely result. It is best to stand by until directed to act by the Haz Mat team.

B. *+1* The water supply is well away from the annex. The line is also stretched dry from the windward side of the incident. Students must be evacuated via the safest area. The gym exit on the east side of the building is best. However, evacuation of the students is best left to the rescue and the ladder companies.

C. *-2* Any use of water in the area of the product or its vapors will result in serious injuries to those unprotected. The best strategy for the fire department is one of non-intervention.

D. *+2* Engine company personnel can best be utilized in a standby mode in a safe area. No water should be used here.

E. *0* The protection and evaluation of the senior citizens' building is not the responsibility of the first-arriving engine company. Evaluation of this area may be accomplished by the battalion chief (the deputy will be in command). He can best decide what protection and/or evacuation strategy to take. The decision should be made utilizing input from the Haz Mat team officer.

6. Engine 2 should get no more involved in this incident than Engine 1.

A. *-2* Two wrongs do not make a right. Water applied to chlorine gas will result in numerous injuries and potential death.

B. *+1* Hydrant #2 is on the windward side and out of the area of the leak. However, if you chose the proper tactics for Engine 1, that hydrant would already be taken as it is the best hydrant given the location of the incident and should be the water source for the first-arriving engine company.

C. *+2* Hydrant #1 is the second best hydrant and should be the water source used by Engine 2. As in answer choice "B," the tactics of stretching to Engine 1's location and standing by is the safest action to take to prevent endangerment of the crews.

D. *-2* The hydrant is downwind. The tactics of applying water to this vapor cloud have already been well established as dangerous and improper.

E. *+1* If you chose student evacuation, you got the point in this and the previous question. Technically, this is the responsibility of the ladder and rescue companies. If the tactics are proper and conditions warrant, life safety will usually always get you into the positive scoring area.

7.

A. *-2* The ladder company is likely to have no training other than the operational level. They should not approach the annex for any reason. To operate on the west side of the annex means definite exposure to the vapor cloud. Structural firefighting gear is insufficient for this task.

B. *+2* These tactics will isolate the spill and vapor by negating its ability to permeate the HVAC system. Using the direction and velocity of the wind, a favorable venting direction can be established. This is akin to PPV without the fans and will clear the school in the quickest manner.

C. *-1* It is best to shut the system down in this case. Recirculating the atmosphere via the HVAC system is uncertain and may spread the fumes into uninvolved areas.

D. *-1* This is just plain weird and does nothing more than jeopardize personnel to potential exposure. First, the companies are not trained to don product-specific gear. Just wearing the gear is no guarantee that personnel will know what to do when they get into the Hot Zone. Second, to place fans on the west, leeward side of the annex will cause exposure.

In addition, the fans are a potential ignition source. While chlorine is not flammable, it may act as an oxidizer, causing a malfunction of the internal combustion motor used to power the fans. What problems may result from this is anybody's guess. Do yourself a favor. Keep the PPV fans on the rig.

E. *+1* This is short and sweet. Simply shutting down the HVAC system from a safe area will eliminate any problems caused by the HVAC system. The controls are stated as being in the gymnasium. This area is remote and windward of the Hot Zone.

8. The Haz Mat officer is a pivotal resource in the proper and safe mitigation of this incident. It is imperative that the incident commander uses this resource to its fullest advantage. Remember that a proper incident management system does not recognize rank or ego. The best action is to find the right person for the job. This may mean that if the Haz Mat officer is a captain or lieutenant and is assigned hazard group supervisor, it may be necessary to give orders to a battalion chief assigned to a subordinate support position in the incident management hierarchy. This is perfectly acceptable, as all players should be working toward the common goal of mitigating the incident in the quickest and safest manner possible.

A. *-1* It is not necessary or advantageous to transfer command to the Haz Mat officer. As incident commander, your knowledge of the personnel and resources available, as well as what has taken place thus far, is critical to the operation. It is better to assign the officer to a specific function such as hazard group supervisor or as a command partner (unified command). Either way, the objective must be the same: to work together under a single, coordinated action plan to safely and rapidly stabilize the incident.

   To take this a step further, the incident commander of the jurisdiction where the incident is taking place is legally bound to retain responsibility for the emergency at hand. Authority for your own jurisdiction cannot be transferred to an outside agency.

B. *-2* Choosing this answer denotes that you have sent your own, untrained crews into the middle of the problem. They should not be involved.

C. *-1* Taking the Haz Mat officer into the school would require you to leave the command post. That is almost never a good idea, and certainly not at this incident. The incident commander will always be most effective at the command post. A layout of the school should be made available to the command post as soon as possible. This will usually be in the form of blueprints. As a matter of fact, as the school should be a target hazard, a set should be carried in the command vehicle.

D. *+1* A unified command structure would be a most advantageous way of arriving at a proper and safe course of action to take at this incident. Unified command is a management process that allows the appropriate responding agencies to share responsibility for the incident as well as cooperatively develop a strategy to safely mitigate the incident. Unified command allows different agencies the opportunity to run the incident without giving up authority, responsibility, or accountability. Under a unified command, there is one action plan developed, a single command post, and a single operations officer who puts the jointly developed plan into action. Another area where an operation may benefit from a unified command structure is for an incident that overlaps jurisdictional boundaries. Here, the highest ranking officer from each jurisdiction, rather than running "competing" incident command operations, work together to arrive at the best approach and solution to the problem.

E. **+2** Delegating the authority to the Haz Mat officer to direct the most hazardous portion of the incident is a safe and effective way of maintaining span of control, while allowing an "expert" do the job for which he was requested. This may be a more effective way to use this officer than the unified command option, as it will allow him to focus on just those operations which are designed to address the main hazard, the chlorine leak. Having this officer at the point of operation may be the safest course of action, as direct supervision may be required. The job of the incident commander would be one of support, providing reinforcement as per the direction of the hazard sector.

9. The senior citizens' building is in the direct path of the vapor cloud. It is the most significant exposure at this incident.

A. **-2** Anyone who attempts to evacuate a six-story building filled with senior citizens in this instance should have his head examined. The multitude of problems that may be encountered is staggering. Residents may be non-ambulatory, or ambulatory requiring assistance. Others may be in that old-age-permanently-confused state. This may lead to a virtual "kittens-out-of-the-box" situation. Once they are out, how do you get them to the firehouse on the other side of the school? The leak is between them and the firehouse. Once they are there, how do you provide care for them? Imagine sending in two companies to try to tackle this task? Protection-in-place along with building system isolation tactics is a better strategy here.

B. **+1** Shutting down the HVAC system will keep any vapors that may have entered the building from circulating into living areas. Closing dampers will serve to isolate the building further. Opening windows on the leeward side will allow any vapors to exit the building. These isolation tactics are easier to accomplish than evacuation and will cause less panic in the building.

C. **-1** Just like in the annex, it is best to shut down any air-handling systems. Even after the dampers are closed, isolating the building fumes from the leak that had entered prior to building isolation may be allowed to circulate into previously uninvolved areas.

D. **+2** These are basically the same tactics as answer choice "B." However, the difference is the place in which the protection-in-place is occurring. If the occupants on each floor are moved temporarily to the western, unexposed side of the building, they will be safer. This slight inconvenience is a small risk and one worth taking. (Fig. 11–13)

E. **-2** The actions of placing PPV fans in front of the building, between the vapor cloud and the building will not only endanger the fire personnel and provide a potential ignition source, but will pull the fumes right into the structure. Leave the fans on the ladder truck, close the windows, and shut down the HVAC system.

Fig. 11–13 Protection-in-place may be the best alternative when confronted with a fire resistive building exposed to vapors. Evacuating may create more problems than it solves. Moving occupants to an unexposed area of the building may be all that is required. Note the air conditioners in the windows. This building has no HVAC system, simplifying defensive measures.

10. Assignment of a safety officer is one of the most important decisions the incident commander has to make. This officer must be both knowledgeable and reliable. This is not a job to be taken lightly. Thus, it is not a job given to someone who routinely shows a disregard for safety. It is imperative that he knows the scope of his duties.

   A. *+1* Establishment and disciplined maintenance of control zones, especially the isolation of the hot zone, must be enforced at all times. The safety officer should see to it that all on-scene personnel are aware of zone boundaries as well as any exotic hazards presented by the product. Clear and easily identified zone boundary markers along with a gatekeeper/passport zone entry system are the best way of maintaining zone integrity. The safety officer should not, however, limit his concern to the hot zone, but should address the entire operational area. At large incidents, it might be efficient to establish a safety division accountable to the assigned safety officer. The establishment of a safety division will allow for a manageable span of control, allowing coverage of a large area, and decentralizing the duties and responsibilities for safety of responders to a closely coordinated team whose sole purpose is incident safety.

   B. *0* The mitigation of the hazard is not the duty of the safety officer. The safety of this operation should be the main concern of the safety officer. He should operate in such a manner that ensures personnel working in all areas of the incident, both in direct involvement and support, are operating in the safest manner possible.

   C. *+2* This blanket statement in regard to the safety of all operating personnel is the true purpose of the safety officer. It must be stated again that at large incidents or incidents spread over a large area, the safety officer can be most effective and efficient if a staff is provided by the incident commander. This and many other type incidents are fraught with dangers, both seen and unseen. It is the responsibility of the incident commander to make the safety officer's assignment as smooth as possible. This, when taken care of, will, in turn, make the incident commander's job more palatable.

   D. *-1* The question seeks the most important duty of the safety officer. Determining the severity of the threat posed by the product may be highly technical and is more the domain of the hazard sector. If appropriate, the safety officer should give input where required, but should be careful not to overstep the limitations of his training.

   E. *+1* These acts take answer choice "A" a step further, but do not address the big picture. The big picture is the entire incident and operation. While danger is at its greatest in the hot and warm zones, the cold and public zones must not be ignored.

Passing Score for Scenario 11–3 = 14 Points